IP A to Z

IP
A to Z

Nathan J. Muller

McGraw-Hill

New York Chicago San Francisco Lisbon London Madrid
Mexico City Milan New Delhi San Juan Seoul
Singapore Sydney Toronto

The McGraw-Hill Companies

Cataloging-in-Publication Data is on file with the Library of Congress

1 2 3 4 5 6 7 8 9 0 DOC/DOC 0 9 8 7 6 5 4 3 2

ISBN 0-07-141086-4

The sponsoring editor for this book was Steve Chapman, the editing supervisor was Steven Melvin, and the production supervisor was Pamela Pelton. It was set in New Century Schoolbook following the AZ01 design by Deirdre Sheean of McGraw-Hill's Professional's composition unit, Hightstown, N.J.

RR Donnelley was printer and binder.

To my son Matthew
Commemorating his bicycle
ride across America. . .

CONTENTS

PREFACE

The Internet is a system of packet-switched data networks based on open standards protocols. It spans more than 200 countries and is used by over 700 million people worldwide on a daily basis for research, education, commerce, and entertainment. It supports a variety of communications methods, including electronic mail, paging, telephone calls, facsimile, videoconferencing, and collaborative computing. Even the range of radio and television broadcasts can be extended to other countries over the Internet.

The value of the Internet is appreciated by governments around the world to such an extent that tax dollars are being used to wire schools, libraries, and rural healthcare clinics for Internet access at a rapid pace to provide all citizens with equal access to its information resources and services. Such efforts are particularly robust in the United States, and similar efforts are underway in other industrialized countries. Many schools and universities offer distance-learning programs over the Internet. There are even university programs that exist only on the Internet, giving busy professionals the chance to earn degrees, regardless of their location or work schedule.

Businesses large and small have also come to see the value of the Internet. Using the same protocols and technologies as the Internet, they have built private networks called intranets to improve internal communication, facilitate information distribution, broaden access to corporate resources, enable group scheduling, and provide a browser front-end to various corporate databases. With links to the public Internet, intranets allow employees to engage in a range of activities, including performing research, staying informed of market behavior, and engaging in electronic commerce. Intranets have become the means through which businesses can reinvent themselves, perhaps to establish an international presence that allows them to participate fully in the global economy.

From its obscure beginning as a military research project over 30 years ago, the Internet has blossomed into an engine for technological innovation and national economic prosperity, so much so that the Federal government in the United States has adopted a hands-off policy with regard to regulating and taxing the Internet—at least for the immediate future. Even the Federal courts have been reluctant to impose their will on the Internet, except in narrow cases involving fraud, gambling, and pornography. Law enforcement, too, sees the value of the Internet, but is limited to using it as an investigative tool out of concern that privacy rights might be violated. Accordingly, a court order must be secured before any person's Internet connection or computer can be monitored for suspected criminal activity.

The reason the Internet merits such special attention from government, education, and business is that information is one of the nation's most critical economic resources—for service industries as well as manufacturing, for economic as well as national security. It is estimated that over 70 percent of U.S. workers are in information-related jobs, and the rest are in industries that rely heavily on information. In an era of global markets and global competition, the technologies to create, manipulate, manage, and use information are of strategic importance. The Internet is the enabling medium over which the free flow of information takes place.

The Internet is considered so important that a growing number of users want to take it with them wherever they go. Already, Internet connectivity is now available with all major mobile phone services from which users can send and receive electronic mail, check stock prices and sports scores, keep track of bank accounts, or browse Web pages. Trains and planes have or are being equipped with wireless technology to allow Internet access on the go. When travelers arrive at their hotel, the room is equipped with a data port to which they can plug in a notebook computer and have access to all corporate information on their intranet and anything they need on the Internet. In a growing number of cities, wireless "hot spots" are being created to allow Internet access in parks, airport terminals, and cafes. Communities not served by cable or digital subscriber line (DSL), too, are setting up wireless networks to share the expense of a high-speed T1 connection to the Internet.

The Internet has become so popular worldwide and sufficiently sophisticated and complex as to merit dozens of books on the topic that are published every year. This encyclopedia is a quick reference that clearly explains the essential concepts of the Internet, including services, applications, protocols, access methods, development tools, administration and management, standards, and regulation. It is designed as a companion to other books you may want to read about the Internet, providing clarification of concepts that may not be fully covered elsewhere.

The information contained in this book, especially as it relates to specific vendors and products, is believed to be accurate at the time it was written and is, of course, subject to change with continued advancements in technology and shifts in market forces. Mention of specific products and services is for illustration purposes only and does not constitute an endorsement of any kind by either the author or the publisher.

Nathan J. Muller

A

AMERICAN REGISTRY FOR INTERNET NUMBERS

The American Registry for Internet Numbers (ARIN) is one of three regional Internet registries that collectively manage Internet Protocol (IP) address assignments. The IP addresses originate from International Corporation for Assigned Names and Numbers (ICANN). ARIN is responsible for North America, South America, the Caribbean, and sub-Saharan Africa. Previously, the management of IP address space had been a responsibility of the Internet Network Information Center (InterNIC) operated by Network Solutions, Inc. (NSI) under the authority of the Internet Assigned Numbers Authority (IANA).

Like the Reseaux IP Européens (RIPE) and Asia Pacific Network Information Center (APNIC), the other two regional IP registries, ARIN is a nonprofit organization. Its funding structure is similar to that of RIPE and APNIC, coming from membership dues and fees for registration and maintenance. Registration fees apply only to those IP number allocations received directly from ARIN. Members are responsible for determining the organization's goals and how best to meet those goals. ARIN's business affairs and finances are the responsibility of its Board of Trustees.

Services

ARIN provides several services that are necessary for the continued stability of the Internet, including the allocation of IP numbers, management of autonomous system numbers (ASNs), maintenance of inverse address mapping and a routing registry, and providing help desk support. In addition, ARIN performs database maintenance and verifies registrants' reassignment information.

IP Allocations ARIN issues IP addresses in either of two formats: IP version 4 (IPv4) or IPv6. The former has been in use for over 20 years, while the later is relatively new. The difference is that IPv4 addresses are only 32 bits in length, which makes them a scarce resource that must be carefully managed, given the explosive growth of the Internet. IPv6 addresses are 128 bits in length, providing a virtually inexhaustible supply of unique IP addresses.[1]

Most organizations receive IP addresses from their upstream provider—those Internet service providers (ISPs) from which they obtain their Internet service. If a customer changes service providers or terminates a contract with its upstream ISP, the customer is asked to return its network addresses and must then use its new address space to renumber its network.

Downstream organizations would prefer to keep the IP addresses assigned to them by their upstream provider even when they choose to contract with another ISP. In this way, they would not have to go through the trouble of renumbering into new addresses. However, blocks of IP addresses allocated to any first-tier or upstream ISP must remain contiguous so that addresses can be aggregated. If these blocks were to become fragmented, routing table overload could result. ARIN and the other regional Internet registries monitor for this potential problem so that it can be addressed effectively.

---

[1] IPv4 address space equates to 2 to the 32nd power, yielding a maximum of over 4 billion IP addresses. The exact number of IPv4 addresses is 4,294,967,296. By comparison, IPv6 address space equates to 2 to the 128th power, or over 340 undecillion IP addresses, one for every grain of sand on the planet. The exact number of IPv6 addresses is 340,282,366,920,938, 463,463,374,607,431,768,211,456.

Autonomous System Numbers (ASNs) ASNs are globally unique numbers used to identify autonomous systems, which are connected groups of IP networks that have a single and clearly defined routing policy and are under single ownership, trust, and administrative control. ARIN is responsible for assigning these numbers.

Inverse Address Mapping Inverse address mapping is the Domain Name System (DNS) that performs address-to-name resolution, which is the reverse of the originating name-to-address transmission. Inverse address mapping is designed to facilitate queries to locate specific servers on networks that originated the transmission. ARIN maintains the authoritative inverse address information for networks within its region so that this process can be performed.

Routing Registry ARIN maintains a routing registry that provides routing information supplied by commercial entities to organizations sending data across the Internet. It serves as a registration service whereby network operators can submit, maintain, and retrieve router configuration information. The registry serves as a repository for routing policy system information and improves the ability of organizations to configure and manage their networks.

Help Desk Support The Registration Services Group within ARIN maintains a staff of analysts whose responsibility is to review IP requests and to answer questions that the requesting organizations may have.

Summary

Continued operation of the Internet depends in large part on the conservation and efficient use of IPv4 address space until the transition to IPv6 is complete. Toward that end, ARIN is responsible for maintaining a public trust. Not only does ARIN promote the conservation of IP address space, it maintains impartiality while determining the size of address blocks to be allocated or assigned and supports efforts to keep the global routing tables to a manageable size to ensure the efficient routing of information over the Internet.

See also

Domain Name System
International Corporation for Assigned Names and Numbers

AMERICAN STANDARD CODE FOR INFORMATION INTERCHANGE

American Standard Code for Information Interchange (ASCII) refers to the 7-bit binary encoding scheme that is used to assign a number to the most frequently used characters in American English. ASCII, which also includes the encoding for common keyboard functions, is understood by almost all applications used by personal computers, including applications that are run over the Internet.

Although ASCII uses a 7-bit binary encoding scheme, which makes for 128 possible characters and functions, there is provision for an eighth bit, the left-most bit, which is reserved for parity. In ASCII, the capital letter *C*, for example, is assigned the decimal code 67 and assigned 01000011 in binary. The 0 bit in the eighth position to the left is reserved for the parity bit, which is used for error checking when data are sent via modem over phone lines. Table A-1 compares the decimal, octal, hex, and binary encoding schemes.

International ASCII

Although ASCII's 7 bits are enough to encode the common characters used in American English, this is not enough to include the symbols frequently used in other countries, such as the British pound symbol or the German umlaut. There is a version of ASCII standardized by the International Organization for Standardization (ISO) that includes the original 128 characters along with an additional 128 charac-

TABLE A-1 Comparison of Common Encoding Schemes

Decimal	Octal	Hex	Binary	Value	Definition
000	000	000	00000000	NUL	Null character
001	001	001	00000001	SOH	Start of header
002	002	002	00000010	STX	Start of text
003	003	003	00000011	ETX	End of text
004	004	004	00000100	EOT	End of transmission
005	005	005	00000101	ENQ	Enquiry
006	006	006	00000110	ACK	Acknowledgment
007	007	007	00000111	BEL	Bell
008	010	008	00001000	BS	Backspace
009	011	009	00001001	HT	Horizontal tab
010	012	00A	00001010	LF	Line feed
011	011	00B	00001011	VT	Vertical tab
012	012	00C	00001100	FF	Form feed
013	013	00D	00001101	CR	Carriage return
014	014	00E	00001110	SO	Serial in/shift out
015	015	00F	00001111	SI	Serial out/shift out
016	016	010	00010000	DLE	Data link escape
017	017	011	00010001	DC1/XON	Device control 1
018	018	012	00010010	DC2	Device control 2
019	019	013	00010011	DC3/XOFF	Device control 3
020	020	014	00010100	DC4	Device control 4
021	021	015	00010101	NAK	Negative acknowledgement
022	022	016	00010110	SYN	Synchronous idle
023	023	017	00010111	ETB	End of transmission block

TABLE A-1 *(Continued)*

Decimal	Octal	Hex	Binary	Value	Definition
024	024	018	00011000	CAN	Cancel
025	025	019	00011001	EM	End of medium
026	026	01A	00011010	SUB	Substitute
027	027	01B	00011011	ESC	Escape
028	028	01C	00011100	FS	File separator
029	029	01D	00011101	GS	Group separator
030	030	01E	00011110	RS	Request to send/record separator
031	031	01F	00011111	US	Unit separator
032	032	020	00100000	SP	Space
033	033	021	00100001		!
034	034	022	00100010		“”””
035	035	023	00100011		#
036	036	024	00100100		$
037	037	025	00100101		%
038	038	026	00100110		&
039	039	027	00100111		ë
040	040	028	00101000		
041	041	029	00101001		
042	042	02A	00101010		*
043	043	02B	00101011		+
044	044	02C	00101100		“,”
045	045	02D	00101101		-
046	046	02E	00101110		.
047	047	02F	00101111		/

048	048	030	00110000	0
049	049	031	00110001	1
050	050	032	00110010	2
051	051	033	00110011	3
052	052	034	00110100	4
053	053	035	00110101	5
054	054	036	00110110	6
055	055	037	00110111	7
056	056	038	00111000	8
057	057	039	00111001	9
058	058	03A	00111010	:
059	059	03B	00111011	;
060	060	03C	00111100	
061	061	03D	00111101	=
062	062	03E	00111110	>
063	063	03F	00111111	?
064	064	040	01000000	@
065	065	041	01000001	A
066	066	042	01000010	B
067	067	043	01000011	C
068	068	044	01000100	D
069	069	045	01000101	E
070	070	046	01000110	F
071	071	047	01000111	G
072	072	048	01001000	H
073	073	049	01001001	I
074	074	04A	01001010	J
075	075	04B	01001011	K

TABLE A-1 *(Continued)*

Decimal	Octal	Hex	Binary	Value	Definition
076	076	04C	01001100	L	
077	077	04D	01001101	M	
078	078	04E	01001110	N	
079	079	04F	01001111	O	
080	080	050	01010000	P	
081	081	051	01010001	Q	
082	082	052	01010010	R	
083	083	053	01010011	S	
084	084	054	01010100	T	
085	085	055	01010101	U	
086	086	056	01010110	V	
087	087	057	01011111	W	
088	088	058	01011000	X	
089	089	059	01011001	Y	
090	090	05A	01011010	Z	
091	091	05B	01011011	[	
092	092	05C	01011100	\	
093	093	05D	01011101	]	
094	094	05E	01011110	^	
095	095	05F	01011111	_	
096	096	060	01100000	`	
097	097	061	01100001	a	
098	098	062	01100010	b	
099	099	063	01100011	c	
100	100	064	01100100	d	

101	101	065	01100101	e
102	102	066	01100110	f
103	103	067	01100111	g
104	104	068	01101000	h
105	105	069	01101001	i
106	106	06A	01101010	j
107	107	06B	01101011	k
108	108	06C	01101100	l
109	109	06D	01101101	m
110	110	06E	01101110	n
111	111	06F	01101111	o
112	112	070	01110000	p
113	113	071	01110001	q
114	114	072	01110010	r
115	115	073	01110011	s
116	116	074	01110100	t
117	117	075	01110101	u
118	118	076	01110110	v
119	119	077	01110111	w
120	120	078	01111000	x
121	121	079	01111001	y
122	122	07A	01111010	z
123	123	07B	01111011	{
124	124	07C	01111100	|
125	125	07D	01111101	}
126	126	07E	01111110	~
127	127	07F	01111111	DEL

ters, such as the British pound symbol and the American cent symbol. This increase to 256 characters is achieved by dispensing with the need for a parity bit so that the full 8 bits can be used for encoding characters. There are several variations of this ISO-8859 standard that can be applied to different language families:

- Latin-1 (western European languages)
- Latin-2 (non-Cyrillic central and eastern European languages)
- Latin-3 (southern European languages and Esperanto)
- Latin-5 (Turkish)
- Latin-6 (northern European and Baltic languages)
- 8859-5 (Cyrillic)
- 8859-6 (Arabic)
- 8859-7 (Greek)
- 8859-8 (Hebrew)

EBCDIC

Extended Binary-Coded Decimal Interchange Code (EBCDIC) is IBM's 8-bit scheme for representing 256 possible characters as numbers. The operating system of the AS/400 midrange computer, for example, processes data internally in EBCDIC format. Although EBCDIC is used widely on large IBM computers, most other computers, including PCs and Macintoshes, use ASCII codes or Unicode.

Unicode

While ISO ASCII uses 8 bits to encode characters, Unicode uses 16 bits, which means that it can represent more than 65,000 characters, the first 256 characters of which are identical to Latin-1. Thus Unicode provides a unique number for every character, no matter what the platform, program, or language. While this is hardly necessary for English and western European languages, it is necessary for Chinese,

Japanese, Korean, and other languages with a large number of ideographic characters.

An extension mechanism allows for the encoding of as many as 1 million additional characters. This capacity is sufficient for all known character encoding requirements, including full coverage of all the world's historic scripts. There is even a proposal to accommodate within Unicode Sumero-Akkadian cuneiform, the ancient Near Eastern writing system used for a number of languages from the end of the fourth millennium B.C. until the first century B.C. The purpose of encoding cuneiform in Unicode is to broaden the study and translation of ancient texts via computers and networks.

Work on Unicode began in 1988, but the project was not given wide exposure until the Unicode Consortium was formed in 1991. Also known as ISO-10646, Unicode has been adopted for use in the products of such industry leaders as Apple, Hewlett-Packard, IBM, Microsoft, Oracle, PeopleSoft, Sun, Sybase, Unisys, and many others. With business becoming more global and national economies becoming more interdependent, many operating systems, databases, and programming languages now use Unicode as the character set instead of ASCII.

Summary

ASCII is an encoding scheme that allows computers to read, process, store, and move information between computers. However, the inability of ASCII to correctly represent the characters of a variety of languages has resulted in a proliferation of encoding schemes worldwide. Unicode is the ultimate encoding standard. It can potentially encode the characters of every language, past and present. The current edition of the Unicode Standard, version 3.0, contains over 49,000 characters, covering the principal written languages of the Americas, Europe, the Middle East, Africa, Asia, and the Pacific Rim. The Unicode Standard is fully synchronized with the ISO-10646, providing a formal, internationally recognized basis for its character encoding.

APPLICATION SERVICE PROVIDERS

Application service providers (ASPs) host business-class applications from their data centers and make them available to customers on a subscription basis over the network.[2] Among the business functions commonly outsourced in this way are customer relationship management (CRM), financial management, human resources, procurement, and enterprise resource planning (ERP). The ASP owns the applications, and subscribers are charged a fixed monthly fee for use of the applications over secure network connections.

An ASP enables customers to avoid many of the significant and unpredictable ongoing application management challenges and costs. Following the implementation of software applications, performed for a fixed fee or on a time and materials basis, customers pay a monthly service fee based largely on the number of applications used, total users, the level of service required, and other factors. By providing application implementation, integration, management, and various upgrade services and related hardware and network infrastructure, the ASP reduces IT burdens of its customers, enabling them to focus on their core businesses and react quickly to dynamic market conditions.

Traditionally, organizations have installed, operated, and maintained enterprise software applications internally. The implementation of enterprise software applications often takes twice as long as planned. Moreover, the ongoing costs of operating these applications, including patching, upgrading, training, and management expenses, are often significant, unpredictable, and inconsistent and may increase over time. The emergence of the Internet, the increased communications bandwidth, and the rewriting of enterprise soft-

[2]The difference between Web hosting or e-commerce hosting and the kind of hosting performed by ASPs is that in the former case the customer owns the applications and merely runs them on the shared or dedicated servers of an Internet service provider. In the latter case the ASP has strategic alliances with third-party software providers for licensed use of the applications over the network. The ASP pays fees to the software firms based on factors such as the number and type of customers' users.

ware to be delivered over IP networks are transforming the way enterprise software applications are being provided to companies.

Instead of in-house installations, these applications are beginning to be hosted by third parties, in which the hosting company maintains the applications on an off-site server, typically in a data center, and delivers the applications to customers over the Internet as a service. In addition, competitive pressures have led to a renewed focus on core competencies, with many businesses concluding that building and maintaining IT capabilities across their entire set of applications are not core competencies. In response to these factors, companies are adopting hosted applications rather than managing them in-house.

An ASP typically can complete a standard implementation of its services in 2 to 14 weeks. This allows customers to avoid the longer implementation times frequently experienced with installing and integrating customized, sophisticated applications. This enables customers to achieve the desired benefits quickly by reducing the time required to establish or augment their IT capabilities.

To address this market, many types of companies are setting themselves up as ASPs in this relatively new market, including long-distance carriers, telephone companies, computer firms, ISPs, software vendors, integrators, and business management consultants. Intel Corp., for example, has built data centers around the world to be ready to host electronic business sites for millions of businesses that will embrace the Internet within 5 years.

Another ASP, Corio, enables businesses to obtain best-of-breed applications at an affordable cost. Corio is responsible for maintaining and managing the applications and ensuring their availability to its customers from its data centers. For a fixed monthly fee for the suite of integrated business applications and services, businesses can achieve a 70 percent reduction on average of total cost of ownership (TCO) in the first year versus traditional models and a 30 to 50 percent TCO reduction over a 5-year period.

Growing the Market

Application outsourcing has been around for nearly 30 years under the concept of the service bureau. In the service bureau arrangement, business users rented applications running the gamut from rudimentary data processing to high-end proprietary payroll. Companies such as EDS and IBM hosted the applications at centralized sites for a monthly fee and typically provided access via low-speed private-line connections.

In an early 1990s incarnation of the service-bureau model, AT&T rolled out hosted Lotus Notes and Novell NetWare services, complete with 24 × 7 monitoring and management. Users typically accessed the applications over a frame relay service or dedicated private lines. AT&T's Notes hosting effort failed and was discontinued in early 1996. The carrier lacked the expertise needed to provide application-focused services and did not offer broad enough access to these applications. The lesson: Large telecommunications companies are focused on building networks, which is quite different from implementing and managing enterprise applications.

In 1998, there emerged renewed interest in this type of service with a new twist—that of providing an array of standardized services to numerous business customers. Economies of scale could be achieved in this "one to many" model by cost reductions incurred in service delivery—specifically by relying on managed IP networks. Further cost reductions could be achieved by developing implementation templates, innovative application management tools, and integration models that can be used for numerous applications across a variety of companies and industries.

To help sell the benefits of applications outsourcing, 25 companies have formed the Applications Service Provider Industry Consortium. The consortium includes a wide range of companies, including AT&T and UUNET, a WorldCom company, on the service-provider side. Compaq Computer, IBM, and Sun Microsystems are representative of the systems and software vendors. Interconnect companies such as Cisco Systems are also members. The consortium's goals include education, common definitions, research, standards, and best practices.

Summary

Several trends have come together to rekindle the market for applications outsourcing. The rise of the Internet as an essential business tool, the increasing complexity of enterprise software programs, and the shortage of IT expertise have created a ready environment for carriers and other companies entering the applications hosting business. The economics of outsourcing are compelling, and new companies are being created to deal with customers' emerging outsourcing requirements. A true ASP, however, supports a range of enterprise applications. When companies outsource, they want a service provider to manage all their corporate applications, not just one. This gives ASPs a competitive advantage over firms that specialize in providing only their own applications or a limited number of third-party applications.

See also

Collocation Arrangements for Businesses
Electronic Software Distribution
Outsourcing
Service Level Agreements

ASYNCHRONOUS TRANSFER MODE

Asynchronous Transfer Mode (ATM) is a cell switching technology that offers low-latency transmission with quality-of-service (QoS) guarantees in support of data, voice, video, and Internet traffic at multimegabit-per-second speeds. ATM is also highly scalable, making it equally suited for interconnecting legacy systems and local area networks (LANs) and for building wide area networks (WANs) over today's high-performance optical fiber infrastructures. ATM-based networks may be accessed through a variety of standard interfaces, including frame relay.

Applications

There are many applications that are particularly well suited for ATM networks, including

- *High-speed Internet access* Many national carriers use ATM to move large volumes of Internet traffic through their fiber backbone networks. Internet traffic from corporations, small and regional ISPs, and government agencies is aggregated at various points within the carrier's ATM network and delivered to a network access point (NAP) on the Internet. At the NAPs, traffic is sorted out among interconnected carrier networks, which send it over the most efficient path to the proper destination.
- *LAN internetworking* ATM can be used to interconnect LANs over the WAN. Special protocols make the connection-oriented ATM network appear as a connectionless Ethernet or token ring LAN segment.
- *Videoconferencing or broadcasting* ATM can be provisioned for interactive videoconferencing between two or more locations or to support point-to-multipoint video broadcasts.
- *Telemedicine* With ATM, large amounts of bandwidth can be provisioned to support the rapid exchange of high-resolution diagnostic images and multimedia patient records while permitting interactive consultations among medical specialists at different locations.
- *Private-line connectivity* An ATM virtual circuit can be used to provide a more economical way to provision leased lines on the WAN. ATM protocols can emulate $N \times 64$ kbps DS0 transport.
- *PBX voice trunking* An ATM virtual circuit can be used to interconnect PBXs and maintain full PBX feature sup-

port, call routing, and switching. Voice trunking combines multiple calls onto a single virtual circuit for further bandwidth optimization, reduced delay, and lower cost. PBX voice trunking requires an integrated access device at the customer premises, between the PBX and ATM switch, that performs the protocol conversions necessary to extend feature signaling across the ATM network.

ATM also offers a consolidation solution for any company that maintains separate networks for voice, video, and data. The reason for separate networks is to provide appropriate bandwidth and preserve performance standards for the different applications. But ATM can eliminate the need for separate networks, providing a unified platform for multiservice networking that meets the bandwidth and QoS needs of all applications. Although the startup cost for ATM is high, the economics of network consolidation mean that companies do not have to wait very long to realize return on their investment.

Quality of Service

ATM serves a broad range of applications very efficiently by allowing an appropriate QoS to be specified for each application. Various categories have been developed to help characterize network traffic, each of which has its own QoS requirements. These categories and QoS requirements are summarized in Table A-2.

CBR is intended for applications where the PVC requires special network timing requirements (i.e., strict PVC cell loss, cell delay, and cell delay variation performance). For example, CBR would be used for applications requiring circuit emulation (i.e., a continuously operating logical channel) at transmission speeds comparable with DS1 and DS3. Such applications would include private-line-like service or voice-type service, where delays in transmission cannot be tolerated.

TABLE A-2 ATM Quality of Service (QoS) Categories

		Quality of Service Requirements			
Category	Application	Bandwidth Guarantee	Delay Variation Guarantee	Throughput Congestion Guarantee	Feedback
Constant bit rate (CBR)	Provides a fixed virtual circuit for applications that require a steady supply of bandwidth, such as voice, video, and multimedia traffic.	Yes	Yes	Yes	No
Variable bit rate (VBR)	Provides enough bandwidth for bursty traffic such as transaction processing and LAN interconnection, as long as rates do not exceed a specified average.	Yes	Yes	Yes	No
Available bit rate (ABR)	Makes use of available bandwidth and minimizes data loss through congestion notification. Applications include e-mail and file transfers.	Yes	No	Yes	Yes
Unspecified bit rate (UBR)	Makes use of any available bandwidth for routine communications between computers but does not guarantee when or if data will arrive at their destination.	No	No	No	No

Variable bit rate–real time (VBR-rt) is intended for applications where the PVC requires low cell delay variation. For example, VBR-rt would be used for applications such as variable bit rate video compression and packet voice and video, which are somewhat tolerant of delay. Variable bit rate–nrt (VBR-nrt) is intended for applications where the PVC can tolerate larger cell delay variations than VBR-rt. For example, VBR-nrt would be used for applications such as data file transfers.

Available bit rate (ABR) is offered as a low-cost method of transporting applications traffic that can tolerate delay variations. The first application that offers traffic to the network gets to use the available bandwidth. Other applications that attempt to offer traffic to the network must wait until the bandwidth becomes free. If congestion builds up in the network, ABR traffic is held back to relieve the congestion condition.

Unspecified bit rate (UBR) handles traffic on a best-effort basis with no guarantee of delivery. Newsgroup updates, network monitoring messages, and file transfers are examples of nonessential traffic that is highly tolerant of delay and cell loss. If congestion starts building in the network, the ATM cells for such traffic are the first to be discarded to relieve the congestion.

Operation

QoS enables ATM to admit a CBR voice connection while protecting a VBR connection for a transaction processing application and allowing an ABR or UBR data transfer to proceed over the same network. Each virtual circuit will have its own QoS contract, which is established at the time of connection setup at the user-to-network interface (UNI). The network will not allow any new QoS contracts to be established if they will adversely affect its ability to meet existing contracts. In such cases, the application will not be able to get on the network until the network is fully capable of meeting the new contract.

When the QoS is negotiated with the network, there are performance guarantees that go along with it: maximum cell rate, available cell rate, cell transfer delay, and cell loss ratio. The network reserves the resources needed to meet the performance guarantees, and the user is required to honor the contract by not exceeding the negotiated parameters. Several methods are available to enforce the contract. Among them are traffic policing and traffic shaping.

Traffic policing is a management function performed by switches or routers on the ATM network. To police traffic, the switches or routers use a buffering technique referred to as a “leaky bucket.” This technique entails traffic flowing (leaking) out of the buffer (bucket) at a constant rate (the negotiated rate) regardless of how fast it flows into the buffer. If the traffic flows into the buffer too fast, the cells will be allowed onto the network only if enough capacity is available. If there is not enough capacity, the cells are discarded and must be retransmitted by the sending device.

Traffic shaping is a management function performed at the UNI of the ATM network. It ensures that traffic matches the contract negotiated between the user and network during connection setup. Traffic shaping helps guard against cell loss in the network. If too many cells are sent at once, cell discards can result, which will disrupt time-sensitive applications. Because traffic shaping regulates the data transfer rate by evenly spacing the cells, discards are prevented.

Cell Structure

Voice, video, and data traffic is usually composed of bytes, packets, or frames. These larger payloads are chopped up into smaller fixed-length cells by the customer’s router or the carrier’s ATM switch. ATM cells are fixed at 53 octets[3] and consist of a 5-octet header and 48-octet payload (Figure A-1).

[3] This odd cell size was the result of a compromise among international standards bodies. The United States wanted the cell’s data payload size to be 64 bytes, and Europe wanted a data payload of 32 bytes. The compromise was simply to average the two, which equals 48 bytes. The cell’s header required 5 bytes, providing an overall cell size of 53 bytes.

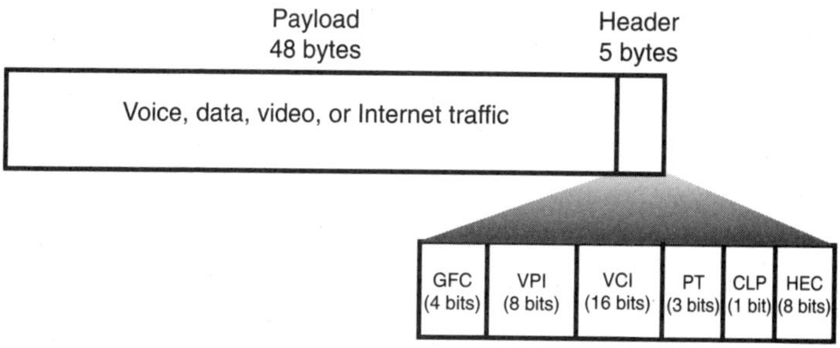

Figure A-1 ATM cell structure.

The cell header contains the information needed to route the information field through the ATM network. The header has several fields, which add up to 40 bits (5 bytes) as follows:

- *Generic Flow Control (GFC)* This 4-bit field has only local significance; it enables customer premises equipment at the UNI to regulate the flow of traffic for different grades of service.
- *Addressing* An 8-bit Virtual Path Identifier (VPI) is used in conjunction with the VCI to identify the next destination of a cell as it passes through a series of switches on the ATM network. The Virtual Channel Identifier (VCI) is a 16-bit field used to identify the virtual channel on a particular virtual path.
- *Payload Type (PT)* This 3-bit field is used to indicate whether the cell contains user information or connection management information. This field also provides for network congestion notification.
- *Cell Loss Priority (CLP)* This 1-bit field, when set to a 1, indicates that the cell may be discarded in the event of congestion. When set to 0, it indicates that the cell is of higher priority and should not be discarded.
- *Header Error Check (HEC)* This 8-bit field is used by the physical layer for detection and correction of bit errors in the cell header. The header carries its own error check to validate the VPIs and VCIs and prevent delivery of cells

to the wrong UNI at the remote end. Cells received with header errors are discarded. Higher-layer protocols are responsible for initiating lost cell recovery procedures.

Initially, there was concern about the high overhead of cell relay, with its ratio of 5 header octets to 48 data octets. However, with innovations in Wave Division Multiplexing (WDM) to increase fiber's already high capacity, ATM's overhead is no longer a serious issue. Instead, the focus is on ATM's unique ability to provide a QoS in support of all applications on the network.

Virtual Circuits

ATM virtual circuits (VCs) can be bidirectional or unidirectional, meaning that each VC can be configured for one-way or two-way operation. The VCs can be configured as point-to-point (i.e., permanent virtual circuit), switched, or multipoint. They also can be symmetric or asymmetric in nature. In other words, each bidirectional virtual channel can be configured for symmetric operation (same speed in both directions) or asymmetric operation (different speeds in each direction).

A VC has two components: a virtual path and a virtual channel. In this simplified view of an ATM network (Figure A-2), the customer has two locations connected together by a virtual path, which contains a bundle of virtual channels. Each of the three virtual channels is assigned to a particular end system, such as a PBX, server, or router. The individual connections between the end devices at each location are identified by ATM addresses, which consist of a Virtual Channel Identifier and the Virtual Path Identifier (VCI/VPI).

In this example, an integrated access device (IAD) is used to consolidate the VCs and to deliver them to the ATM switch via a dedicated line. VC-1 provides LAN users with access to a mainframe. VC-2 provides trunking between the PBXs. VC-3 provides LAN users with access to a remote server.

In a large network, there may be hundreds of virtual paths. The ATM standards allow up to 65,000 virtual channels to

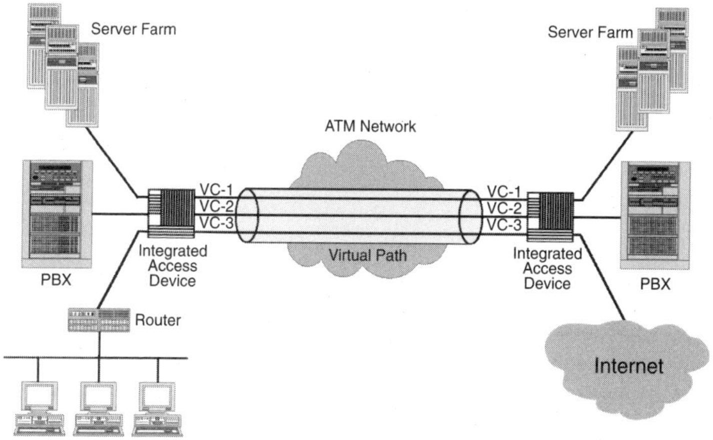

Figure A-2 A simplified view of virtual circuits through an ATM network.

share the same virtual path. This scheme simplifies network management and network recovery. When a virtual path must be reconfigured to bypass a failed port on an ATM switch, for example, all its associated virtual connections go with it, eliminating the need to reconfigure each VC individually.

Inverse Multiplexing over ATM

Today even midsized companies with multiple traffic types and three or more distributed locations can benefit from ATM's sustained throughput, low latency, and adept traffic handling via appropriate QoS mechanisms. The availability of ATM-based inverse multiplexers and $N \times$ T1 access makes ATM suitable for mainstream use, particularly for companies who appreciate the benefits of ATM but have been locked out of the service because of its high cost of implementation.

In the past, T3 links were the minimum bandwidth required to access ATM networks, making the cost prohibitive for the vast majority of companies. Inverse Multiplexing over ATM (IMA) solves the bandwidth gap problem. With IMA, companies can aggregate multiple DS1 circuits to achieve just the right amount of bandwidth they need for their applications and pay for only that amount on an $N \times$ T1 basis. The advantage of IMA is that such companies can scale up to the bandwidth they need, starting with a single T1, and then add links as more bandwidth is justified.

For example, when the bandwidth of four T1s is bonded by the IMA device, the virtual connection through the service provider's network is provisioned at 6 Mbps. When the bandwidth of eight T1s is bonded by the IMA device, the virtual connection through the service provider's network is provisioned at 12 Mbps. Regardless of the number of T1 access links in place, the IMA device bonds them together, combining the bandwidth into a fatter logical pipe that can support mixed-media applications, including IP traffic, running over interconnected LANs (Figure A-3).

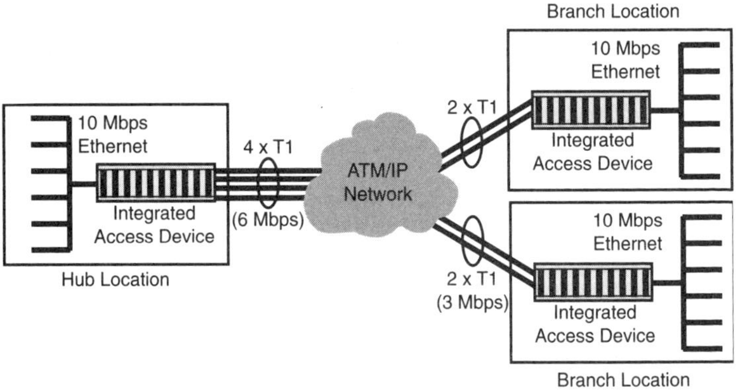

Figure A-3 Inverse Multiplexing over ATM (IMA) allows the use of bonded multiple T1 access lines into and out of the service provider's ATM network instead of forcing companies to use more expensive T3 access lines at each location. This makes it cost-effective for midsized companies to take advantage of ATM services to support a variety of applications, including broadband IP traffic.

Summary

A solid base of standards now exists to allow equipment vendors, service providers, and end users to implement a wide range of applications via ATM. The standards will continue to evolve as new applications emerge. The rapid growth of the Internet is one area where ATM can have a significant impact. With the Internet forced to handle a growing number of multimedia applications—telephony, videoconferencing, faxes, and collaborative computing, to name a few—congestion and delays are becoming ever more frequent and prolonged. ATM backbones will play a key role in alleviating these conditions, enabling next-generation networks to be used to their full potential.

See also

Frame Relay

Integrated Access Devices

Inverse Multiplexers

B

BANDWIDTH MANAGEMENT SYSTEMS

The stringent performance demands of multimedia and priority traffic on corporate intranets and virtual private networks (VPNs) has led to development of a new class of tools, usually referred to as bandwidth managers or packet shapers, that permit allocation of bandwidth according to traffic type. These tools, combined with load balancing servers and network caching strategies, reduce congestion on IP networks, ensure delivery of priority messages, and support real-time traffic, including voice.

To accomplish all this, bandwidth managers employ a variety of traffic control techniques—including TCP rate control, queuing, and policy definition—to ensure that essential traffic makes it to its destination in a timely manner, especially during periods of network congestion. The need for these capabilities depends on several factors, such as

- The variety of traffic types that are run over an intranet or TCP/IP WAN service.
- Whether certain traffic is of more value than other traffic.
- If any traffic has delay characteristics that require priority handling.

If an organization is running traffic that is all of equal value, these capabilities will not be needed. But if the organization finds itself running more multimedia applications on its intranet—such as streaming audio or video, IP telephony, and collaborative computing—then adding such capabilities usually will improve network performance more economically than upgrading the network with more bandwidth.

Operation

Bandwidth managers are positioned at the edge of corporate nets, deciding which traffic gets preferential treatment before reaching the wide area network (WAN) router. They are usually deployed between the WAN and the local area network (LAN) but may sit on a LAN that is connected to a WAN-linked router. Policies define how various types of traffic are handled.

Bandwidth managers identify and manipulate traffic classes by looking at the type-of-service (TOS) bit in the IP header or the IP address, the TCP or UDP port number, the DNS, the application, or the URL. Traffic shapers, using the TOS bit, can identify traffic without adding to the IP header. This eliminates the need to change routers and switches. To prioritize traffic, they use a variety of different strategies, including queuing, changing the size of the TCP/IP window (TCP rate control), or a combination of both. Some also can handle protocols other than IP—such as IPX—and some offer automatic traffic identification.

Queuing techniques may be used separately or with TC/IP rate control. Queuing types include priority, weighted, and class-based. Priority queuing sets queues for high- and low-priority traffic and empties high-priority queues first. Cisco's weighted fair queuing (WFQ), for example, assigns traffic to priority queues and also assigns an appropriate amount of bandwidth. Its class-based queuing (CBQ) guarantees a transmission rate to a queue, and other queues can borrow from unused bandwidth to support their traffic.

While queuing technology helps prioritize traffic, it also has some drawbacks. For example, it works only on outbound traffic, and packets delayed in queues either get dropped or time out, requiring retransmissions that cause a significant reduction in network efficiency. TCP rate control takes a more efficient and precise approach to bandwidth allocation: It applies TCP rate-based flow-control policies to both individual traffic flows and classes of flows. This results in predictable service level control for true IP quality of service (QoS). Since it does not rely on a queue and thus provides bidirectional bandwidth management for both inbound and outbound traffic, TCP rate control increases network efficiency by avoiding retransmissions and packet loss.

Types of Solutions

Bandwidth management products come in the form of hardware- and software-only solutions or a combination of hardware and software. Hardware-based bandwidth managers tend to offer the best performance because they rely on application-specific integrated circuits (ASICs) and dedicated memory to implement bandwidth management. They can be difficult to upgrade, however. Software systems provide considerable flexibility and are easy to upgrade, but with a performance penalty. Combined hardware-software systems have a performance and feature set somewhere between the two. Bandwidth management software and devices are also starting to be integrated with other network management components such as firewalls and integrated access devices.

Policy-based, application-adaptive bandwidth management hardware solutions control the use of WAN bandwidth and deliver end-to-end quality of service on a per-application-flow basis. Among the techniques used for ensuring the performance of business-critical applications are automatic traffic discovery and classification, in-depth performance analysis, rate-based traffic control, and application service-level management. A TCP rate control mechanism lets network administrators go beyond simple traffic priorities to set kilobit-per-second partitions for each classified traffic flow. It also can show a graphical representation of the "top ten" types of traffic running through a WAN.

Software bandwidth management solutions typically include a policy server, which communicates with proxy agents that are distributed throughout the network. The proxy agents place the QoS policies into the installed base of routers and other edge devices via appropriate drivers. These device drivers act as translators that interpret the abstract policies into actual commands that can be acted on by equipment throughout the network, such as routers, access points, and packet switches. A database is used to store the policy information.

Problems of sluggish performance on IP networks are addressed by designating performance classes tailored to the needs of specific applications. For example, the network administrator might want to guarantee a minimum bandwidth service class for mission-critical applications while designating a low-latency performance class for delay-sensitive applications such as voice over IP (VoIP). At the same time, the network administrator can set a standard class for applications that can tolerate some delay, such as e-mail and intranet access. Under this classification scheme, the key traffic in the minimum bandwidth class will get network service even during times of congestion.

The software automatically discovers and classifies network devices and lays out a network's topology. Network managers can point and click to check the settings on different devices in the network. The bandwidth management system uses these data to construct rules that assign QoS priorities to specific applications. The software then translates those rules into the actual configuration commands that network devices can interpret to appropriately support the applications. The rule-based interface allows network administrators to set up dedicated classes on an IP WAN for different applications and to configure devices individually or as a group in one operation.

Devices are classified as edge, gateway, or core and then grouped by geographic location or other parameters. These capabilities allow network administrators to create generic rules that apply to all members that perform the same function and then drill down for specific site requirements. For example, access routers at the edge of the network are responsible for classifying traffic, controlling access, and enforcing queuing mechanisms. Gateway and core routers can then be relieved of having to set the priority bits themselves and need only to enforce the queuing policies.

Load Balancing

Another way to improve handling of IP traffic on corporate intranets is to make use of load-balancing systems. While bandwidth management tools enable allocation of portions of available bandwidth to different users, load balancers operate on the server side, routing traffic to the best server available to handle a job. In a load-balanced network, incoming traffic is distributed among replicated servers, thus permitting server clusters to share the processing load, provide fail-back capability, and speed response time for users.

Advanced load-balancing policies can be defined that reflect the capabilities of individual servers on the network. Policies are defined that redirect traffic based on the capabilities of the servers. For example, all video can be redirected to the video server, all Web traffic to the Web server, and all employees in marketing to the marketing server. The policy manager continuously adjusts both the flow and priority of applications through the network and the distribution of those applications to servers. Traffic can be balanced between available servers using algorithms such as

- *Round-robin*, whereby each server will be treated with equal priority.
- *Weighted round-robin*, whereby each server can be given an individual weight or priority based on its ability to deliver specific applications.

- *Maintenance rerouting,* whereby traffic can be rerouted to another server when originally targeted server becomes unavailable.

Network Caching

Network caching offers an effective and economical way to relieve some of the demand for bandwidth while improving response time. Corporations with their own intranets maintain an active cache of the most-often-visited Web sites so that when these pages are requested again, the download occurs from the locally maintained cache server instead of the request being routed over the backbone network to the actual server. The result is a faster download speed with minimal use of network bandwidth.

Caches can reside at various points in the network. For enterprises, caches can be deployed on servers throughout campus networks and in remote and branch offices. Within enterprise networks, caches are on the way to becoming as ubiquitous as IP routers. Just about every large company now depends on Web caches to keep its intranets running smoothly.

There are two types of cache techniques: passive and active. With the former, the cache waits until a user requests the object again and then sends a refresh request to the server. If the object has not changed, the cached object is served to the requesting user. If the object has changed, the cache retrieves the new object and serves it to the requesting user. However, this approach forces the end user to wait for the refresh request, which can take as long as the object retrieval itself. It also consumes bandwidth for unnecessary refresh requests.

With active caching, the cache performs the refresh request before the next user request—if the object is likely to be requested again and the object is likely to have changed on the server. This automatic and selective approach keeps the cache up to date so that the next request can be served immediately. Network traffic does not increase because an object in cache is refreshed only if it is likely to be requested again and only if there is a statistically high probability that it has changed on the source server.

For example, the Web page of a major broadcast network might contain a logo object that never changes, while the "Breaking News" object changes often. If this page is popular among corporate users, the "Breaking News" object will be refreshed prior to the next user's request. Because only content that is likely to change is refreshed, users are served with the most updated information without putting unnecessary traffic on the network.

By contrast, previous generations of cache technology do not accommodate the individual nature of cached objects. They rely on global settings, which treat all objects equally, thereby severely limiting the hit ratio. Since the passive cache requires frequent, redundant refresh traffic, it induces significant response-time delays.

Active caches can achieve hit ratios of up to 75 percent, meaning a greater percentage of user requests can be served by the cache. If the requested data are in the cache and are up to date, the cache can serve them to the user immediately on request. If not, the user must wait while the cache retrieves the requested data from the network. Passive caches, on the other hand, typically achieve hit rates of only 30 percent, forcing users to go to the network 2.5 times more often to get the information they need.

Some kinds of objects in a Web page cannot be cached and are individually marked by their Web server as such. One object of this type is a database-driven object, such as a real-time stock quote. While this particular object is not cacheable, the rest of the objects in the page usually are. For example, a Web page that delivers stock quotes may contain 30 other objects; only one of those objects—the stock ticker—may not be cacheable. If all the remaining objects can be cached, a significant performance benefit will result.

Summary

Business-critical applications often struggle to compete with less important traffic for the finite amount of bandwidth available on corporate networks. The result is that end-user response times can fall below acceptable levels, which decreases productivity. If this condition persists, business processes and revenues may become negatively impacted. This has led to the availability of bandwidth management tools from a variety of vendors, which offer the means to set QoS policies once and deploy them selectively or globally throughout the network. Ideally, the selected solution will integrate seamlessly into the existing network without requiring new protocols, standards, topologies, or hardware changes. Along with bandwidth management, users should give consideration to implementing load-balancing and network caching solutions. These solutions can very economically provide dramatic improvements in response time and free up scarce bandwidth to keep vital applications running smoothly.

See also

Content Delivery Networks

Quality of Service

BLUETOOTH

Bluetooth is an omnidirectional wireless technology that provides limited-range voice and data transmission over the unlicensed 2.4-GHz frequency band, allowing connections with a wide variety of fixed and portable devices that normally would have to be cabled together. Up to eight devices—one master and seven slaves—can communicate with one another in a so-called piconet at distances of up to 30 feet. Table B-1 summarizes the performance characteristics of Bluetooth products that operate at 1 Mbps in the 2.4-GHz range.

Applications

Among the many things users can do with Bluetooth is swap data and synchronize files merely by having the devices come within range of one another. Images captured with a digital camera, for example, can be dropped off at a PC for editing or a color printer for output on photo-quality paper—all without having to connect cables, load files, open applications, or click buttons.

The technology is a combination of circuit switching and packet switching, making it suitable for voice as well as data. Instead of fumbling with a cell phone while driving, for example, the user can wear a lightweight headset to answer a call and engage in a conversation even if the phone is tucked away in a briefcase or purse.

While useful in minimizing the need for cables, wireless LANs are not intended for interconnecting the range of mobile devices people carry around everyday between home and office. For this, Bluetooth is needed. And in the office, a Bluetooth portable device can be in motion while connected to the LAN access point as long as the user stays within the 30-foot range.

TABLE B-1 Performance Characteristics of Bluetooth Products

Feature/Function	Performance
Connection type	Spread spectrum (frequency hopping)
Spectrum	2.4-GHz ISM (industrial, scientific, and medical) band
Transmission power	1 milliwatt (mW)
Aggregate data rate	1 Mbps using frequency hopping
Range	Up to 30 feet (9 meters)
Supported stations	Up to eight devices per piconet
Voice channels	Up to three synchronous channels
Data security	For authentication, a 128-bit key; for encryption, the key size is configurable between 8 and 128 bits
Addressing	Each device has a 48-bit MAC address that is used to establish a connection with another device

Bluetooth can be combined with other technologies to offer wholly new capabilities, such as automatically lowering the ring volume of cell phones or shutting them off as users enter quiet zones such as churches, restaurants, theaters, and classrooms. On leaving the quiet zone, the cell phones are returned to their original settings.

Topology

The devices within a piconet play one of two roles: that of master or slave. The master is the device in a piconet whose clock and hopping sequence are used to synchronize all other devices (i.e., slaves) in the piconet. The unit that carries out the paging procedure and establishes a connection is by default the master of the connection. The slaves are the units within a piconet that are synchronized to the master via its clock and hopping sequence.

The Bluetooth topology is best described as a multiple-piconet structure. Since Bluetooth supports both point-to-point and point-to-multipoint connections, several piconets can be established and linked together in a topology called a "scatternet" whenever the need arises (Figure B-1).

Piconets are uncoordinated, with frequency hopping occurring independently. Several piconets can be established and linked together ad hoc, where each piconet is identified by a different frequency-hopping sequence. All users participating on the same piconet are synchronized to this hopping sequence. Although synchronization of different piconets is not permitted in the unlicensed ISM band, Bluetooth units may participate in different piconets through Time Division Multiplexing (TDM). This enables a unit to sequentially participate in different piconets by being active in only one piconet at a time.

With its service discovery protocol, Bluetooth enables a much broader vision of networking, including the creation of personal area networks (PANs), where all the devices in a person's life can communicate and work together. Technical

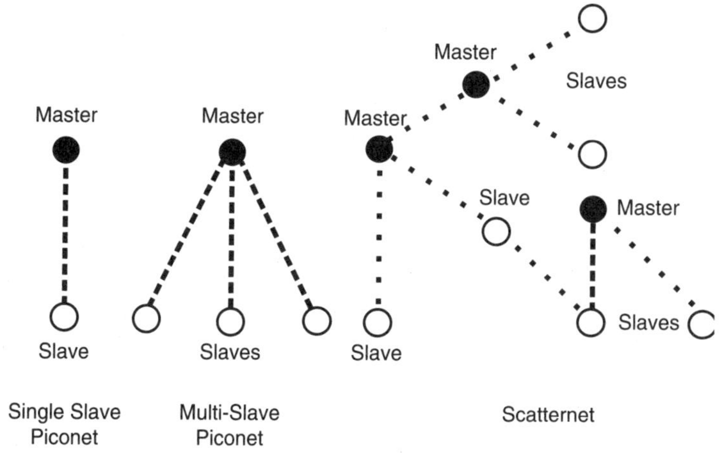

Figure B-1 Possible topologies of networked Bluetooth devices, where each is either a master or slave.

safeguards ensure that a cluster of Bluetooth devices in public places, such as an airport lounge or train terminal, would not suddenly start talking to one another.

Technology

Two types of links have been defined for Bluetooth in support of voice and data applications: an asynchronous connectionless (ACL) link and a synchronous connection-oriented (SCO) link. ACL links support data traffic on a best-effort basis. The information carried can be user data or control data. SCO links support real-time voice and multimedia traffic using reserved bandwidth. Both data and voice are carried in the form of packets, and Bluetooth devices can support active ACL and SCO links at the same time.

ACL links support symmetric or asymmetric, packet-switched, point-to-multipoint connections, which are typically used for data. For symmetric connections, the maximum data rate is 433.9 kbps in both directions, send and receive. For asymmetric connections, the maximum data rate is 723.2 kbps in one direction and 57.6 kbps in the reverse direction. If errors are detected at the receiving device, a notification is sent in the header of the return packet so that only lost or corrupt packets need to be retransmitted.

SCO links provide symmetric, circuit-switched, point-to-point connections, which are typically used for voice. Three synchronous channels of 64 kbps each are available for voice. The channels are derived through the use of either Pulse Code Modulation (PCM) or Continuous Variable Slope Delta (CVSD) modulation. PCM is the standard for encoding speech in analog form into the digital format of ones and zeros. CVSD is another standard for analog-to-digital encoding but offers more immunity to interference and therefore is better suited than PCM for voice communication over a wireless link. Bluetooth supports both PCM and CVSD; the appropriate voice-coding scheme is selected after negotiations between the link managers of each Bluetooth device before the call takes place.

Voice and data are sent as packets. Communication is handled with Time Division Duplexing (TDD), which divides the channel into time slots, each 625 microseconds (μs) in length. The time slots are numbered according to the clock of the piconet master. In the time slots, master and slave can transmit packets. In the TDD scheme, master and slave alternatively transmit (Figure B-2). The master starts its transmission in even-numbered time slots only, and the slave starts its transmission in odd-numbered time slots only. The start of the packet is aligned with the slot start. Packets transmitted by the master or the slave may extend over as many as five time slots.

With TDD, bandwidth can be allocated on an as-needed basis, changing the makeup of the traffic flow as demand warrants. For example, if the user wants to download a large data file, as much bandwidth as is needed will be allocated to the transfer. Then, at the next moment, if a file is being

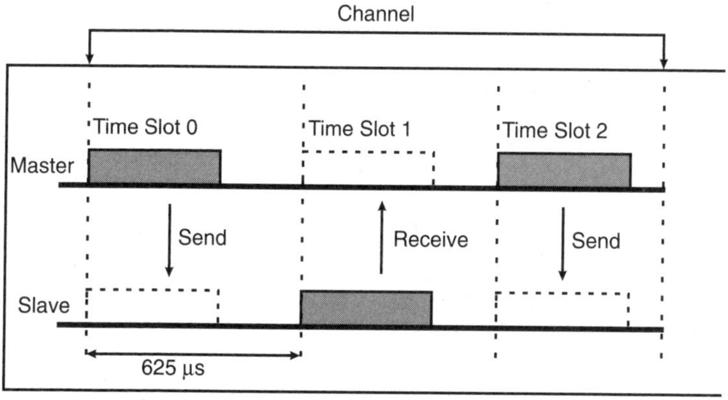

Figure B-2 With the TDD scheme used in Bluetooth, packets are sent over time slots of 625 microseconds (µs) in length between the master and slave units within a piconet.

uploaded, that same amount of bandwidth can be allocated to that transfer.

No matter what the application—voice or data—making connections between Bluetooth devices is as easy as powering them up. In fact, one advantage of Bluetooth is that it does not need to be set up—it is always on, running in the background, and looking for other devices that it can communicate with.

When Bluetooth devices come within range of one another, they engage in a service discovery procedure, which entails the exchange of messages to become aware of each another's service and feature capabilities. Having located available services within the vicinity, the user may select from any of them. After that, a connection between two or more Bluetooth devices can be established.

The radio link itself is very robust, using frequency-hopping spread spectrum technology to overcome interference and fading. Spread spectrum is a digital coding technique in which the signal is taken apart or "spread" so that it sounds more like noise as it is sent through the air. With the addition of frequency hopping—having the signals skip from one frequency to another—wireless transmissions are made even more secure. Bluetooth specifies a rate of 1600 hops per second among 79 frequencies. Since only the sender and receiver know the hopping sequence for coding and decoding the signal, eavesdropping is virtually impossible. For enhanced security, Bluetooth also supports device authentication and encryption.

Other frequency-hopping transmitters in the vicinity will be using different hopping patterns and much slower hop rates than Bluetooth devices. Although the chance of Bluetooth devices interfering with non-Bluetooth devices that share the same 2.4-GHz band is minimal, should non-Bluetooth transmitters and Bluetooth transmitters coincidentally attempt to use the same frequency at the same moment, the data packet transmitted by one or both devices will become garbled in the collision, and a retransmission of the affected data packets will be required. A new data packet will be sent again on the next hopping cycle of each transmitter. Voice packets, because of their sensitivity to delay, are never retransmitted.

Bluetooth complements infrared's point-and-shoot ease of use with omnidirectional signaling, longer-distance communications, and capacity to penetrate walls. For some users, having both Bluetooth and infrared will provide the optimal short-range wireless solution. For others, the choice of adding Bluetooth or infrared will be based on the applications and intended usage.

Summary

Communicator platforms of the future will combine a number of technologies and features in one device, including mobile Internet browsing, messaging, imaging, location-based applications and services, mobile telephony, personal information management, and enterprise applications. Bluetooth will be a key component of these platforms. Since Bluetooth radio transceivers operate in the globally available ISM (industrial, scientific, and medical) radio band of 2.4 GHz, products do not require an operator license from a regulatory agency, such as the Federal Communications Commission (FCC) in the United States. The use of a generally available frequency band means that Bluetooth-enabled devices can be used virtually anywhere in the world and link up with one another for ad hoc networking when they come within range.

See also

Wireless LANs

BRIDGES

Bridges are used to extend or interconnect LAN segments. At one level, they are used to create an extended network that greatly expands the number of devices and services available to each user. At another level, bridges can be used for segmenting LANs into smaller subnets to improve performance, control access, and facilitate fault isolation and testing without impacting the overall user population.

The bridge does this by monitoring all traffic on the subnets that it links. It reads both the source and destination addresses of all the packets sent through it. If the bridge encounters a source address that is not already contained in its address table, it assumes that a new device has been added to the local network. The bridge then adds the new address to its table.

In examining all packets for their source and destination addresses, bridges build a table containing all local addresses. The table is updated as new packets are encountered and as addresses that have not been used for a specified period of time are deleted. This self-learning capability permits bridges to keep up with changes on the network without requiring that their tables be updated manually.

The bridge isolates traffic by examining the destination address of each packet. If the destination address matches any of the source addresses in its table, the packet is not allowed to pass over the bridge because the traffic is local. If the destination address does not match any of the source addresses in the table, the packet is discarded onto an adjacent network. This filtering process is repeated at each bridge on the internetwork until the packet eventually reaches its destination. Not only does this process prevent unnecessary traffic from leaking onto the internetwork, it acts as a simple security mechanism that can screen unauthorized packets from accessing various corporate resources.

Bridges also can be used to interconnect LANs that use different media, such as twisted pair, coaxial, and fiberoptic cabling and various types of wireless links. In office environments that use wireless communications technologies such as spread spectrum and infrared, bridges can function as an access point to wired LANs. On the WAN, bridges even switch traffic to a secondary port if the primary port fails. For example, a full-time wireless bridging system can establish a modem connection on the public network if the primary wire or wireless link is lost because of environmental interference.

In reference to the OSI model, a bridge operates at Layer 2; specifically, it operates at the Media Access Control (MAC) sublayer of the Data Link layer. It routes by means of the Logical Link Control (LLC), the upper sublayer of the Data Link layer (Figure B-3).

Because the bridge connects LANs at a relatively low level, throughput often exceeds 30,000 packets per second (pps). Multiprotocol routers and gateways, which provide LAN interconnection over the WAN, operate at higher levels of the OSI model and provide more functionality. In performing more protocol conversions and delivering more functionality, routers and gateways are generally more processing-intensive and, consequently, slower than bridges.

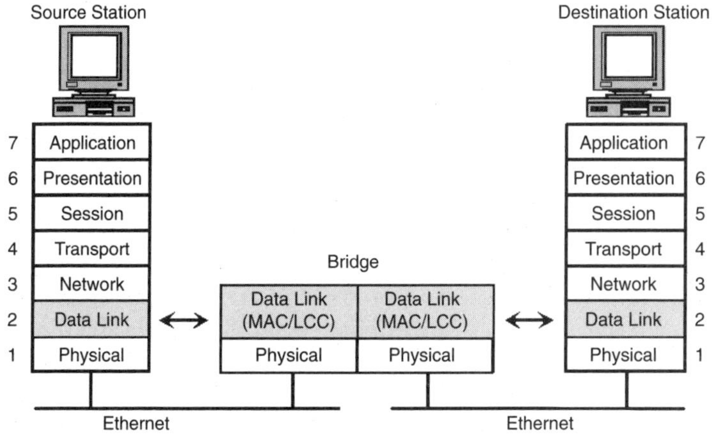

Figure B-3 Bridge functionality in reference to the OSI model.

See also

Gateways

Repeaters

Routers

BUILDING LOCAL EXCHANGE CARRIERS

A building local exchange carrier (BLEC) designs, constructs, deploys, and manages high-speed broadband networks inside commercial office buildings to meet the communications needs of tenants, which typically are small and midsized businesses. Among the services the BLEC may offer are high-speed Internet access, enhanced conference calling services, Web hosting, managed network security, remote access, and information technology services. The BLEC has the capability to provide services using wireless, optical, and copper-based technologies. The BLEC may seek to enhance existing customer relationships by also offering similar broadband services to customers' branch offices and other businesses located in buildings in which it does not have an installed network.

The in-building broadband data network transmits data to and from each customer at a variety of speeds. Using information on the demand characteristics of customers in a particular building, the BLEC installs broadband data equipment by connecting each building network to a central facility in each metropolitan area, usually over lines leased from other carriers. At this metropolitan hub, traffic is aggregated and distributed to the appropriate locations.

The BLEC partners with large-scale real estate owners and secures the right to install its broadband data network inside office buildings that meet its criteria. It generally targets buildings with more than 100,000 rentable square feet and 10 or more tenants. The BLEC pays real estate owners either a modest portion of the gross revenue generated from tenants in their buildings or a fixed rental fee. In having their services available in the building, BLECs believe they assist the real estate owner with their tenant leasing and retention efforts. A typical lease or license agreement with a real estate owner is for a term of 10 or more years.

Once an agreement is reached with the building owner, the BLEC's sales and marketing efforts focus primarily on tenants. The BLEC develops building-specific marketing and promotional techniques, such as lobby events and advertising in landlord newsletters. In many cases, BLECs work with building managers to demonstrate their services to tenants and prospective tenants.

The BLEC provides affordable services and products with a range of choices for tenants. In addition to pricing based on the number of desktops connected to the network, the BLEC may offer tenants data communication service bundles that combine a broad range of high-speed connectivity, business communication applications, and professional network management services, including Internet connections, multiple customer-branded e-mail accounts, Web hosting, remote access, a desktop business portal with a customer-specific design, and professional information technology services. This approach permits customers to obtain the services they need for a low monthly cost and add on as the BLEC demonstrates the ability of its services to enhance productivity.

The BLEC supports its services through a national customer care center and a national operations control center, which are staffed 24 × 7 and continuously monitor the network to detect disruptions in service, remotely resolve problems, configure networks, and compile data on customer service levels. In addition, field operations personnel augment the customer care center by providing on-site support.

Network Architecture

Inside the building, the BLEC installs a broadband data infrastructure that typically runs from the basement of the building to the top floor inside the building's vertical utility shaft. This broadband data infrastructure is designed to carry data and voice traffic for all the building's tenants for the foreseeable future. Tenants receive services through a data cable from their LAN to the BLEC's infrastructure in the vertical utility shaft.

Inside the building, usually in the basement, the BLEC establishes a point of presence (POP). In each building POP, the BLEC connects the broadband data cable to routers and other equipment that enables the transmission of data and video traffic and the aggregation and dissemination of traffic to and from those cables. The BLEC typically negotiates with the building owner for the right to use a small amount of space in the basement to establish the POP.

Within each market the BLEC serves, it has a metropolitan POP where traffic is aggregated and distributed to and from all the building networks via broadband data circuits. These broadband data lines typically are leased from carriers that have previously installed local transport capacity in that market. The POP contains all the equipment necessary to provide services in the metropolitan area and may include network computer servers and traffic routers. Through a collocation arrangement, each metropolitan POP is connected to multiple major service providers that provide Internet connectivity.

Summary

Ever since passage of the Telecommunications Act of 1996, which was aimed at deregulating the industry, the FCC has been playing referee. The FCC has mostly sided with the competitors, forcing the Bell companies to lease key components of their networks to their rivals. FCC officials see the battle over building access in the same context. When FCC staff took up the issue in late 2000, it considered imposing rules that would force landlords to provide access to the rival companies in exchange for fair compensation. But some commissioners voiced concerns that the FCC would be on shaky legal ground if it sought to regulate landlords. Members of Congress, under intense lobbying from building owners, have since echoed this concern. Although FCC staff remains convinced that the FCC has the authority to regulate on this issue as part of its mandate to foster telecommunications competition, a negotiated resolution between private parties has become the more palatable approach.

See also

Competitive Local Exchange Carriers

Incumbent Local Exchange Carriers

C

CABLE MODEMS

One of the services most in demand from cable companies is broadband Internet access. To deliver data services over a cable network, one television channel in the 50- to 750-MHz range is typically allocated for downstream traffic to homes and businesses, while another channel in the 5- to 42-MHz band is used to carry upstream signals. A head-end cable modem termination system (CMTS) communicates through these channels with cable modems to create a virtual local area network (VLAN) connection. Most cable modems are external devices that connect to a personal computer (PC) through a standard 10Base-T Ethernet card or Universal Serial Bus (USB) connection, although internal PCI modem cards are also available.

A single downstream 6-MHz television channel supports up to 27 Mbps of downstream data throughput from the cable head-end using 64 QAM (quadrature amplitude modulation) transmission technology. This downstream bandwidth is shared by all subscribers on the cable segment, giving users between 1 and 3 Mbps of bandwidth at any given time, assuming that not all will access the bandwidth at the same time. The speed can be boosted to 36 Mbps using

256 QAM. Upstream channels may deliver 500 kbps to 10 Mbps, depending on the amount of spectrum allocated for the service. This upstream and downstream bandwidth is shared by the active data subscribers connected to a given cable network segment, which typically range from 250 to 500 on a modern HFC network.

Most cable systems are not yet equipped for two-way capability. To get broadband Internet access, subscribers must still use a dial-up modem and telephone line for the upstream data path. Simple requests for Web pages are issued over the low-speed modem connection, and the rich content is returned over the high-speed cable connection (Figure C-1). Performance suffers when users want to send large files upstream, in which case the data are sent at the speed of the modem—no more than 56 kbps—over the dial-up line.

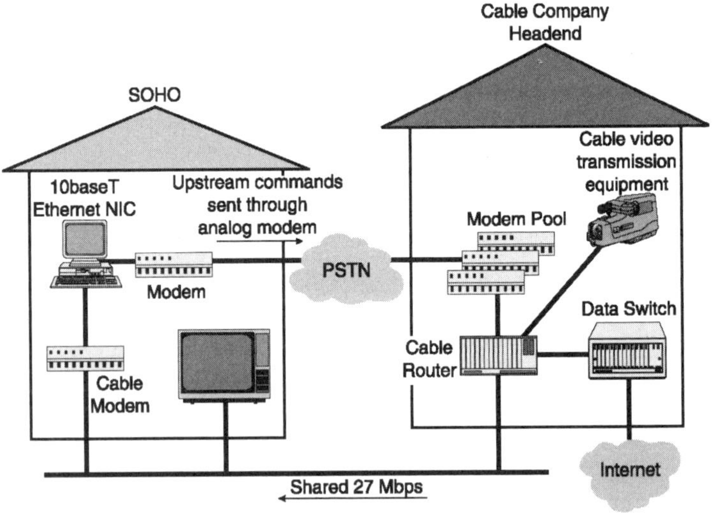

Figure C-1 One-way cable systems require Internet subscribers to use a separate phone line and modem for the upstream path. Sometimes the cable modem contains an integral dial-up modem.

The key standard for cable modems is the Data over Cable Service Interface Specification (DOCSIS) developed by CableLabs and approved by the International Telecommunication Union (ITU) in 1998. DOCSIS specifies downstream traffic transfer rates between 27 and 36 Mbps over a radiofrequency (RF) path in the 50- to 750+-MHz range and upstream traffic transfer rates between 320 kbps and 10 Mbps over an RF path of between 5 and 42 MHz. It also defines interface standards for cable modems and supporting equipment. With certification from CableLabs, manufacturers can produce cable modems for retail sale, so consumers no longer have to depend on leased cable modems from their cable providers.

Cable operators are typically charging between $20 and $60 per month for the broadband Internet service, which includes cable modem rental and unlimited Internet access. To qualify for the lower price, the subscriber usually must choose a premium package of entertainment services. For the amount of bandwidth, even the higher price is cheaper than the Digital Subscriber Line (DSL) services offered by telephone companies.

Cable versus DSL

When facing competition from cable operators, DSL service providers like to point out that cable is a shared service. Actual performance of the Internet connection deteriorates as new customers are brought online to share the bandwidth. This point is exaggerated because cable provides much greater bandwidth than DSL, and most cable operators limit the number of subscribers on a segment. In any case, the same argument can be applied to DSL service, since the fiber on the network side of the DSL Access Multiplexer (DSLAM) is routinely overprovisioned and becomes a bottleneck during peak demand.

Cable operators point out that DSL has distance limitations from customer premises to the DSLAM—the farther away the customer is from the DSLAM, the less bandwidth is available. Since online prequalification is only 70 percent accurate at best, customers really do not know what speed their DSL service will operate at until after installation. Furthermore, since DSL is provisioned over an analog line, it will not work if load coils are attached to the line, if there is a bridge tap along the line, or if the ILEC has attached the line to pair-gain equipment.

Summary

To survive in the new competitive climate ushered in by the Telecommunications Act of 1996, cable companies are investing billions of dollars to upgrade their networks for full-duplex operation to support advanced services, such as local and long-distance voice services in competition with the telephone companies, as well as broadband Internet access, video-on-demand, and interactive television. Among the technology choices for upgrading CATV networks for advanced services are hybrid fiber/coax (HFC), fiber-in-the-loop (FITL) systems, and SONET rings.

See also

Cable Telephony

Digital Subscriber Line Technologies

Hybrid Fiber/Coax

CABLE TELEPHONY

Telephone service can be implemented over two-way cable networks in two ways: circuit switching and packet switching. In the former case, digitized voice signals are delivered in the traditional time-division mode of circuit switching over 6-MHz RF channels provisioned over HFC networks. Anywhere from a handful to a maximum of 500 to 2000 households share access to the service stream in any one coaxially connected service area. The trouble with circuit-switched voice over cable is that the platforms for implementing the service are proprietary.

Another way to implement telephone service over cable is via packet switching. Like the circuit-switched version of cable telephony, Internet Protocol (IP) uses shared 6-MHz channels. While IP over cable offers the advantage of standardization, the migration path to IP is so long that it often justifies the incremental cost of starting out on a proprietary system. Vendors of proprietary circuit-switched solutions support migration by offering gateways that connect the local "cablephone" host digital terminal at the head end with IP backbones. This allows long-haul traffic to be transferred via IP while retaining the circuit-switched mode for local connections within and outside the cable network.

Voice-Enabled Cable Modems

Although IP over cable has been possible for a long time, the relatively new wrinkle is the ability to support data and voice through the same cable modem. The modem, in turn, is connected to the same coaxial cable that delivers TV service. Previously, to put voice and data on the cable network required a modem and a separate voice gateway, which is more complex and expensive.

The voice-enabled cable modem, also known an integrated multimedia terminal adapter (MTA), has a telephone port and an integral four-port Ethernet hub. Data speeds on the cable can be 10 Mbps or higher, depending on the cable system and how many active users are on the same shared-cable subnet. The integrated MTA uses proprietary traffic prioritization techniques to ensure that there is enough bandwidth to prevent voice packets from getting delayed.

The MTA sits on the customer premises (Figure C-2), where it converts phone and PC traffic into IP-formatted packets for transport over the cable network. The MTA sends

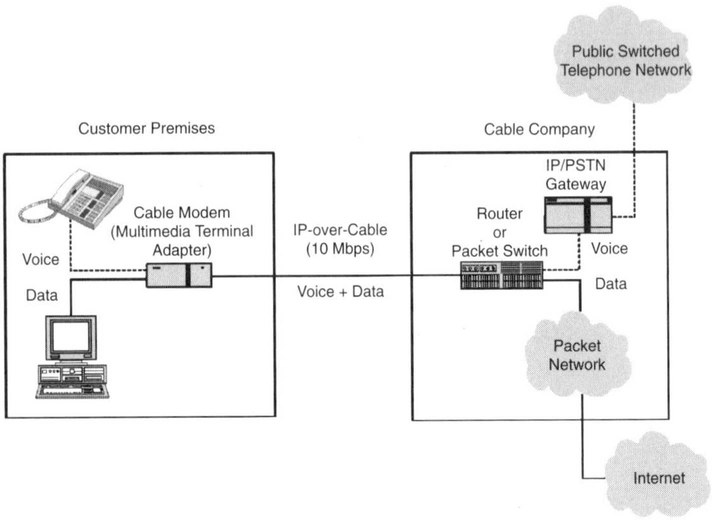

Figure C-2 The route of a telephone call over a cable network.

the packets to a router or packet switch on the cable network, where voice and data are sorted out. Data are routed to the cable operator's packet network, while voice is routed to an IP-PSTN gateway. The gateway decompresses the packets and returns them to analog form so that voice can be received at a phone on the public network or corporate PBX.

With many corporations devising ways of implementing voice over IP internally, the cable option allows them to extend it out to telecommuters. Cable modems already support data at the Ethernet speed of 10 Mbps, which gives telecommuters a viable way to retrieve information from the corporate local area network (LAN) via the Internet access connection. Although ISDN's Basic Rate Interface (BRI) is more flexible in terms of bandwidth allocation and call-handling features, it provides only 128 kbps for user data, which is not enough capacity for frequent LAN access, making it far more expensive than cable.

Convenience Features

There is virtually no difference in the convenience features of telephone service over cable and that over the PSTN. Depending on cable operator, the following convenience features may be offered:

- *Call forwarding* Forwards incoming calls to a designated number. "Busy" call forwarding forwards calls when the phone is in use. "No answer" forwards calls when no one picks up the phone. "Selective" forwards calls only from specified numbers. "Variable" forwards all calls.
- *Call forwarding select/preferred call forwarding* Forwards only the calls from people placed on an automated list.
- *Call return/last call return* Provides the phone number of the last person who called and gives the option to call him or her back.
- *Call waiting* Provides notification of an incoming call and allows click over to take the call if desired.
- *Custom ring/distinctive ring* Assigns two telephone numbers to one line, each with a unique ring type.
- *Directory assistance* Dial 411 for local and long distance phone numbers. Charges apply.
- *Priority ring/distinctive ring* Assigns specific numbers a special ring to know when and if someone specific is calling.
- *Remote access to call forwarding* Activates or deactivates call forwarding even when away from home.
- *Repeat dialing/continuous dialing* Automatically calls back a busy number for up to 30 minutes, if desired.
- *Speed dial/speed calling* Speed dial 8 stores up to eight dial numbers for single-digit dialing. Speed dial 30 stores up to 30 numbers for two-digit dialing.
- *Toll restriction/long-distance blocking* Ensures that only local calls can be made from the phone.

- *Three-way calling* Adds a third party to an existing phone call.
- *911* Provides instant access to emergency, police, fire, and medical services.

Privacy Features

Phone service over cable offers the following privacy features, which are comparable to those offered by conventional phone service:

- *Anonymous call rejection* Rejects calls from unfamiliar numbers. The caller will receive a message that the call has been blocked.
- *Call blocking* Rejects specific types of calls: specific area codes, long-distance calls, or operator-assisted calls.
- *Caller ID* Identifies incoming calls with names and phone numbers. (Customer-provided Caller ID equipment required.)
- *Caller ID with call waiting/call waiting caller ID* Indicates who is calling, even when already busy with another call. (Requires a specially equipped call waiting box.)
- *Caller ID blocking/caller ID "per call" blocking* When a call is being made, permanently blocks the display of the person's name on caller ID units.
- *Call screening/selective call blocking* Selects up to 12 numbers from which calls will not be accepted, and the caller will hear an automated message.
- *Call trace/customer-originated trace* Dials a special code to trace and handle annoying or harassing calls.
- *900/976 Blocking/custom code restriction* Account comes with 900 and 976 numbers automatically blocked. Subscribers may choose to have these numbers available if they prefer.

In addition to convenience and privacy features, cable operators offer voice mail. Subscribers also can add pager notification so that they will be informed of new voice mail messages.

Summary

Convergent services that combine entertainment programming, broadband Internet access, and telephone services present a challenge that could cost the voice-oriented ILECs and CLECs substantial market share in the not-too-distant future. The early market success of cable-telephone services is pressuring more cable operators to enter the market. For consumers, this convergence can mean lower monthly costs, simplified billing, and in some cases, higher satisfaction with the services they purchase.

See also

Digital Subscriber Line Technologies

Internet Telephony

LAN Telephony

Voice-Data Convergence

Voice over IP

CALL CENTERS

Call centers are specialized work environments that are equipped, staffed, and managed to handle a large volume of inbound or outbound calls. An inbound call center specializes in taking calls via a toll-free number to take sales orders or provide customer assistance. An outbound call center is staffed with salespeople who make calls to sell a product or service. Of course, the distinction between inbound and outbound call centers is arbitrary; the same call center can do both. On the inbound side, calls can come in to place orders for computers, for example. On the outbound side, calls can be placed to customers to confirm hardware and software specifications, notify customers of the shipping date, and determine satisfaction with sales assistance. And when inbound call volume drops, agents can do outbound calling.

Integration with the Internet

One of the newest developments in electronic commerce is the integration of the Internet with traditional call centers. The Internet-enabled call center allows companies to personalize relationships with Web site visitors by providing access to a customer service agent during a critical moment—when the visitor has a question, the answer to which will influence the decision to buy. With the ability to intervene in the online purchase decision and influence the outcome, a company can realize several benefits:

- *Competitive differentiation* By providing a convenient value-added service that improves customers' Web experiences via live agent assistance.
- *Sales generation* By removing obstacles in the buying process with immediate interaction between customers and knowledgeable call center agents.
- *Increased customer satisfaction* By delivering technical support and responding to customer needs quickly with personalized one-on-one service.

A company may use its Web page to allow consumers to buy a product or service online. Customers can buy a standard desktop computer, for example, or configure their own system with desired features and options. Selected items are added to a virtual shopping cart, which keeps a tally of the purchases. Visitors can make changes until the configuration meets a budget target. Configuration conflicts are even pointed out, giving the customer an opportunity to resolve the problem from a list of possible choices.

As the customer makes changes, the new purchase price is displayed, along with the monthly lease cost, in case the customer wants to consider this finance option. When the customer is ready to buy, the shopping cart adds shipping and handling charges and applicable state sales tax. The customer completes the transaction by entering contact and payment information and hitting the "Submit" button to send the purchase order to the company. Customers can check the status of a recent purchase by entering their order number into an online search field.

The company can make it easy for customers to request online assistance. At any time, customers can ask questions by selecting a preferred method of online communication. A question can be asked by entering it into an online form, in which case a sales representative will respond via e-mail, phone, or fax at a time convenient to the customer. Customers also can initiate an interactive text chat session with a sales representative or place a telephone call over the Internet to talk to a representative. If the customer wants to initiate a voice call but is not familiar with the procedure, a help window can be requested that provides information on system requirements and step-by-step instructions for placing the call.

AT&T provides a service called Interactive Answers (Figure C-3), which sets up telephone calls over the Internet when a customer clicks the "Call" button on a Web page. With a click of the button, customers are connected by phone to a customer service rep in a call center. If all call center agents are busy, the customer receives an online message stating that the call will be returned as soon as possible. Through dynamic queuing, the return call is made automatically as soon as an agent is free.

Regardless of what form of communication is selected, all go to the company's call center for handling by the next available agent or the agent who can most effectively respond to a customer's request (Figure C-4). While having a real-time conversation with the Web site visitor, the call

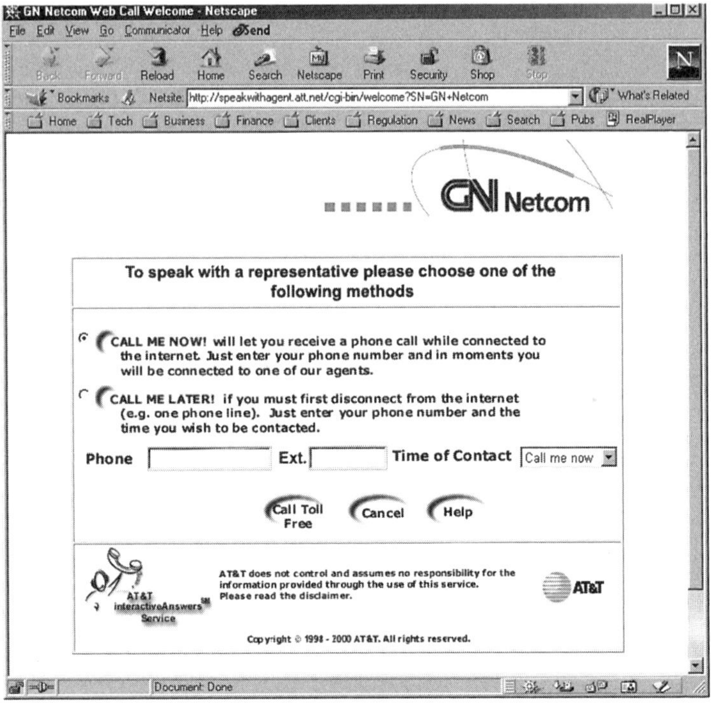

Figure C-3 Companies can subscribe to AT&T's Interactive Answers service, which allows them to place a click-to-call link from their Web site, allowing customers to talk with a call center agent over the Internet.

center agent can "push" Web pages to the customer's computer with appropriate text and images that help answer complex questions or illustrate examples.

In implementing an Internet call center—plus offering an online configuration and pricing tool, securing payment information with encryption, and providing an order status checker—the company has not only provided online shoppers with a new level of convenience, it also has removed key potential barriers to online sales. These barriers include customer uncertainty due to lack of decision-making information and doubt about the safety of electronic commerce.

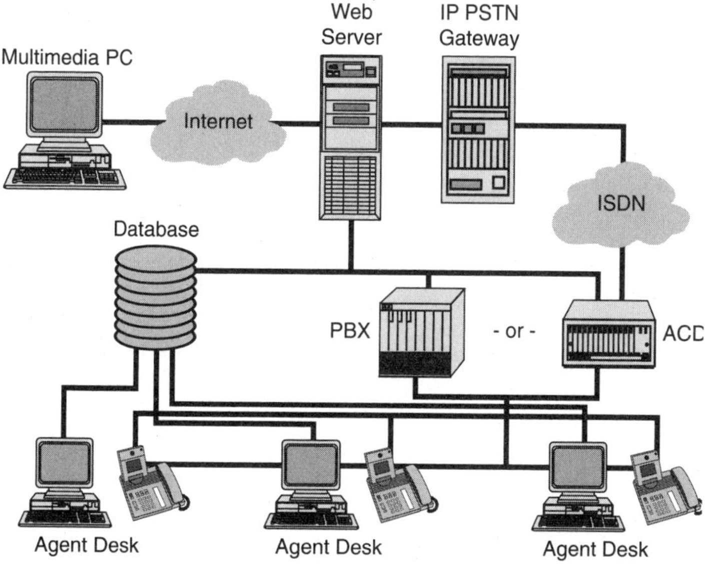

Figure C-4 Calls launched from a Web page are directed to an IP-PSTN gateway, which translates voice over IP (VoIP) into the digital ISDN format. The call is held in queue at the ACD for distribution to the next available agent. Instead of a screen pop-up of a customer record, the agent sees the Web page the caller is viewing.

Summary

A call center can consist of only two or three agents or as many as several thousand agents at distributed locations around the world. Sometimes calls will be transferred automatically across time zones so that the organization can provide customers with 24-hour service. The size and distribution of the call center will determine what kind of ACD system, dialing equipment, lines, and services it will need. Management tools will be used to track and monitor customer interaction and agent performance in real time, providing current information and updates to agents and managers during outbound call campaigns.

See also

Electronic Commerce
Internet Telephony
Private Branch Exchanges

CARNIVORE

Carnivore is an FBI surveillance tool that tracks suspected criminals online by sifting through data from Internet service providers (ISPs) to find the senders and recipients of a suspect's e-mail. The Carnivore system is actually a Windows 2000 application with custom software that plugs into a network hub to monitor traffic in a passive listening mode. If the traffic meets specific filtering criteria, the information is collected and stored on disk.

Architecture

When placed on an ISP's network (Figure C-5), the collection computer receives all packets on the Ethernet segment to which it is connected and records packets or packet segments that match Carnivore filter settings. The one-way tap ensures that Carnivore cannot transmit data on the network, and the absence of an installed IP stack ensures that Carnivore cannot process any packets other than to filter and optionally record them. Carnivore can neither alter packets destined for other systems on the network nor initiate any packets.

Control computers are located at law enforcement sites. When connected by modem to the collection computer, a control computer operator can set and change filter settings, start and stop collection, and retrieve the collected information. Using off-the-shelf tools from Packeteer and CoolMiner, the operator can reconstruct target activity

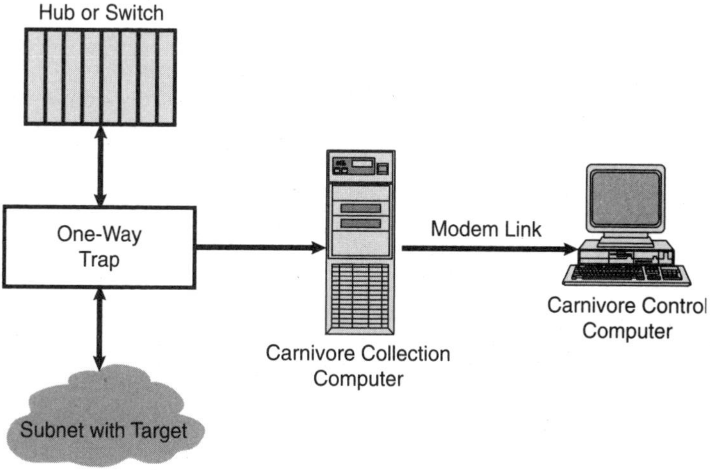

Figure C-5 Architecture of the Internet surveillance tool Carnivore used by the FBI.

from the collected IP packets. In pen mode, the operator can see the TO and FROM e-mail addresses and the IP addresses of computers involved in File Transfer Protocol (FTP) and Hypertext Transfer Protocol (HTTP) sessions. In full-collection mode, the operator can view the content of e-mail messages, HTTP pages, and FTP sessions. Carnivore operators are anonymous to the system. All users are logged in as "administrator," and no audit trail of actions is maintained.

Carnivore has proven useless against suspects clever enough to encrypt their files. Accordingly, the FBI has developed an enhancement called "Magic Lantern" that allows investigators to secretly break encryption keys. The software is actually a Trojan virus that is transmitted to the target computer. Once installed, it captures keystrokes, which contain critical encryption key information, and transmits them to the FBI, allowing computer files to be read remotely by investigators with court authorization.

Privacy Concerns

The intended use of Carnivore is to monitor allegedly criminal activity under a court order, much as telephone wiretaps do. But the system sparked debate in Congress and objections from privacy groups who were worried that Carnivore could be used to track the e-mail of innocent people.

Under pressure from Congress, as well as privacy and civil rights groups, the Justice Department agreed to have Carnivore's source code reviewed by an independent evaluation team to see if it performed as described by the FBI. Illinois Institute of Technology Research Institute (IITRI), which verified Carnivore's operation, issued the following conclusions in November 2000:

- When Carnivore is used in accordance with a court order, it provides investigators with no more information than permitted by that order. When Carnivore is used under pen trap authorization, it collects TO and FROM information and also indicates the lengths of messages and the lengths of individual fields within those messages, possibly exceeding court-permitted collection.
- Carnivore operation introduces no operational or security risks to the ISP network where it is installed, unless the ISP must make changes to its network to accommodate Carnivore. Such changes may introduce unexpected network behavior.
- Carnivore reduces, but does not eliminate, the risk of both intentional and unintentional unauthorized acquisition of electronic information by FBI personnel but introduces little additional risk of acquisition by persons other than FBI personnel.
- While operational procedures or practices appear sound, Carnivore does not provide protections, especially audit functions, commensurate with the level of the risks. While the system was designed to and can perform fine-tuned searches, it is also capable of broad sweeps. Incorrectly

configured, Carnivore can record any traffic it monitors.

- Carnivore does not have nearly enough power to spy on almost everyone with an e-mail account. In order to work effectively, it must reject the majority of packets it monitors. It also monitors only the packets traversing the wire to which it is connected. Typically, this wire is a network segment handling only a subset of a particular ISP's traffic.
- There is inadequate provision (e.g., audit trails) for establishing individual accountability for actions taken during the use of Carnivore.
- The current implementation of Carnivore, version 1.3.4 SP3, has significant deficiencies in protection for the integrity of the information it collects.

Although IITRI was not specifically asked to address questions of constitutionality and of illegal activity by the FBI, it voiced its concern that the presence of Carnivore without safeguards fuels the concerns of privacy advocates and reduces the expectations of privacy by citizens at large and increases public concern about the potential unauthorized activity of law enforcement agents.

Summary

Carnivore is now known by the more innocuous designation DCS1000. Despite numerous challenges from privacy groups and threats from Congress, Carnivore is in use today by the FBI, which notes that electronic surveillance has been extremely effective in securing the conviction of more than 25,600 dangerous felons over the past 13 years. According to the FBI, in many cases there is no substitute for electronic surveillance, since the evidence cannot be obtained through other traditional investigative techniques. Carnivore serves to limit the messages viewable by human eyes to those that are strictly included within the court order. ISP knowledge and assistance, as directed by court order, are required to install the software on the network.

See also

Network Security

CELLULAR DATA COMMUNICATIONS

One of the oldest services for sending data over a cellular communications network is Wireless IP, also known as Cellular Digital Packet Data (CDPD). Wireless IP provides a way of passing data packets over analog cellular voice networks at speeds of up to 19.2 kbps. Although CDPD employs digital modulation and signal processing techniques, the underlying service is still analog. The medium used to transport data consists of the idle radio channels typically used for Advanced Mobile Phone System (AMPS) cellular service.

Channel hopping automatically searches out idle channel times between cellular voice calls. Packets of data select available cellular channels and go out in short bursts without interfering with voice communications. Alternatively, cellular carriers also may dedicate voice channels for CDPD traffic to meet high traffic demand. This situation is common in dense urban environments where cellular traffic is heaviest.

Once the user logs on to the network, the connection stays in place to send or receive data. In accordance with the IP, the data are packaged into discrete packets of information for transmission over the CDPD network, which consists of routers and digital radios installed in current cell sites. In addition to addressing information, each IP packet includes information that allows the data to be reassembled in the proper order via the Transmission Control Protocol (TCP) at the receiving end. The transmissions are encrypted over the air link for security purposes.

Although CDPD piggybacks on top of the cellular voice infrastructure, it does not suffer from the 3-kHz limit on voice transmissions. Instead, it uses the entire 30-kHz RF channel during idle times between voice calls. Using the entire channel contributes to CDPD's faster data transmission rate. Forward error correction ensures a high level of wireless communications accuracy. With encryption and authentication procedures built into the specification, CDPD offers more robust security than any other native wireless data transmission method. As with wireline networks, CDPD users also can customize their own end-to-end security.

To take advantage of CDPD, the user should have an integrated mobile device that operates as a fully functional cellular phone and Internet appliance. For example, the AT&T PocketNet Phone contains both a circuit-switched cellular modem and a CDPD modem to provide users with fast and convenient access to two-way wireless messaging services and Internet information. GTE provides a similar service through its Wireless Data Services. Both companies have negotiated intercarrier agreements, which enable their customers to enjoy seamless CDPD service in virtually all markets across the country. AT&T's Wireless IP service, for example, is available in 3000 cities in the United States.

Among the applications for CDPD are access to the Internet for e-mail and to retrieve certain Web-based content. AT&T PocketNet Phone users, for example, have access to two-way messaging, airline flight information, financial information, show times, restaurant reviews, and door-to-door travel directions. AT&T provides unlimited access to featured sites on the wireless Internet, which means there are no per-minute charges for surfing wireless Web sites.

Companies also can use CDPD to monitor alarms remotely, send/receive faxes, verify credit cards, and dispatch vehicles. Although CDPD services might prove too expensive for heavy database access, the use of intelligent agents can cut costs by minimizing connection time. Intelligent agents gather requested information and report back only the results the next time the user logs on to the network.

Summary

Wireless IP is an appealing method of transporting data over cellular voice networks because it is flexible, fast, widely available, and compatible with a vast installed base of computers and has security features not offered with other wireless data services. One caveat: The carrier's wireless data network is different from its wireless voice network. Therefore, users of AT&T Digital PocketNet service, for example, will not be able to access that service everywhere voice calls can be made. It is important to look at coverage maps and compare service plans before subscribing to this type of service.

See also

Voice over IP

CELLULAR TELEPHONES

Bell Labs built the first cellular telephone in 1924. After decades of development, cellular telephones have emerged as a "must have" item among mobile professionals and consumers alike, growing in popularity every year since they became commercially available in 1983. Their widespread use for both voice and data communications has resulted from significant progress made in their functionality, portability, the availability of network services, and the declining cost for equipment and services. Functionality has been increased with optional Internet access.

Features and Options

Cellular telephones offer many features and options, including

- *Voice activation* Sometimes called "hands-free operation," this feature allows the user to establish and answer calls by issuing verbal commands. This safety feature enables a driver to control the unit without becoming visually distracted.

- *Memory functions* Allow storage of frequently called numbers to simplify dialing. Units may offer as few as 10 memory locations or in excess of 100, depending on model and manufacturer.
- *Multimode* Allows the phone to be used with multiple carriers. The phone can be used to access digital service where it is available and then switch to an analog service of another wireless carrier when roaming.
- *Multiband* Allows the phone to be used with multiple networks using different frequency bands. For example, the cell phone can be used to access the 1900-MHz band when it is available and then switch to the 800-MHz band when roaming.
- *Visual status display* Conveys information on numbers dialed, state of battery charge, call duration, roaming indication, and signal strength. Cell phones differ widely in the number of characters and lines of alphanumeric information they can display. The use of icons enhances ease of use by visually identifying the phone's features.
- *Programmable ring tones* Some cellular phones allow the user to select the phone's ring tone. Multiple ring tones can be selected, each assigned to a different caller. A variety of ring tones may be downloaded from the Web.
- *Silent call alert* Features include visual or vibrating notification in lieu of an audible ring tone. This can be particularly useful in locations where the sound of a ringing phone would constitute an annoyance.
- *Security features* Includes password access via the keypad to prevent unauthorized use of the cell phone as well as features to help prevent access to the phone's telephone number in the event of theft.
- *Voice messaging* Allows the phone to act as an answering machine. A limited amount of recording time (about 4 minutes) is available on some cell phones. However, carri-

ers also offer voice-messaging services that are not dependent on the phone's memory capacity. While the phone is in standby mode, callers can leave messages on the integral answering device. While the phone is off, callers can leave messages on the carrier's voice-mail system. Users are not billed for airtime charges when retrieving their messages.

- *Call restriction* Enables the user to allow use of the phone by others to call selected numbers, local numbers, or emergency numbers without permitting them to dial the world at large and rack up airtime charges.
- *Call timers* Provide the user with information as to the length of the current call and a running total of airtime for all calls. These features make it easier for users to keep track of call charges.
- *User-defined ring tones* Offer users the option to compose or download ring tones of their choice to replace the standard ring tone that comes with the cell phone.
- *Data transfer kit* For cell phones that are equipped with a serial interface, there is software for the desktop PC, which allows users to enter directory information via keyboard, rather than the cell phone keypad. The information is transferred via the kit's serial cable. Through the software and cable connection, information can be synchronized between the PC and cell phone, ensuring that both devices have the most recent copy of the same information.

Internet-Enabled Mobile Phones

Internet-enabled mobile phones potentially represent an important communications milestone, providing users with access to Web content and applications, including the ability to participate in electronic commerce transactions. The Wireless Application Protocol (WAP), an internationally accepted specification, allows wireless devices to retrieve content from the Internet, such as general news, weather, airline schedules, traffic reports, restaurant guides, sports scores, and stock prices.

Users also can personalize these services by creating a profile, which might request updated stock quotes every half-hour or specify tastes in music and food. A user also could set up predefined locations, such as home, main office, or transit, so that the information is relevant for that time and location. With access to real-time traffic information, for example, users can obtain route guidance on their cell phone screens via the Internet. Up-to-the-minute road conditions are displayed directly on the cell phone screen. Street-by-street guidance is provided for navigating by car, subway, or simply walking, taking into account traffic congestion to work out the best itinerary. Such services can even locate, and guide users to, the nearest facilities such as free parking lots or open gas stations using either an address entered on the phone keypad or information supplied by an automatic location identification (ALI) service.

One vendor that has been particularly active in developing WAP-compliant Internet-enabled mobile phones is Nokia, the world's biggest maker of mobile phones. The company's model 7110 works only on GSM 900 and GSM 1800 in Europe and Asia but is indicative of the types of new mobile phones that about 70 other manufacturers are targeting at the world's 200 million cellular subscribers. It displays Internet-based information on the same screen used for voice functions. It also supports SMS and e-mail and includes a calendar and phonebook as well.

The phone's memory also can save up to 500 messages—SMS or e-mail—sorted in various folders such as the inbox, outbox, or user-defined folders. The phonebook has enough memory for up to 1000 names, with up to five phone and fax numbers and two addresses for each entry. The user can mark each number and name with a different icon to signify home or office phone, fax number, or e-mail address, for example. The phone's built-in calendar can be viewed by day, week, or month, showing details of the user's schedule and calendar notes for the day. The week view shows icons for the jobs the user has to do each day. Up to 660 notes in the calendar can be stored in the phone's memory.

Nokia has developed several innovative features to make it faster and easier to access Internet information using a mobile phone:

- *Large display* The screen has 65 rows of 96 pixels (Figure C-6), allowing it to show large and small fonts, bold or regular, as well as full graphics.

Figure C-6 Display screen of the Nokia 7110.

- *Microbrowser* Like a browser on the Internet, the microbrowser feature enables the user to find information by entering a few words to launch a search. When a site of interest is found, its address can be saved in a "Favorites" folder or input using the keypad.
- *Navi Roller* This built-in mouse looks like a roller (Figure C-7) that is manipulated up and down with a finger to scroll and select items from an application menu. In each situation, the Navi Roller knows what to do when it is clicked—select, save, or send.
- *Predictive text input* As the user presses various keys to spell words, a built-in dictionary continually compares the word in progress with the words in the database. It selects the most likely word to minimize the need to con-

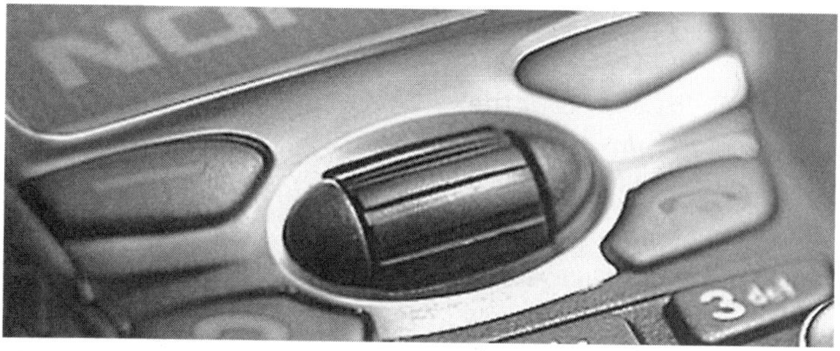

Figure C-7 Close-up of the Navi Roller on the Nokia 7110.

tinue spelling out the word. If there are several word possibilities, the user selects the right one using the Navi Roller. New names and words can be input into the phone's dictionary.

However, the Nokia phone cannot be used to access just any Web site. It can access only Web sites that have been developed using WAP-compliant tools. The WAP standard includes its own Wireless Markup Language (WML), which is a simple version of the HyperText Markup Language (HTML) that is widely used for developing Web content. The strength of WAP is that it is supported by multiple airlink standards and, in true Internet tradition, allows content publishers and application developers to be unconcerned about the specific delivery mechanism.

Third-Generation Phones

The world is moving toward third-generation (3G) mobile communications systems that are capable of bringing high-quality mobile multimedia services to a mass market. The International Telecommunication Union (ITU) has put together a 3G framework known as International Mobile Telecommunications 2000 (IMT-2000). This framework encompasses a small number of frequency bands, available on a globally harmonized basis, that make use of existing national and regional mobile and mobile-satellite frequency allocations.

Along the way toward 3G is a 2.5G service known as General Packet Radio Service (GPRS), which offers true packet data connectivity to digital cell phone users. GPRS leverages IP technologies, adding convenience and immediacy to mobile data services. GPRS is ideal for wireless data applications with bursty data, especially WAP-based information retrieval and database access.

GPRS enables wireless users to have an "always on" data connection, as well as high data transfer speeds of up to 144 kbps, versus 19.6 kbps with Cellular Digital Packet Data (CDPD) service, which is layered over analog cellular networks.

GPRS packet-based service should cost users less than circuit-switched services, since communication channels are shared rather than dedicated only to one user at a time. It also should be easier to make applications available to mobile users because the faster data rate means that middleware currently needed to adapt applications to the slower speed of wireless systems will no longer be needed. To take advantage of GPRS, however, mobile users will have to buy new cell phones that specifically support the data service.

Summary

Cellular phones are getting more intelligent, as evidenced by the combination of cellular phone, personal digital assistant (PDA), Web browser, and always-on GPRS connection into one unit. These devices not only support data communications, but they also support voice messaging, e-mail, fax, and micropayments over the Internet as well. Third-party software provides the operating system and such applications as calendaring, card file, and to-do lists. With more cellular phones supporting data communications, cellular phones are available that provide connectivity to PC desktop and databases via Bluetooth, infrared, or serial RS-232 connections. Information can even be synchronized between cell phones and desktop computers to ensure that the user is always accessing the most up-to-date information.

See also

Bluetooth

Cellular Data Communications

Personal Digital Assistants

CENTRAL OFFICE SWITCHES

Central office switches in North America are categorized into several types. Class 5 central office switches process local calls and offer Custom Local Area Signaling Services (CLASS) such as call block, call return, call trace, caller ID, distinctive ring, and speed dial. Class 4 central office switches process long-distance calls. Tandem office switches provide high-speed trunks that distribute traffic between Class 5 and Class 4 switches. A cluster of Class 5 switches can be connected to a tandem switch (Figure C-8), which also provides a connecting point for interexchange carriers (IXCs). The tandem switch stores routing data to access a particular switch in another service provider's network and vice versa. This eliminates the need for each end office to store routing information, which provides faster call transfer and reduces administrative costs.

The switches themselves are modular, making possible various service and feature enhancements via software upgrades, card additions, and adjunct system connections. They can be equipped to provide access to centrex/business group features and can function as a database gateway, connecting telephony functions with online databases or interactive services and packet networks. They also can be equipped to support LAN interconnection over the wide area

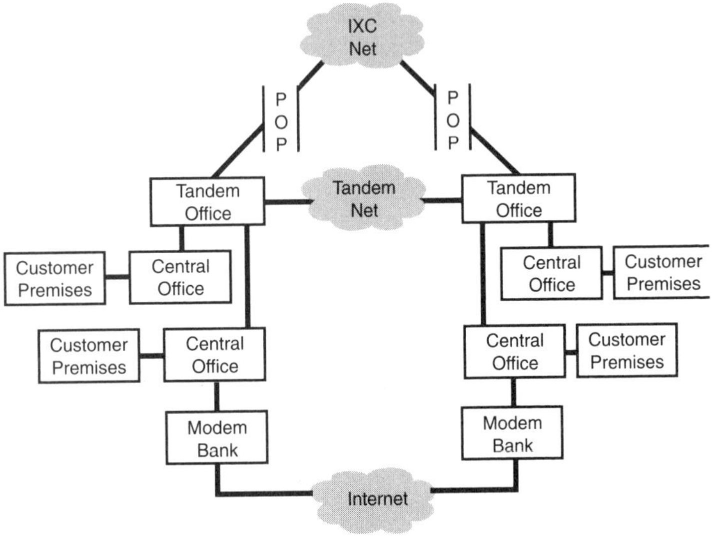

Figure C-8 A simple tandem office configuration. Dial-up calls to the Internet are diverted to alternative resources connected to the PSTN, such as modem banks operated by Internet service providers.

network (WAN). In recent years, support for a variety of broadband technologies has been added, including frame relay, ATM, and SONET.

Classless Switches

A new breed of "classless" central office switch, sometimes referred to as a "soft switch," has become commercially available to Competitive Exchange Carriers (CLECs). Since many of these carriers provide customers with local and long-distance services, they could do so more efficiently and economically with a single platform that could handle both types of traffic rather than purchase and provision separate Class 4 and Class 5 switches.

The benefits of a classless platform can be extended even further, since more voice calls are handled by IP-based packet data networks (PDNs). With IP telephony, there is no need for differently equipped switches in the network. Such switches, also known as "soft switches," can interface with both the PSTN and PDN, supporting traffic in a variety of formats, including TDM and IP. And since they support IP, they also facilitate the deployment of virtual private networks (VPNs). These classless switches can scale up to 100,000 ports, handling up to 50,000 calls simultaneously.

Feature management software allows any combination of service features to be directed to any port, eliminating the rigidity of the traditional Class 4/5 hierarchy of network switching. For example, for IP-based carriers, the soft switch can function as a Class 4 switch. This means that it will perform the toll/tandem switching functions needed to interconnect IP networks with local exchange telephone clients. At the same time, however, the soft switch can support the incremental addition of new advanced service features that are not traditionally associated with Class 4 switches, such as call redirection on busy and voice mail—features normally associated with Class 5 switches. This flexible architecture permits any telephone switch feature to be distributed to any switch interface on demand.

Summary

Today's all-digital, highly modular central office switches allow carriers to build different types of switching centers using various hardware combinations. The switch can function as an end office, an access tandem, or a remote unit capable of serving rural communities. The modular design allows software to be updated and new processors and subsystems to be added with relative ease, creating a ready migration path for future technologies and services. Classless switches are even more flexible, allowing them to assume the functions traditionally provided by separate Class 4 and Class 5 switches via changes to software. This makes it easier for IP-based carriers to launch competitive services without unnecessary infrastructure expenses.

See also

Centrex

Integrated Services Digital Network

CENTREX

Centrex—short for central office exchange—is a service that handles business calls at the telephone company's switch rather than through a customer-owned, premises-based Private Branch Exchange (PBX). Centrex provides a full complement of station features, remote switching, and network interfaces that provides an economical alternative to owning a PBX. Centrex offers remote options for businesses with multiple locations, providing features that appear to users and the outside world as if the remote sites and the host switch are one system.

Centrex users have access to direct inward dialing (DID) features, as well as station identification on outgoing calls. Each station has a unique line appearance in the central office, in a manner similar to residential telecommunications subscriber connections. A centrex call to an outside line exits the switch in the same manner as a toll call exits a local exchange. Users dial a four- or five-digit number without a prefix to call internal extensions and dial a prefix (usually 9) to access outside numbers.

The telephone company operates, administers, and maintains all centrex switching equipment for the customers. It also supplies the necessary operating power for the switching equipment, including backup power to ensure uninterrupted service during commercial power failures.

Centrex may be offered under different brand names. BellSouth calls it Essex and SBC Communications calls it Plexar, while Verizon calls it CentraNet. Centrex is also offered through resellers that buy centrex lines in bulk from the local exchange carrier. Using its own or commercially purchased software, the reseller packages an offering of centrex and perhaps other basic and enhanced telecommunications services to meet the needs of a particular business. The customer gets a single bill for all local, long-distance, 800, 900, and calling-card services at a fee that is less than the customer would otherwise pay.

IP Centrex

Centrex can be enhanced to support telephone calls over IP packet networks. The advantage of integrating centrex with IP packet-based nets is to enable the telephone company to reduce facility costs by using the inherent efficiencies of IP networks. It also enables the telephone company to offer value-added services to its centrex customers, such as virtual private networks (VPNs), telecommuter access, and virtual call centers. Telephone companies can also use the IP network to extend the market reach for their centrex offerings by serving corporate locations out of their local serving area.

The integration of centrex with IP is accomplished with a gateway that gives the existing Class 5 central office switch the ability to offer the full spectrum of voice, data, and centrex features over an IP network (Figure C-9). The gateway connects to the Class 5 switch using the Telcordia Technologies (formerly Bellcore) Standard GR-303 interface. The gateway uses the existing stable and secure Class 5 infrastructure for billing; Operations, Administration, Maintenance, and Provisioning (OAM&P); signaling (SS7); and trunking. The telephone company can increase revenues with new IP applications that can be deployed immediately and minimize expenditures by providing these new services using existing equipment.

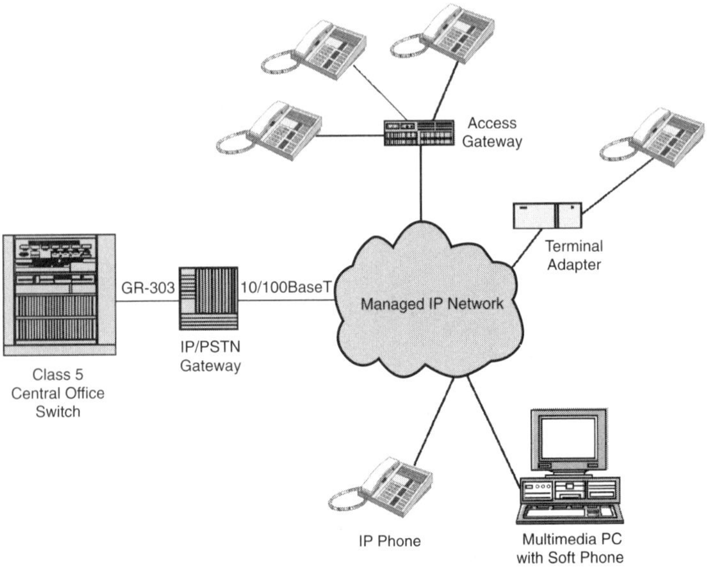

Figure C-9 An IP gateway connects to the Class 5 central office switch using the industry standard GR-303 interface to economically with centrex features.

At the customer premises, a variety of equipment can be used to get voice and data traffic through the IP network to the gateway and then to the centrex switch. A multiport access gateway is a concentration device that connects multiple legacy key sets to an IP network. Standard phone wiring is needed only between desktop key sets and the gateway.

Terminal adapters connect a single legacy phone to an IP network. They perform coding/decoding and packetization using Digital Signal Processors (DSPs) for high performance. Some terminal adapters come bundled with a PC user interface. Others have a PSTN connection as a backup.

Soft phones are telephony applications that run on a PC. Soft phones typically use the PC's main processor to perform coding/decoding and packetization in addition to any other work being done on the PC. There are also IP phones that replace conventional phones and connect directly to the IP network, usually via an Ethernet jack. The coding/decoding and packetization functions are performed in the key set. There are also wireless IP phones, which perform the coding/decoding function in the handset.

Summary

Centrex offers high-quality, dependable, feature-rich telephone service that supports a variety of applications. For many organizations, centrex offers distinct advantages over on-premises PBX or key/hybrid systems. Centrex can save money over the short term because there is no outlay of cash for an on-premises system. If the service is leased on a month-to-month basis, there is little commitment and no penalty for discontinuing the service. A company can pick up and move without worrying about reinstalling the system, which may not be right for the new location.

Centrex systems are easily expanded to accommodate customer growth by adding communication paths, memory, intercom lines, tie trunks, and CO lines as needed. Now, with IP connectivity, centrex can be used for VPNs, virtual call centers, and economical telecommuter access. If there is a centrex problem, repair is immediate and inexpensive. There is no need for a company to invest in a spare parts inventory, test equipment, or technical staff to take care of a PBX—the telephone company is responsible for all that. And as the CO switching equipment is updated, the centrex services are also updated.

See also

Call Centers

Central Office Switches

Private Branch Exchanges

Voice over IP

COMPETITIVE LOCAL EXCHANGE CARRIERS

Competitive Local Exchange Carriers (CLECs) offer voice, data services, and value-added services at significantly lower prices than the Incumbent Local Exchange Carrier (ILEC), enabling residential and business users to save money on such things as local calls, call handling features, lines, and Internet access. Typically, CLECs offer service in major cities, where traffic volumes are greatest and, consequently, users are hardest hit with high local exchange charges from the incumbent carrier. Some CLECs call themselves integrated communications providers (ICPs) because their networks are designed from the outset to support voice and data services as well as Internet access. Others call themselves data local exchange carriers (DLECs) because they specialize in data services such as DSL, which is used primarily for Internet access.

As of 2001, the ILECs still controlled 97 percent of the market for local services, according to the FCC, which means that the CLECs are trying to sustain themselves on the remaining 3 percent as they attempt to take market share from the ILECs. To deal with this situation, the CLECs have adopted different strategies based on resale and facilities ownership.

Resale versus Ownership

CLECs may compete in the market for local services by setting up their own networks or by reselling lines and services purchased from the ILEC. They may have hybrid arrangements for a time, which are part resale and part facilities ownership. Most CLECs prefer to have their own networks because the profit margins are higher than for resale. However, many CLECs start out in new markets as resellers. This enables them to establish a local presence, build brand awareness, and begin building a customer base while they assemble their own facilities-based network.

Although this strategy is used by many CLECs, many fail to carry it out properly. They get into financial trouble by using their capital to expand resale arrangements to capture even more market share instead of using that capital to quickly build their own networks and migrate customers to the high-margin facilities. Depending on the service, it could take a carrier 3 to 4 years to break even on a pure resale customer versus only 6 to 9 months on a pure facilities-based customer. With capital markets drying up for telecom companies and customers deferring product and services purchases, prolonged dependence on resale could set the stage for bankruptcy.

CLECs employ different technologies for competing in the local services market. Some set up their own Class 5 central office switches, enabling them to offer "dial tone" and the usual voice services, including ISDN, as well as features such as caller ID and voice messaging. The larger CLECs build their own fiber rings to serve their metropolitan customers with high-speed data services. Some CLECs have chosen to specialize in broadband data services by leveraging existing copper-based local loops, offering DSL services for Internet access. Others bypass the local loop entirely through the use of broadband wireless technologies, such as Local Multipoint Distribution Service (LMDS), enabling them to feed customer traffic to their nationwide fiber backbone networks without the incumbent carrier's involvement.

Despite the risks, some CLECs view resale as a viable long-term strategy. It not only allows them to enter into new markets more quickly than if they had initially deployed their own network, but it also reduces initial capital requirements in each market, allowing them to focus capital resources initially on the critical areas of sales, marketing, and operations support systems (OSS). In addition, the strategy allows them to avoid deployment of conventional circuit switches and maintain design flexibility for the next generation of telecommunications technology.

Unfortunately, the resale strategy also results in lower margins for services than for facilities-based services. This means the CLEC must pass much of its customer revenues back to the ILEC to pay the monthly fees for access lines. When investors stopped stressing market growth over profits in 2000, these CLECs found that capital was hard to get. By then, many had no money to invest in their own facilities where margins are greater. Most financial analysts doubt that CLECs can rely strictly on resale and survive. Although the ILECs have a vested interest in the survival of some resale CLECs in order to receive regulatory approval to provide in-region long distance, once that approval is gained, some analysts believe that the ILECs may have no further interest in cooperating with the CLECs.

Summary

With the Telecommunications Act of 1996, CLECs and other types of carriers are allowed to compete in the offering of local exchange services and must be able to obtain the same service and feature connections as the ILECs have for themselves—and on an unbundled basis. If the ILEC does not meet the requirements of a 14-point checklist to open up its network in this and other ways, it cannot get permission from the FCC to compete in the market for long distance services.

See also

Building Local Exchange Carriers
Interexchange Carriers

CONTENT DELIVERY NETWORKS

Despite efforts to improve the performance of public and corporate networking infrastructures, expediting the delivery of popular digital content remains operationally difficult for most companies. The strain on backbone bandwidth and server capacity, already evident with today's Internet connections, only gets worse as the user population grows and broadband access becomes available. Content Delivery Networks (CDNs) represent the latest attempt to deal with these problems.

Simply, this "edge network" solution acts as a distribution mechanism for Web content, which is replicated on cache servers at many points of presence (PoPs) on different backbone providers' networks so that content can be delivered directly from these servers without needing to traverse the frequently congested Internet backbone. Users no longer have to go to origin sites across the Internet backbone to access specific content. Consequently, content is delivered faster and more reliably, greatly improving the user's Web experience.

CDNs employ various technologies to improve the performance of Web sites, reduce hardware and bandwidth costs, and boost reliability by mirroring a Web site's content on distributed servers. With infrastructure technologies like caching, CDNs push replicated content closer to the network edge to minimize delay. Global load balancing ensures that users are transparently routed to the nearest content source.

Typically, large enterprise Web site owners subscribe to the CDN and determine the content it will serve. This can be done by selectively reassigning URLs to embedded objects. That way, dynamic or localized content—such as banner ads, Java applets, and graphics, which represent 70 percent of a typical Web page—can be served up locally by the company's own Web site, avoiding the CDN, while static and easily distributed content can be retrieved from the nearest CDN server. Although such services are currently used mostly for Internet applications, there is great potential for the technology in intranets and extranets as well.

Enabling Technologies

Several technologies are being implemented by CDNs—among them network caching. The idea behind caching is to have frequently accessed content delivered from a server located as close as possible to the user so that requests do not have to be sent to the origin Web server for processing. Despite growing use of caching, success has been limited for delivering digital content simultaneously to tens or hundreds of thousands of users, let alone the millions of consumers served by other one-to-many content distribution methods, such as television and radio.

In an effort to improve caching performance, adaptive prioritization software and refresh algorithms have been implemented to deliver geographically relevant content quickly and securely to within one router hop from the end user. Such systems are designed around master and local caches. The master cache is located at a key Internet access point and acts as the primary gatherer of content in response to real-time user requests. Once gathered, data are sent via the broadband content delivery network to local caches at the edge nodes of the Internet—typically in an ISP's local PoP or in an enterprise's branch office. Special software facilitates communication between the master cache and each local cache. In some cases, the use of point-to-multipoint satellite links to deliver content bypasses the congested Internet, significantly enhancing the end users' experience by reducing download times.

Another innovative method of expediting content distribution is through intelligent wide area traffic management, which adds intelligence to the standard Domain Name System (DNS). Adding an intelligent wide area traffic manager improves DNS over standard services by directing content requests to the most optimal site. The determination of the site's location is based on criteria such as geographic proximity, network topology, and endpoint (cache and Web server) availability. The traffic manager also allows specific rules to be built into the DNS process, such that if a request comes from anywhere in a particular region, it will first be sent to data center "X". The second decision would be to determine whether there's a data center available to handle that user. If the one in New York is too busy for a French user, perhaps the data center in Boston might provide the next best experience.

Another technology, IP multicast, also offers the prospect of significantly better high-volume digital content delivery, especially for streaming content. Multicasting relies on a centralized server and unique streams for each client requesting data from the server. To avoid network and server congestion under heavy user loads, stream replication is pushed to the edges of the network, where it has less impact. Traditionally, however, IP multicast has provided only a best-effort service that does not guarantee content delivery. To gain reliability with multicast, users normally would have to implement a back-channel connection to the server that indicates whether the recipient is receiving content properly so that adjustments can be made in response to poor performance. But too many back-channel requests also can slow down a server, limiting the scalability of multicast.

Recent technological advances are improving IP multicast so that it can more reliably support digital content distribution. Digital Fountain, for example, uses "randomized algorithms" to continuously multicast data—including audio and video—across a network such that every packet sent is unique. Recipients need only pick up a certain number of packets, in any order, to replicate the entire file locally. Each packet contains random bits of information about the original file. Using the special algorithms, the decoder can recreate the data. No back channels are needed to ensure reliable delivery, even during times of high packet loss.

Since no one packet is critical, requests for retransmission do not go back to the server when packets are missing or out of sequence. With this technique, it also does not matter when recipients join the multicast group—they will still receive the entire file. Users merely click an HTML link that sends 150 bytes of information about the file's encoding parameters to the receiver, and the rest is done behind the scenes. All that is needed is about 4 percent more bandwidth above what it would normally take to send the stream without encoding. For instance, a file that normally requires 200 kbps of bandwidth would need 208 kbps under Digital Fountain's system.

Another technique for improving Internet performance is cookie-based switching that allows traffic to be routed to designated servers, helping to eliminate bottlenecks for users on the basis of the nature of the request. Such switches typically sit between a server farm and a router or backbone switch on the network.

Since security functions are computationally intensive, overall performance can be improved by offloading such processing to dedicated servers. Normally, computation-intensive Secure Sockets Layer (SSL) requests are delivered from centralized servers located far away from Internet users. This delay can be overcome with a switch that identifies inbound SSL-encrypted traffic and sends it to the security server, which handles the SSL handshake, key exchange, and decryption and conversion of the session between HTTPS (HTTP Secure) and HTTP. The traffic is then sent back to the Web server. By offloading the security processing, a Web server could be freed up to handle content requests more quickly instead of getting bogged down handling encryption algorithms.

Advantages of CDNs

CDNs offer a number of compelling benefits:

- *Speed-boosting performance* Web businesses can deliver more engaging sites and achieve more page views, more advertising revenue, and more transactions without the risk of alienating visitors with mind-numbing delay.
- *Performance guarantees* Companies typically see performance improvements of 2 to 10 times or more. If the service provider fails to perform and is not meeting its obligations as stated in the Service Level Agreement, appropriate credits are applied to the next invoice.

- *Rich content* Web site designers can offer a custom experience to users without regard to possible performance delays. Targeted banner ads, individual layouts, and fresh content provide for a more stimulating experience to visitors, which keeps them coming back.
- *Peak traffic operation* The CDNs, which usually are much larger than any single Web site, have the network capacity to support an on-rush of visitors, even during peak periods.
- *Easy implementation* From the Web site owner's perspective, there is no new capital investment or systems to install, such as switches or redirectors. From the visitor's point of view, there is nothing new to install and no PC or browser changes are required.
- *No process interruptions* Uninterrupted interaction with users, counting hits, placing cookies, and dynamically generating pages are essential to Internet businesses. Depending on the service provider, CDNs do not get in the way of these and other personalized user interactions. Businesses can continue to track visitors while still delivering customized content.
- *Reduced network costs* CDNs reduce site infrastructure and bandwidth requirements, saving monthly fees and equipment and labor expenses. Web site owners pay only for the amount of content delivered.
- *Instant feedback* Traffic reports provide information about how much and where content is delivered. Historical information is also available for running custom analyses.
- *Enhance the end-user experience* By expediting information flow, CDNs can help businesses reduce subscriber churn, especially among broadband users.

For companies that want to put together their own infrastructure for content delivery for security and performance reasons, management becomes a key issue that must be addressed. Network professionals must know how to design the infrastructure for optimal efficiency, which entails setting up and managing network caches. If the company intends to provide a commercial service, it must have the tools in place to measure CDN usage for billing purposes. If the CDN is for internal use, accounting software should be considered if usage by departments and groups is charged back.

Summary

As more businesses increasingly rely on the Internet to offer new services, improve customer satisfaction, and reach new markets, information flow is bogged down by lack of bandwidth and heavy server loads. What is needed is a way to expedite the delivery of data, images, and streaming content to users that does not require huge investments in bandwidth and processing capacity at the Web servers. The emergence of CDNs has improved the performance of Web sites by distributing content in a more efficient manner than has ever been possible before. By offloading certain types of content to these overlay networks, businesses can realize substantial performance gains with little or no investment in additional bandwidth or processing capacity. Some companies using CDN services are seeing a 20 to 50 percent decrease in download times from their Web sites.

See also

Application Service Providers

Bandwidth Management Systems

Internet Service Providers

CYBER CRIME

Cyber crime refers to any criminal activity conducted with a computer attached to a network, usually the Internet. According to the FBI, cyber crime is one of the fastest evolving areas of criminal behavior and represents a significant threat to U.S. national and economic security. PriceWaterhouseCoopers LLC says businesses spent $300 billion in 2000 fighting hackers and computer viruses.

To counter the mounting cyber threat, the FBI uses its National Infrastructure Protection and Computer Intrusion squads located in 16 field offices. It also has developed baseline computer intrusion team capabilities in nonsquad field offices. Further, it has established partnerships with state and local law enforcement through cyber crime task forces. As of mid-2001, the U.S. Department of Justice had 48 prosecutors working on cyber crime in U.S. attorneys' offices nationwide to prosecute hacking and copyright violations.

Types of Cyber Threats

Over the past several years there have been a range of computer crimes, ranging from simple hacking by juveniles to sophisticated intrusions that may be sponsored by organized crime and foreign powers. Some intrusions result in the theft of credit card numbers or proprietary information or the loss of sensitive government information and can threaten national security and undermine confidence in e-commerce. A denial-of-service attack that can knock e-commerce sites offline can have significant consequences not only for victim companies but also for consumers and the economy as a whole.

Insider Threat The disgruntled insider is a principal source of computer crimes. Insiders do not need a great deal of knowledge about computer intrusions because their knowledge of victim systems often allows them to gain unrestricted access to cause damage to the system or to steal system data. The FBI estimates that insiders are responsible for 55 percent of reported malicious activity cases.

Hackers Hackers are also a common threat. They sometimes crack into networks simply for the thrill of the challenge or for bragging rights in the hacker community. More recently, however, there are more cases of hacking for illicit financial gain or other malicious purposes. While remote cracking once required a fair amount of skill or computer knowledge, hackers can now download attack scripts and protocols from the Web and launch them against victim sites. Attack tools have become more sophisticated, available, and easier to use.

Hactivism There is a rise in so-called hacktivism—politically motivated attacks on publicly accessible Web sites or e-mail servers. These groups and individuals overload e-mail servers and hack into Web sites to send a political message. While these attacks generally have not altered operating systems or networks, they still damage services and deny the public access to information and infringe on others' rights to communicate.

Virus Writers Virus writers are posing an increasingly serious threat to networks and systems worldwide. When unleashed onto the Internet and opened by computer users, viruses can destroy files, applications, or operating systems.

Criminal Groups The FBI is also seeing the increased use of cyber intrusions by criminal groups who attack systems for purposes of monetary gain. One method of penetrating secure systems is "dumpster diving" to gather old phone books and technical manuals for systems. This information is used to call employees and trick them into giving up log-on and password information, which is then used to break into targeted systems.

Distributed Denial-of-Service Attacks This method of attack entails hackers planting tools on a number of unwitting victim systems. Then, when the hacker sends the command, the victim systems in turn begin sending messages against a target system. The target system is overwhelmed with the traffic and is unable to function. Users trying to access that system are denied its services.

Terrorists The FBI says terrorists have been known to use information technology and the Internet to formulate plans, raise funds, spread propaganda, and communicate securely. Some groups have already used cyber attacks to inflict damage on their enemies' information systems. For example, a group calling itself the Internet Black Tigers conducted a successful denial-of-service attack on servers of Sri Lankan government embassies. Italian sympathizers of the Mexican Zapatista rebels attacked Web pages of Mexican financial institutions. While the FBI has yet to see a significant instance of cyber terrorism with widespread disruption of critical infrastructures, all these facts portend the use of cyber attacks by terrorists to disrupt the critical systems of targeted governments or civilian populations.

Foreign Intelligence Services Foreign intelligence services have started to use cyber tools as part of their information gathering and espionage tradecraft. According to the FBI, foreign intelligence services increasingly view computer intrusions as a useful tool for acquiring sensitive U.S. government and private sector information.

Sensitive Intrusions Intrusions into Department of Defense computer networks as well as networks of other federal agencies, universities, and private-sector entities are on the increase. The FBI claims intruders have successfully accessed U.S. government networks and taken enormous amounts of unclassified but sensitive information, with many of these intrusions appearing to originate in Russia.

Information Warfare One of the greatest potential threats to national security is the prospect of information warfare by foreign militaries against critical infrastructures in the United States. According to the FBI, several nations are developing information warfare doctrine, programs, and capabilities for use against each other or other nations. The reason for these information warfare programs is that they cannot defeat the United States in a head-to-head military encounter, and they believe that information operations are a way to strike at what they perceive as America's dependence on information technology to control critical government and private-sector systems.

Challenges to Law Enforcement

The FBI points out that the growing problem of cyber crime poses unique challenges to law enforcement, which require novel solutions, close teamwork among agencies and with the private sector, and adequate numbers of trained and experienced agents and analysts with sophisticated equipment.

Identifying the Intruder One major difficulty that distinguishes cyber threats from physical threats is determining who is attacking computer systems, why, how, and from where. This difficulty stems from the ease with which individuals can hide or disguise their tracks by manipulating logs and directing their attacks through networks in many countries before hitting their ultimate target.

Jurisdictional Issues Another significant challenge comes from investigating incidents that span multiple states and often many countries. This is the case even when the hacker and victim are both located in the United States, where law enforcement can subpoena records and execute search warrants on suspects' homes, seize evidence, and examine it. Overseas, there is dependence on the local authorities. In some cases, the local police forces simply do not understand or cannot cope with the technology. In other cases, nations simply do not have laws against computer intrusions.

Human and Technical Issues Threat investigations are compounded by human and technical challenges. Once the problem of having enough agents, computer scientists, and analysts to work computer intrusions is overcome, there is the issue of recruiting people to fill these positions, training them in the rapidly changing technology, and retaining them. The FBI's manpower shortage is acute. A distributed denial-of-service attack can draw a tremendous amount of personnel resources. Most of technical analysts are pulled from other work to examine the log files received from victim companies. Tracking down hundreds of leads absorbs the energy of a dozen field offices. A single computer espionage case can consume as much as 6000 worker-hours.

Encryption One of the biggest challenges to FBI computer investigative capabilities lies in the increasingly widespread use of strong encryption and digital telecommunications. technologies, both of which place a tremendous burden on the FBI's electronic surveillance capabilities. Today the most basic communications employ layers of protocols, formatting, compression, and proprietary coding that were nonexistent only a few years ago. New cryptographic systems provide robust security to conventional and cellular telephone conversations, facsimile transmissions, LANs and WANs, Internet communications, PCs, wireless transmissions, electronically stored information, remote keyless entry systems, advanced messaging systems, and RF communications systems.

State and Local Assistance State and local authorities, often the first to arrive on a crime scene, are even less likely to have the expertise to investigate computer intrusions by gathering and examining cyber media and evidence. The challenge for the federal government is to provide the training and backup resources to the state and local levels so that they can successfully conduct investigations and prosecutions in their jurisdictions.

Building Investigative Capacity The FBI must identify, recruit, and train personnel who have the technical, analytical, investigative, and intelligence skills for engaging in cyber investigations. This includes personnel to provide early warnings of attacks, read and analyze log files, write analytic reports and products for the field and the private sector, and support other investigations with cyber components.

Developing Forensic and Technical Capabilities Improving technical capabilities to access plain text communications is a critical challenge, according to the FBI. The ultimate objective is to provide field investigators with an integrated suite of automated data collection systems, operating in a low-cost and readily available personal computer environment, that will be capable of identifying, intercepting, and collecting targeted data of interest from a broad spectrum of data telecommunications transmissions media and networks. The most technically complex component of electronic surveillance has been and always will be the deciphering of encrypted signals and data. The ability to gather evidence from FBI electronic surveillance and seized electronic data will significantly depend on the development of and deployment of signal analysis and decryption capabilities.

Summary

The cyber crime scene is dynamic—it grows, contracts, and can change shape instantly. Determining whether an intrusion is even occurring often can be difficult in the cyber world, and usually a determination cannot be made until a suspicion is raised and an investigation initiated. The reason the FBI pushed hard for the Communications Assistance for Law Enforcement Act (CALEA) was to force the telecommunications industry to proactively address law enforcement's need and authority to conduct lawfully authorized electronic surveillance. The FBI has had less success in garnering acceptance of Carnivore, a surveillance tool that tracks suspected criminals online by sifting through all the data from an Internet service provider to track a suspect's e-mail.

See also

Carnivore

Network Security

D

DATA COMPRESSION

Data compression, a standard feature of most bridges and routers, as well as modems, improves throughput by capitalizing on the redundancies found in the data to reduce frame size and thereby allow more data to be transmitted over a link. There are also compression programs available for the desktop, such as WinZIP, which allows files to be compressed for transmission over an Internet connection when attached to an email message, for example. An algorithm detects repeating characters or strings of characters and represents them as a symbol or token. At the receiving end, the process works in reverse to restore the original data.

There are many different algorithms available to compress data, which are designed for specific types of data sources and the redundancies found in them but do a poor job when applied to other sources of data. For example, the Moving Pictures Experts Group (MPEG) compression standards were designed to take advantage of the relatively small difference from one frame to another in a video stream and so do an excellent job of compressing motion pictures. On the other hand, MPEG would not be effective if applied to still images. For this data source, the Joint Photographic Experts Group (JPEG) compression standards would be applied.

JPEG is "lossy," meaning that the decompressed image is not quite the same as the original compressed image—there is some degradation. JPEG is designed to exploit known limitations of the human eye, notably that small color details are not perceived as well as small details of light and dark. JPEG eliminates the unnecessary details to greatly reduce the size of image files, allowing them to be transmitted faster and take up less space in a storage server.

On wide area network (WAN) links, the compression ratio tends to differ by application. The compression ratio can be as high as 6 to 1 when the traffic consists of heavy-duty file transfers. The compression ratio is less than 4 to 1 when the traffic is mostly database queries. When there are only "keep alive" signals or sporadic query traffic on a T1 line, the compression ratio can dip below 2 to 1.

Encrypted data exhibit little or no compression because the encryption process expands the data and uses more bandwidth. However, if data expansion is detected and compression is withheld until the encrypted data are completely transmitted, the need for more bandwidth can be avoided.

Types of Data Compression

There are several different data compression methods in use today over WANs—among them are TCP/IP header compression, link compression, and multichannel payload compression. Depending on the method used, there can be a significant tradeoff between lower bandwidth consumption and increased packet delay.

TCP/IP Header Compression With TCP/IP header compression, the packet headers are compressed, but the data payload remains unchanged. Since the TCP/IP header must be replaced at each node for IP routing to be possible, this compression method requires hop-by-hop compression and decompression processing. This adds delay to each compressed/decompressed packet and puts an added burden on the router's CPU at each network node.

TCP/IP header compression was designed for use on slow serial links of 32 kbps or less and to produce a significant performance impact. It needs highly interactive traffic with small packet sizes. In such traffic, the ratio of layer 3 and 4 headers to payload is relatively high, so just shrinking the headers can result in a substantial performance improvement.

Payload Compression Payload compression entails the compression of the payload of a layer 2 WAN protocol, such as Point-to-Point Protocol (PPP), Frame Relay, High-Level Data Link Control (HDLC), X.25, and Link Access Procedure, Balanced (LAPB). The layer 2 packet header is not compressed, but the entire contents of the payload, including higher-layer protocol headers (i.e., TCP/IP), are compressed. They are compressed using the industry standard Lemple-Ziv algorithm or some variation of that algorithm.

Layer 2 payload compression applies the compression algorithm to the entire frame payload, including the TCP/IP headers. This method of compression is used on links operating at speeds from 56 to 1.544 Mbps and is useful on all traffic types, as long as the traffic has not been compressed previously by a higher-layer application. TCP/IP header compression and layer 2 payload compression, however, should not be applied at the same time because it is redundant and wasteful and could result in the link not coming up to not passing IP traffic.

Link Compression With link compression, the entire frame—both protocol header and payload—is compressed. This form of compression is typically used in LAN-only or legacy-only environments. However, this method requires error correction and packet sequencing software, which adds to the processing overhead already introduced by link compression and results in increased packet delays. Also, like TCP/IP header compression, link compression requires hop-by-hop compression and decompression, so processor loading and packet delays occur at each router node the data traverses.

With link compression, a single data compression vocabulary dictionary or history buffer is maintained for all virtual circuits compressed over the WAN link. This buffer holds a running history about what data have been transmitted to help make future transmissions more efficient. To obtain optimal compression ratios, the history buffer must be large, requiring a significant amount of memory. The vocabulary dictionary resets at the end of each frame. This technique offers lower compression ratios than multichannel, multihistory buffer (vocabularies) data compression methods. This is particularly true when transmitting mixed local area network (LAN) and serial protocol traffic over the WAN link and frame sizes are 2 kilobytes or less. This translates into higher costs, but if more memory is added to get better ratios, this increases the upfront cost of the solution.

Mixed-Channel Payload Data Compression By using separate history buffers or vocabularies for each virtual circuit, multichannel payload data compression can yield higher compression ratios that require much less memory than other data compression methods. This is particularly true in cases where mixed LAN and serial protocol traffic traverses the network. Higher compression ratios translate into lower WAN bandwidth requirements and greater cost savings.

But performance varies because vendors define payload data compression differently. Some consider it to be compression of everything that follows the IP header. However, the IP header can be a significant number of bytes. For overall compression to be effective, header compression must be applied. This adds to the processing burden of the CPU and increases packet delays.

External Data Compression Solutions Bridges and routers can perform data compression with optional software or add-on hardware modules. While software-based compression capabilities can support fractional T1/E1 rates, hardware-based compression off-loads the bridge/router's main processor to deliver even higher levels of throughput. With a compression ratio of up to 4 to 1, a data compression module can support up to 16 Mbps of compressed data throughput without imposing additional traffic latency. This is enough to keep four T1/E1 circuits full of compressed data in both directions simultaneously.

The use of a separate digital signal processor (DSP) for data compression, instead of the software-only approach, enables the router to perform all its core functions without any performance penalty. This parallel processing approach minimizes the packet delay that can occur when the router's CPU is forced to handle all these tasks by itself.

If there is no vacant slot in the bridge/router for the addition of a data compression module, there are two alternatives: the software-only approach or an external compression device. The software-only approach could bog down the overall performance of the router, since its processor would be used to implement compression in addition to core functions. Although an external data compression device would not bog down the router's core functions, it means that one more device must be provisioned and managed at each remote site.

Summary

Data compression will become increasingly important to most organizations as the volume of data traffic at branch locations begins to exceed the capacity of the wide area links. Multichannel payload solutions provide the highest compression ratios and reduce the number of packets transmitted across the network. Reducing packet latency can be effectively achieved via a dedicated processor like a DSP and by employing end-to-end compression techniques rather than node-to-node compression/decompression. All these factors contribute to reducing WAN circuit and equipment costs as well as improving the network response time and availability for user applications.

See also

Voice Compression

DATA WAREHOUSES

A data warehouse is an extension of the database management system (DBMS) that consolidates information from various sources into a high-level, integrated form used to identify trends and make business decisions. For a large company, the amount of information in a data warehouse could be several trillion bytes, or terabytes (TB). The technologies that are used to build data warehouses include relational databases; powerful, scalable processors; and sophisticated tools to manipulate and analyze large volumes of data for the purpose of identifying previously undetectable patterns and relationships. Benefits include increased revenue and decreased costs due to the more effective handling of great volumes of data.

In addition, the more effective handling of corporate data—identifying patterns of repeat purchases, the frequency and manner with which customers use a company's products, and their propensity to switch vendors when they are offered better prices or more targeted features—can improve customer satisfaction and cement customer loyalty. A change of only a few percentage points of customer retention can equate to hundreds of millions of dollars to a large company.

System Components

Data warehousing frameworks typically comprise four elements: data residing in one or more database systems, software to translate data, connectivity software to transfer data between databases and platforms, and end-user query tools.

Key system components include an information store of historical events (the data warehouse), warehouse administration tools, data manipulation tools, and decision support systems (DSS) that allow strategic analysis of the information.

The effectiveness of the data warehouse architecture depends on how well it addresses the following issues:

- *Warehouse population* A central data repository is built by consolidating data from various sources. The data may consist of transaction-based events and related information needed to isolate and aggregate those events. Volume tends to be high, so performance and cost are key considerations for the warehouse and operational data sources.
- *Warehouse volume* The data in a large warehouse can be made more accessible by arranging them into data marts, or specialized subsets of the data warehouse. It might take days for a query to run through a multiterabyte data warehouse. Data marts emerged to improve system and network performance, since it is not always necessary for everyone in the organization to have direct access to all the data in the warehouse.
- *Warehouse administration* This component focuses on maintaining the metadata (the data about data) that provides analytical derivation, exception recognition, integrity, controls, and security. Metadata, which reside above the warehouse data, define the rules and content of the views provided from the entire domain of available information. They map user queries to the operational data sources needed to satisfy the request.
- *Operational data store (ODS)* The ODS draws its data from the various operational systems in the corporation, but it also can add information derived from data keys in a data mart. As such, it not only adds the value of a

common view of enterprise data, but it also can add data derived from trend analyses done on the data marts.

Web-Enabled Data Warehouses

Data warehouses can be integrated with the corporate intranet to extend access to remote users through their Web browsers. They can access canned reports formatted in HyperText Markup Language (HTML) that can be hosted on the corporate Web site. For users who need more than static HTML reports, plug-ins allow queries to the back-end database. With such tools, employees can do such things as drill down into the reports to find specific information.

Making a data warehouse accessible to Web users solves a dilemma faced by many companies. On one hand, they do not want to limit users by providing only predefined HTML reports that cannot be manipulated. On the other, they do not want to overwhelm users with an OLAP tool they are not trained to understand. A Web-based OLAP tool that allows some interactivity with the data warehouse offers a viable alternative for users who are capable of handling simple queries. In turn, this makes corporate information more accessible to a broader range of users, including business analysts, product planners, and salespeople.

Since not everybody has the same information requirements, some companies have implemented multiple reporting options:

- *Canned reports* These are predefined, executive-level reports that can only be viewed through the Web browser. Users need little to no technical expertise, knowledge of the data, or training because they can only view the reports, not interact with them.
- *Ready-to-run reports* For those with some technical expertise and knowledge of the data, report templates are provided that users fill in with their query requirements. While dynamic, these reports are limited to specified field values, fill-in boxes, and queries.

- *Ad-hoc reports* For the technically astute who are familiar with the data to run free-form queries, unlimited access to the data warehouse is provided. They can fill in all field values, choose among multiple fill-in boxes, and run complex queries.

Web-Based Architecture

Many vendors offer Web tools that support a tiered intranet architecture comprising Web browsers, Web servers, application servers, and databases. The Web servers submit user requests to an application server via a gateway such as the Common Gateway Interface (CGI) or server API. The application server translates HTML requests into calls or SQL statements it can submit to the database. The application packages the result and returns it to the Web server in the proper format. The Web server forwards the result to the client.

This model can be enhanced with Java applets or other client-side programs. For example, the query form can be presented as a Java applet rather than the usual CGI (Figure D-1). Among the advantages of a Java-based query form is that error checking can be performed locally rather than at the server. If certain fields are not filled in properly, for example, an appropriate error message can be displayed before the query is allowed to reach the Web server. This helps control the load on the server.

Summary

Increasingly, data warehouses are reaching the terabyte level because businesses and government agencies are not only collecting more data—they are keeping it longer for the purpose of analyzing trends. Moving to a distributed architecture by deploying data marts can make large data warehouses more manageable. A distributed "logical" warehouse can offer greater efficiency and, in the process, provide

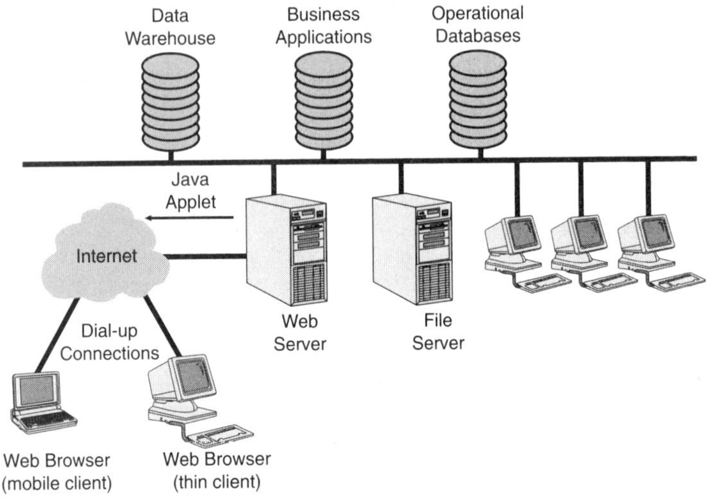

Figure D-1 Through a Web browser, a remote user is presented with a Java applet containing a query field. When the query is launched through the intranet, it goes to the company's Web server, where a canned report is retrieved. If the user requires a custom report, the Web server passes the request to the back-end data warehouse. After processing the request, the results are passed to the Web server and then delivered to the user.

redundancy and load balancing to guard against data loss and improve overall performance. Extending access to the data warehouse over an intranet leverages information assets by making them available to any employee at any location.

See also

Storage Area Networks

DIGITAL DIVIDE

The term *digital divide* refers to the separation of people into two groups according to their level of access to the Internet.

On one side are those who have access to the best information technology that society has to offer in terms of the most powerful computers, the best telephone service, and fastest Internet service, as well as a wealth of content and training relevant to their lives. On the other side are those who for one reason or another do not have access to the newest or best computers, the most reliable telephone service, or the fastest or most convenient Internet services. The difference between these two groups of people is what has been called the *digital divide.*

Social Topology

The private sector builds where there is the potential for high-volume traffic and the means to pay for it. In most communities, the fiberoptic rings circle the business district, not the poor suburban neighborhood or the inner city. Typically, providers that have spent years building their infrastructures do not come back and fill in the underserved neighborhoods. They build out networks only in areas that have a high concentration of the users who can afford the services they offer. Critics charge that while this may be a shrewd financial strategy, the social impact could be devastating.

Neglected communities tend to be the ones struggling hardest, including those with a high concentration of minorities. If the trends do not change, critics argue, these communities will miss out on developing high-speed home uses, such as telemedicine, distance learning, and telecommuting. The result is that economic development will suffer. When inner cities do not get high-speed Internet access, employers are limited in the jobs they can create, and this puts the community further behind the technology curve.

Many forecasters anticipate a huge division between the high-speed haves and have-nots. By the end of 2002, about half of all households will be able to choose between a phone line equipped with Digital Subscriber Line (DSL) and a cable high-speed service. The other half still will not be able to get any high-speed service. The result is a two-tiered system in which some communities enjoy inexpensive state-of-the-art broadband services, while others either pay extravagant amounts or have no service at all.

But the situation is changing. Not only is technology improving, but also the devices are dropping in cost as opposed to increasing. Also important is the $2 billion spent every year since 1997 to wire schools to the Internet, as well as several hundred million dollars a year in projects funded by the departments of education and commerce.

Digital Inclusion

A number of reports, including an August 2000 survey by the Census Bureau, show that what once looked like a divide based on race and income now looks increasingly like one that has more to do with income than anything else.

In December 1998, only 9.2 percent of blacks were connected to the Internet at home versus 26.7 percent of whites, 8.7 percent of Hispanics, and 25.6 percent of Asian Americans and Pacific Islanders.

By August 2000, those numbers had changed radically. Whites stepped up their usage considerably; 50.3 percent of Census respondents said they had an Internet connection at home, an 88 percent increase over the previous 20 months.

But blacks and Latinos flocked to the Internet in even greater numbers. By August 2000, 29.3 percent of blacks and 23.7 percent of Hispanics were online. In other words, blacks increased their usage of the Internet by more than 218 percent, while Hispanics' use was up more than 172 percent.

These changes have come so fast that even those who once backed strong government intervention to close the digital divide concede that much of the problem may besolved more easily than they once thought. The debate has moved from simple access to the Internet to the skills needed to use the Internet effectively as well as the relevance of information they can find once they are connected.

Surveys have revealed that participants want more job classifieds, especially those about entry-level positions. They want better information about housing opportunities and online education. They want more information in their native languages, more sites written at a level that beginning English speakers could understand, and information about health services.

But few sites deliver those things. Just 1 percent of sites surveyed in 2000 were written at a limited-literacy level. Likewise, 1 percent of sites surveyed could direct users to local entry-level jobs, and only 1 percent of sites dealt with low-cost housing.

Nationwide, about 22 percent of adults—44 million people—lack the reading and writing skills necessary to function fully in modern society. Many of those people are primarily non-English speakers. About 32 million adults, or roughly 16 percent of Americans over age 18, fall into this category. Although these people will be online eventually, service providers insist, they claim technology will not likely be of much help to them until the nation reforms its education system first.

Summary

To be on the less fortunate side of the so-called digital divide means that there is less opportunity to take part in the new information-based economy, in which many more jobs will be related to computers. It also means that there is less opportunity to take part in the education, training, shopping, entertainment, and communications opportunities that are available online. Now that a large number of Americans regularly use the Internet to conduct daily activities, people who lack access to those tools are at a growing disadvantage. Therefore, raising the level of digital access by increasing the number of Americans using the technology tools of the digital age is deemed to be a vitally important national goal by policy makers and community activists.

See also

Electronic Commerce

Internet

Telecommuting

DIGITAL SIGNATURES

A digital signature is a method for protecting the integrity of an online transaction. It is based on asymmetric cryptography that entails the use of a unique pair of private and public keys, which are certified by a trusted certificate authority. When a person "signs" a transaction, a unique mathematical code is created with the help of a private key and the actual contents of the document. This "signature," which is bound to the transaction, can identify the signer's identity by its relationship to the digital certificate. If the document is changed in any way, the digital signature is invalidated, as is the transaction. If the document is not changed and passes the validation test, the sender cannot repudiate the document or the transaction.

With the passage of the Electronic Signatures in Global and National Commerce Act (E-SIGN) in June 2000, electronic signatures and online contracts gained legal acceptance in the United States for the first time. The law went into effect on September 30, 2000, making it possible for businesses and individuals to sign enforceable contracts online with just a few mouse clicks. Digital signatures have the same binding legal status as paper and ink.

For individuals, the digital signatures make it easier to do things like buy an insurance policy or open a brokerage account on the Internet or use credit cards for online purchases. For businesses, digital signatures can speed the pace of transactions with suppliers, consultants, service firms, and other partners over the Internet, private intranet, or shared extranet, since it is unnecessary to fax or mail paper documents back and forth. This saves time, money, and labor.

Although the law provides legal recognition to electronic signatures, it does not specify the form they should take; in other words, the law is technology-neutral. Thus an electronic signature can be merely a string of invisible computer code attached to an electronic document. The law also does not make provisions for guaranteeing the authenticity of the documents to which digital signatures are attached. So, for complete protection, digital signatures must be used in conjunction with digital certificates (Figure D-2) from a trusted issuer such as VeriSign. For a small fee, individuals can download a digital certificate from the VeriSign Web site.

Users of Microsoft Outlook e-mail can obtain similar technology for free by clicking on a toolbar and inserting an electronic signature. In both instances, the electronic signature is actually a complex string of invisible computer bits that ties the sender of the message to the message itself.

To make a document tamperproof, however, requires another step: encryption. Basically, encryption works by pro-

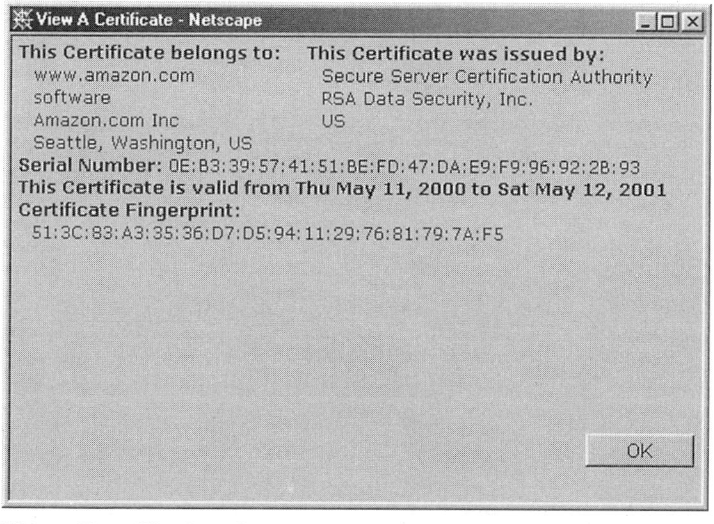

Figure D-2 The digital certificate of Amazon.com issued by RSA Data Security, Inc.

tecting computer files in such a way that they can only be opened with a special code or "key." The encryption process actually entails the use of two keys: a "public key" and a "private key." A person's public key can be used by anyone to encrypt the document, but only the holder of the private key can decrypt the document and make sense of it.

With a private key and the right software, a person can put digital signatures on documents and other data. The software compresses the data into just a few lines by a process called "hashing." These few lines are called a "message digest." The software then encrypts the message digest with a private key. The result is the digital signature. Finally, the software appends the digital signature to the document. All the data that were hashed have been "signed." When the document is sent, the recipient's software decrypts the signature using the sender's public key, changing it back into a message digest. If this works, then it proves the sender actually signed the document, because only the sender has the unique private key. The recipient's software then hashes the document data into a message digest. If the message digest is the same as the message digest created when the signature was decrypted, then the recipient knows that the signed data have not been changed (Figure D-3).

Among the companies that offer digital signature products is Lexign, Inc. (formerly, E-lock Technologies). The company's ProSigner enables users to add electronic signatures to documents easily and quickly. The software also allows multiple signatures on the same document, which can be useful in a number of scenarios. For example, all parties involved in a legal agreement could sign the same document, as could the superior of an employee whose signature had initially granted approval.

For the user, the experience is no more difficult than going through any other step-by-step, wizard-driven process. Once the software is installed, the user downloads a digital certificate, either from Lexign or from another provider such as VeriSign. A company's security administrator also has to

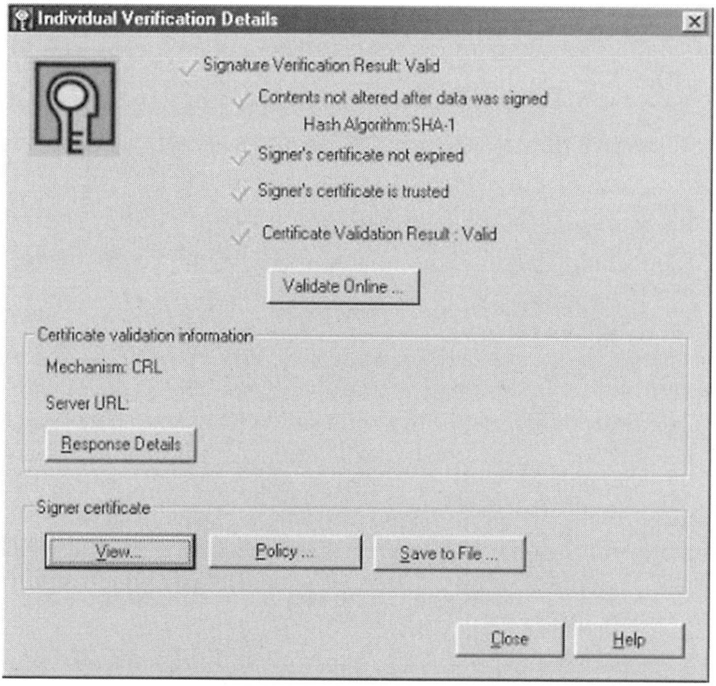

Figure D-3 A digital signature is validated when the recipient opens the document with viewer software such as the E-Lock Reader from E-Lock Technologies, which is available for free download from the company's Web site.

define the software's users and any policies that will be implemented. E-Lock offers two applets for this task.

The Policy Manager lets administrators define and apply rules to documents and certificates. These policies describe the level and strength of security applied. They not only give businesses control over security measures applied to individual documents, but they also make it easier to audit and maintain the uniformity of all digitally signed documents. Using the Profile Manager, end users can assign settings to their profiles that are then applied to documents that they sign. These include what certificate to use and whether users can attach a bitmap to documents. The actual process of document signing can be implemented in one of two ways: by opening Lexign's ProSigner software or by clicking icons on a toolbar that the software installs in Microsoft Office applications.

Summary

A digital signature does not resemble a handwritten signature. It relies on a set of keys to lock and unlock the contents of a document and a digital certificate for authentication. The result is that transactions that once took days or weeks to wrap up can be handled instantly online. The digital signature law does not mandate the use of digital signatures for all transactions; in fact, paper will still be required for many court documents such as eviction notices, wills, and court orders. Still, the digital signature law is expected to take the worry out of doing business on the Internet. With the legal risk out of the way, businesses can now focus on which digital signature systems will best work for them.

See also

Electronic Commerce

Network Security

DIGITAL SUBSCRIBER LINE TECHNOLOGIES

Digital Subscriber Line (DSL) is a category of local loop technologies that turn an existing twisted-pair line, normally used for plain old telephone service (POTS), into a high-speed digital line for Internet access. The electronics at both ends of the local loop compensate for impairments that would normally impede high-speed data transmission over ordinary twisted-pair copper lines. This enables ILECs and their competitors to offer high-speed connections over which a variety of advanced broadband data services and value-added applications can be offered, including Web surfing, news feeds, VPN access, and in some cases, videoconferencing and telephone calls.

DSL provides an economical way to satisfy surging demand among businesses and consumers for huge amounts of bandwidth, which is especially needed for Internet access. In leveraging existing copper local loops, DSL obviates the need for huge capital investment to bring fiber to the customer premises or to the curb in order to offer broadband services. Instead, idle twisted pairs to the customer premises can be provisioned to support high-speed data services. Users are added simply by installing DSL access products at the customer premises and connecting the DSL line to the appropriate voice or data network switch via a DSL concentrator at the central office (CO) or a serving wire center (SWC) where the data and voice are split out for distribution to the appropriate network (Figure D-4).

There are about a dozen DSL technologies currently available—each optimized for a given level of performance relative to the distance of the customer premises from the CO or SWC. The farther away the customer is from the CO or SWC, the lower is the speed of the DSL in both the upstream (toward the network) and downstream (toward the user) directions. The closer the customer location is from the CO or SWC, the greater is the speed in both directions.

Common Characteristics of DSL

Regardless of the specific type of technology used to implement DSL, the different varieties share some common characteristics:

- All DSL services are provisioned over the same unshielded twisted-pair copper wiring that is commonly used for plain old telephone service (POTS). Thus no spe-

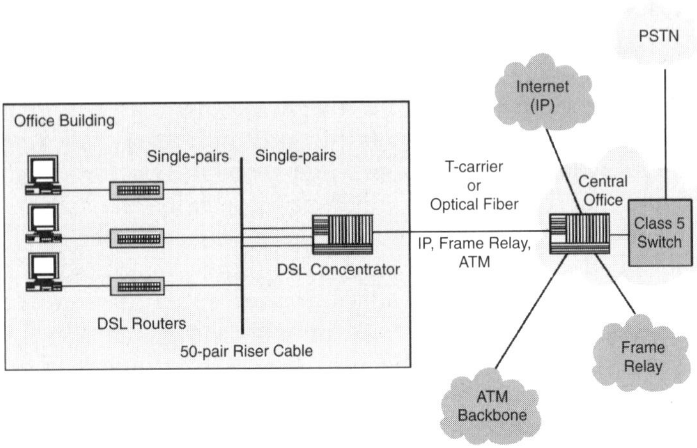

Figure D-4 Each customer location is equipped with one or more DSL modems. DSL lines from multiple customers are concentrated at a DSL access multiplexer (DSLAM) that can be positioned in a building's equipment room or at the CO where voice and data are split out for delivery to appropriate networks.

cial connection must be installed to the customer premises to obtain this type of service.

- All the DSL varieties offer a means to turn a low-quality voice-grade POTS line into a high-quality broadband data line. Electronic equipment at both ends of the POTS lines adapts or compensates for impairments that would normally corrupt high-speed data transmission over ordinary twisted-pair copper lines. The method by which this "line conditioning" is achieved differs by equipment vendor, which helps account for the slight variance in maximum line speeds for the same DSL service at comparable distances. Other factors that account for these variances include the specific gauge of the wire [e.g., 24 American Wire Gauge (AWG) versus 26 AWG] over which the DSL service is provisioned.
- All the DSL technologies conform to a similar configuration of user and service provider equipment. At a telecom-

muter's home, a branch office, or other corporate facility, DSL access requires a copper phone line that is connected to a DSL modem, which is actually a router. At the service provider's CO or SWC, concentration/server equipment connects multiple users and passes the transmissions to their respective voice and data networks.

- Once the DSL modem or router is connected to the DSL and is put into service by the carrier, it remains continuously available to the user without the need to dial up every time access is required—the service is "always on," just like a LAN. To use the service to access the Internet, for example, users merely open their Web browser.
- DSL access concentrators at the CO or SWC help carriers relieve congestion in their voice switching systems. This equipment partitions voice and data traffic, directing data onto a separate packet, frame, or cell-based data network and directing voice onto the Public Switched Telephone Network (PSTN).
- DSL is inherently more secure than other access technologies. DSL provides a dedicated point-to-point connection to the network that cannot be accessed by others, except by physically tapping into the line. On cable TV networks, for instance, many subscribers share the same cable for Internet access, so there is always the possibility of an intrusion by hackers. To guard against this, cable customers must purchase their own firewall software and have the technical expertise to configure and manage it.
- DSL is also very flexible in the types of traffic data formats it can accommodate. In addition to voice, DSL can be used to transport IP, Frame Relay, and ATM traffic from the customer premises through the local loop and to the appropriate network, including the PSTN, the Internet, or a corporate VPN.

When it comes to price, DSL is far more economical than other digital technologies, such as T1. Like T1, DSL is priced at a flat rate per month with unlimited hours of access.

While T1 access circuits cost anywhere from $150 to $700 per month or more, depending on market, DSL can cost much less for comparable bandwidth—as low as $29.95 a month for 256 kbps to $250 a month for 7 Mbps.

Asymmetric versus Symmetric

The bandwidth available over the DSL is carved up in a variety of ways to meet the needs of particular applications. When the upstream and downstream speeds are different, the DSL is referred to as asymmetric; that is, much greater bandwidth is available in the downstream direction than the upstream direction, as in the case of Asymmetric Digital Subscriber Line (ADSL).

ADSL runs at up to 8 Mbps in the downstream direction and up to 640 kbps in the upstream direction, with the actual speeds depending on the distance of the customer location to the CO or SWC. This would meet the needs of Internet users who want to retrieve multimedia Web content very quickly without waiting an indeterminate period for the pages to be loaded to their computer, as is the case with dial-up modem connections. The lower upstream speed of ADSL is more than adequate for issuing information requests from the computer to the network, since simple queries usually do not require more than 16 kbps. This leaves enough bandwidth capacity to handle multiple voice channels as well as data.

Asymmetric operation is fine for such activities as Web surfing, but it may not be appropriate for applications like server mirroring, where huge amounts of data must be able to flow in both directions. Symmetric Digital Subscriber Line (SDSL) service, on the other hand, offers the same amount of bandwidth in both the upstream and downstream directions—160 kbps to about 2 Mbps is available in each direction, depending on the distance of the user's location to the CO or SWC. SDSL and other symmetric DSL services, such as High Bit Rate DSL (HDSL), are more suited to applications that once required a T1/E1 line. Among the popular symmetric applications are videoconferencing, interactive distance learning, and telecommuting.

Two versus Four Wire

Traditionally, voice has been handled in the local loop via two wires (i.e., a "twisted pair"). This continues to be a very economical way to provide millions of residential customers with POTS. Businesses, however, require better-quality local loops for high-speed digital communication. This is achieved by providing them with four-wire connections for such services as T1. Businesses pay a premium price for these connections.

But with the growing popularity of the Internet, even residential customers have a need for more bandwidth than can be obtained with 56-kbps modems and Basic Rate ISDN. Where the infrastructure is almost exclusively two-wire, new DSL technologies have been developed that are capable of providing bandwidth in the multimegabit-per-second range without requiring expensive upgrades of the local loop.

Increasingly, two-wire DSL solutions are moving into the business sector, and their performance now exceeds the previous performance levels of traditional four-wire solutions. This means carriers can offer services to twice the number of businesses or double the transmission speed to the same businesses—without incurring major local loop upgrade costs.

There are other advantages to offering two-wire DSL solutions to businesses. Provisioning T1 service requires the installation of repeaters every 4000 to 6000 feet to boost signal strength. This is an expensive, time-consuming task for the carrier, which inflates the costs of T1 service. Some two-wire DSL technologies, such as High-Speed Digital Subscriber Line Two-Wire (HDSL2), match or exceed the performance of T1 without the need for repeaters.

Rate Adjustment

A version of ADSL is available that adjusts dynamically to varying lengths and qualities of twisted-pair local access lines. Like ADSL, Rate Adaptive DSL (RADSL) delivers a high-capacity downstream channel and a lower-speed upstream channel while simultaneously providing POTS over standard copper loops. Unlike ADSL, which does not tune itself to changing line conditions, RADSL adjusts data rates up or down in much the same way ordinary modems do.

With RADSL, it is also possible to connect over different lines at varying speeds. Connection speed can be determined when the line synchs up, while the connection is active, or as the result of a signal from the CO.

Inverse Multiplexing

Multiple DSL lines can be bonded together to provide users with higher-speed services. For example, two-wire SDSL, which tops out at 2 Mbps, can be bonded with another two-wire SDSL to offer an access speed of up to 4 Mbps. With this much bandwidth, small to medium-sized companies can get the transmission capacity they need without resorting to more expensive four-wire T1 access lines.

The bonding arrangement for DSL requires the user to have multiple telephone lines over which the higher-speed service can be run. No additional hardware is required; a simple software change to the inverse multiplexers in the service provider's network implements the bonding process. Once the DSL lines are bonded together in the service provider's network, the user's data load is balanced across the active lines.

Service Provisioning

Most DSL service providers offer customers the means to check for local DSL availability via forms posted on their Web

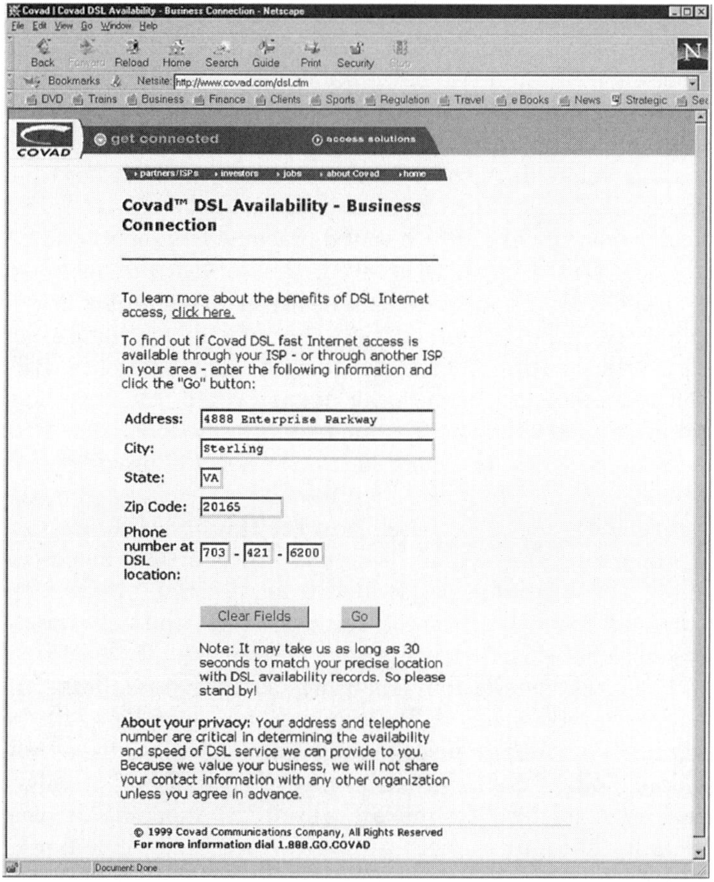

Figure D-5 Covad Communications is among the many companies that provide a Web form to allow consumers to check on the availability and speed of DSL in their service area.

sites, which prompt for the address and phone number of the DSL location (Figure D-5). If DSL service is available, the database application notifies the user of the type of service available as well as the speed of the connection. Since this type of service is only 70 percent accurate, a quirk of DSL services has been that the actual speed of the connection will not be positively known until the service is actually provisioned. Customers in high-rise office buildings have to factor in the vertical distance as part of the total circuit distance. Now there are online services users can call that conduct live tests of their line to determine if it is suitable for DSL and at what speed. These services can even determine the existence of a load coil on the line, which would preclude DSL operation.

Once DSL service is ordered, the service provider will arrange with the local exchange carrier to connect a line to the network interface outside the customer premises or share the existing POTS line for data. Either way, the DSL service provider (or a local agent) visits the customer premises and connects the line to a DSL modem.[1] The DSL modem is typically leased from the service provider, but users may now purchase them from a retail source and configure them through a Web browser (Figure D-6). By connecting the DSL modem to a hub, multiple LAN users can share the available bandwidth to access the Internet for such applications as e-mail, Web browsing, and news group discussions.

For large installations, such as a multitenant building, an appropriately sized DSL concentrator usually will be installed on the premises to aggregate the wires from individual DSL modems in each office. The building manager can have its own technician install the modems at each workstation and connect the wires to the concentrator or have the service provider do it for a fee. The concentrator usually will be scalable in terms of port density to accommodate future growth and support multiple types of DSL to meet the varying needs of users.

Several methods may be used to send aggregated traffic from the on-premises DSL concentrator to the service provider's location. Depending on the amount of traffic

[1]So-called DSL modems are actually routers. They are called modems because most consumers have become familiar with modems for dialing into the Internet or sending faxes from their home computers. Manufacturers stayed with this familiar term rather than risk confusing consumers with the unfamiliar term "router."

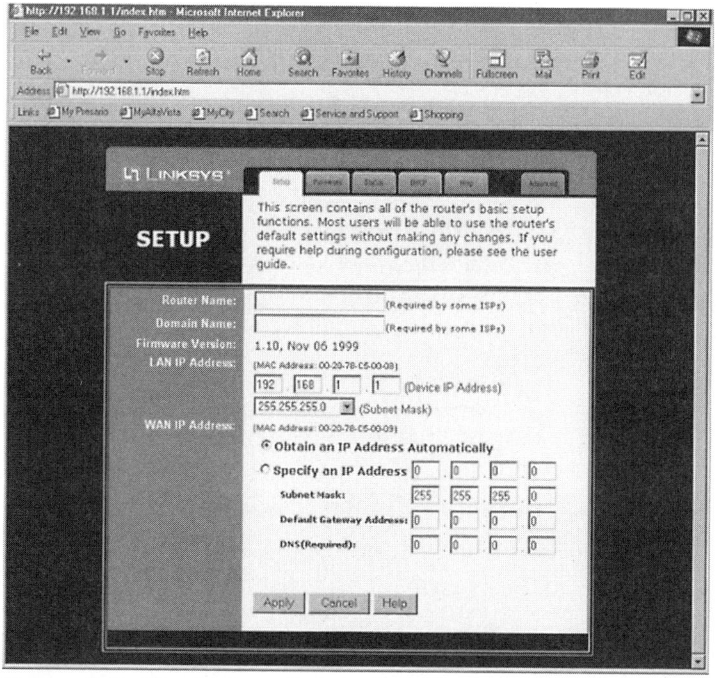

Figure D-6 The Linksys Instant Broadband EtherFast Cable/DSL Router is one of a growing number of off-the-shelf products that is easily configured by the user through a Web browser interface.

involved, leased T1 lines, Frame Relay over DS3, and ATM over SONET may be used. At the service provider's location, another concentrator splits out individual user sessions and tunnels for distribution to other networks.

Depending on the switching platform used, value-added services can be offered over DSL. Cisco's BPX 8650 switching platform, for example, lets service providers offer DSL on an IP-over-ATM backbone. This platform will let the carrier offer quality of service (QoS) through either ATM or multiprotocol label switching (MPLS), thus permitting the delivery of value-added services such as voice over DSL.

Provisioning Obstacles

In addition to distance limitations from CO/SWC to customer premises and the presence of fiber in the loop, there are several other obstacles that may stand in the way of provisioning digital subscriber lines.

Load Coils DSL service cannot be provisioned over lines that have load coils. These are used on persistently noisy phone lines to improve voice quality. The load coils are inserted at 3000-foot intervals along the line (Figure D-7).

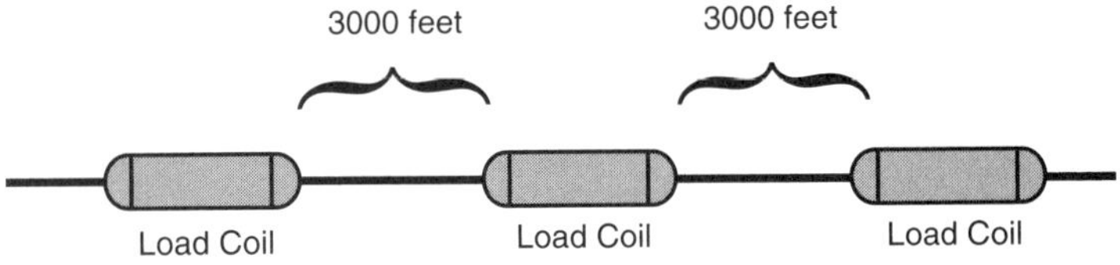

Figure D-7
Load coils improve line quality by limiting the audio band to about 4 kHz to filter out noise, but they also render the line useless for digital services such as DSL, which use the higher ranges of the audio band.

They filter out noise by limiting the usable audio band to about 4 kHz, which is just enough to allow voice conversation. But in limiting the audio band to 4 kHz, the upper frequency ranges of the audio band are no longer available to provision DSL. Although telephone company maps can pinpoint the locations of load coils, the companies will not remove them from the line. Not only would this open the audio band to noise, it also might introduce cross-talk into adjacent pairs. Where load coils are encountered, an unused pair must be selected for DSL.

Bridged Taps A bridged tap and lateral is a circuit that has been used in provisioning analog telephone service for many years. The bridged tap itself is a splicing mechanism for attaching an additional circuit to the normal distribution cable (Figure D-8). One leg allows the normal distribution path to

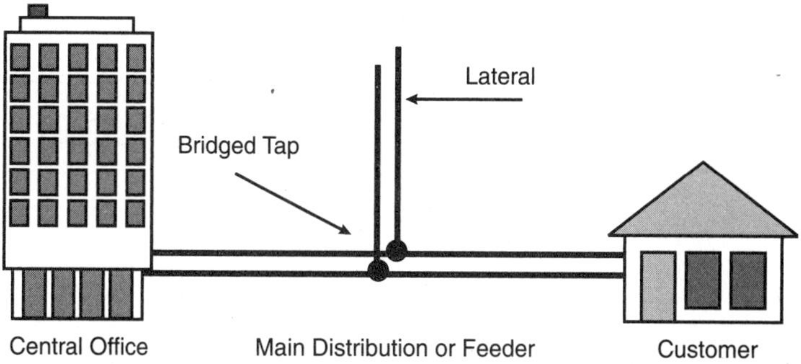

Figure D-8 The bridged tap is a splicing mechanism for attaching an additional circuit to the normal distribution cable. This renders both the lateral and main circuits useless for digital services such as DSL.

continue farther along, and the other is attached to an unused "lateral." A lateral or "spur" is any portion of a cable pair that is not in the direct path between the customer and the CO.

Laterals create problems on voice circuits that have been converted to digital. In addition to the main circuit, a lateral creates a second path for a digital signal, which weakens the signal on both paths. If the digital signal travels down a lateral that is open (i.e., not terminated), it is reflected back into the main circuit, where it can mix with the "good" digital signals. These echoes effectively render the data useless. In order for the digital circuit to operate properly, the bridged taps must be removed.

Pair-Gain Equipment DSL service cannot be provisioned over lines that attach to digital pair-gain systems (Figure D-9). These systems typically use HDSL technology to consolidate four, eight, or twelve 64-kbps channels over a single twisted copper pair, which is connected to a line card of the

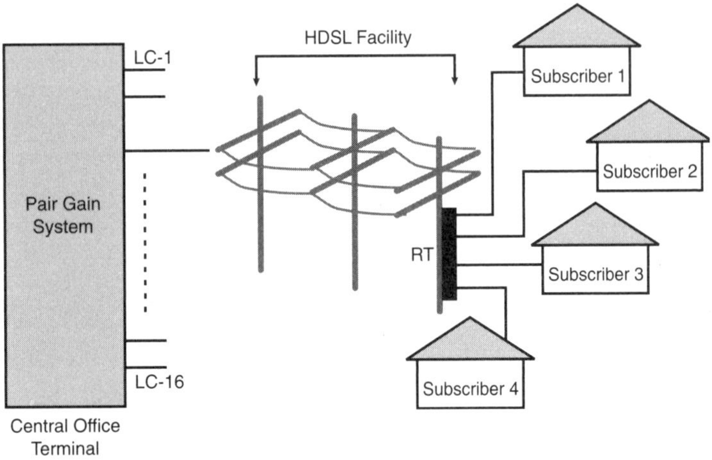

Figure D-9 When a carrier already uses a DSL technology to minimize its remote equipment requirements, the affected subscriber lines cannot be provisioned for DSL services.

pair-gain system. All subscriber circuits are completely independent of each other, and different applications can be mixed within the available channels of the system, including voice, fax, and modem data. These pair-gain systems provide an immediate, low-cost solution to subscriber loop shortages. They also provide the telephone company with flexibility in network planning and conserve existing investments in outside plant. However, DSL services cannot be provisioned over the individual subscriber lines. Instead, HDSL is used by the telephone company to consolidate multiple 64-kbps subscriber lines at a remote terminal for connection to a line card at the pair-gain system.

Digital Loop Carrier Systems (DLCSs) This type of outside voice-centric plant equipment concentrates local loop lines onto a shared T-carrier backbone, which backhauls them to the CO. While the DLCS enables telephone services to be extended to outlying locations that are beyond the normal reach of the

CO, shared backhaul prevents the lines from being provisioned for DSL. This is so because DSL requires a dedicated pair of copper wires from the customer location to the DSL access multiplexer in the CO. About 65 million of the 250 voice lines in the United States are served by DLCSs. Next-generation DLCSs and upgrades to legacy DLCSs overcome this limitation by supporting DSL and voice on the same plug-in cards.

Security

While DSL is certainly more secure than cable or wireless, it is not entirely immune to the problem. Although the access line is certainly more secure than cable because it is dedicated rather than shared, the security problem begins on the other side of the DSLAM, where traffic from multiple copper loops is aggregated for transport over a high-capacity fiber link to the Internet. Since the DSL connection is always on, it is possible for hackers on the Internet to find their way into the DSLAM and, from there, to individual computers on the other side. Since the DSLAM is not equipped with firewall capabilities, the CPE must be equipped to provide security.

Several vendors have addressed the need for security in their DSL products. Netopia, for example, offers a built-in firewall in its SDSL routers. The units come with preconfigured firewalls to disallow all inbound traffic originating from the Internet. They also filter packets on a per-connection profile basis for source/destination address, service, and protocol. Up to 255 rules are available in up to eight filter sets. The routers also support secure VPN access to corporate intranets and extranets via the Point-to-Point Tunneling Protocol (PPTP), plus 56-bit DES extensions for added protection. Security is further enhanced with a Network Address Translation (NAT), which hides all IP addresses on the LAN behind a single statically or dynamically assigned IP address on the Internet.

Management

Like any other service, the management of DSL has carrier and customer components. Initially, DSL services were hard for the carriers to provision, which stalled service delivery. But new tools have become available to enable carriers to streamline the rollout, setup, and ongoing management of DSL. There are now tools that can be used by customers, enabling them to change services, add bandwidth, and monitor carrier performance for compliance with service level agreements (SLAs).

Paradyne's Service Level Agreement Reporter, for example, can be used by both a service provider, to offer SLAs on its DSL service, and corporate network managers, to ensure that such SLAs are being met. The SLA Reporter supports Paradyne's multiservice DSLAM that carriers use to offer a variety of DSL services. The SLA Reporter enables service providers to obtain factual, network-based statistics and operational information to verify the quality of services being provided to their customers.

Graphical charts provide the performance information, which is delivered to customers via a secure Web site. Using familiar Web browser software, customers can view and even interact with the data provided in the SLA Reporter output. This allows corporate network planners, architects, and authorized users to obtain their own local views of network performance, throughput, capacity planning, and line quality.

Another company, Syndesis, offers service-provisioning software for DSL. Its NetProvision Creator streamlines carrier setup of DSL lines and extends service-ordering functions to the customer. The company's NetProvision Activator issues the commands that configure the individual network devices, including the customer modem, the DSL access multiplexer in the carrier switching office, and core switches and routers. By eliminating paper-based provisioning requests, the process is not only speeded up, but the number of order errors is reduced.

NetProvision Creator allows DSL service profiles to be defined in plain language, such as the speed of a line. This feature makes it easy for sales representatives to use the software to take customer orders for DSL services, and it allows customers to modify their DSL-based services via the Web. For example, a customer with a 256-kbps DSL service to an Internet service provider (ISP) could increase that bandwidth to 384 kbps via NetProvision Creator installed on a Web site. NetProvision Activator implements the requested changes.

Service Provider Selection

The choice of DSL service provider typically will hinge on the following key factors:

- Broadness of coverage
- Range of DSL services offered and their price points
- Type of equipment offered and its security features
- Service level agreement
- Availability of technical support
- Field service infrastructure
- Track record of timely installations
- Pace of DSL service rollouts to new areas
- Availability of value-added services
- Financial stability of the service provider

Another factor that enters into the choice of provider is the quality of the business partnerships the DSL service provider has in place. For example, partnering with a big IXC in the launch of enterprise services could help the DSL provider differentiate itself from the competition and accelerate the deployment of value-added services such as voice over DSL and secure VPNs. Partnering with a Web hosting company could allow the DSL provider to further differentiate itself by offering customers a total e-commerce solution that includes site development and trend reporting tools and collocation space for the servers.

Summary

DSL is a local loop technology that provides a high-speed digital service over an unused portion of the audio band on an ordinary twisted-pair copper line. This turns the POTS line into an economical platform for multiservice networking, enabling incumbent local exchange carriers (ILECs) and their competitors (i.e., CLECs, DLECs, and ISPs) to offer high-speed connections over which a variety of broadband data services can be offered, including Internet access, corporate connectivity via VPNs, and extra voice channels. Although fiber is faster and more reliable, it does not extend into enough customer locations. Twisted-pair copper lines, however, reach into every home and business and can be leveraged to provide advanced services and new revenue streams. By some estimates, the U.S. market for DSL is expected to reach $5 billion with 20 million subscribers in 2005. Although some major industry players are struggling financially, even declaring bankruptcy, demand for DSL remains strong.

See also

Cable Television Networks

Local Loop

T-Carrier Facilities

DIRECT BROADCAST SATELLITE

Direct broadcast satellite (DBS) operators use satellites to transmit video programming to subscribers, who must buy or rent a small parabolic dish antenna and pay a subscription fee to receive the programming service. DBS meets consumer demand for entertainment programming, Internet connectivity, and multimedia applications. DBS offers more programming choices for consumers and a platform for the development of new services, including video on demand, interactive TV, Internet messaging services, and personalized on-demand stock quotes. Much of the growing popularity of DBS is attributable to the programming choices available to consumers, as well as the picture quality provided by digital technology. And like cable television systems, DBS offers programming in the high-definition television (HDTV) format.

Operation

DBS operates in the K_u band, the group of frequencies from 12 to 18 GHz. TV shows and movies are stored on tape or in digital form at a video server, while live events are broadcast directly to a satellite. Stored programs are sent to the uplink (ground-to-satellite) center manually via tape or electronically from the video server over fiberoptic cable. Live events also pass through the uplink center. There, all programs—whether live or stored—are digitized (or redigitized) and compressed before they are uplinked to the satellites.

All DBS systems use the MPEG-2 compression scheme because it supports a wide range of compression ratios and data rates. It is capable of delivering a clean, high-resolution video signal and CD-quality sound. The satellites broadcast over 200 channels simultaneously via the downlink. The home satellite dish picks up all the channels and sends them via a cable to a set-top decoder. The set-top decoder tunes one channel, decodes the video, and sends an analog signal to the TV. One of the most popular DBS services is DirecTV, a unit of Hughes Electronics, which markets the service worldwide. First introduced in the United States in 1994, DirecTV offers over 225 channels and has over 10 million customers. The satellite service requires the user to have an 18-inch dish, a digital set-top decoder box, and a remote control. The system features an on-screen guide that lets users scan and select programming choices using the remote. Customers also can use the remote control to instantly order pay-per-view movies, as well as set parental controls and spending limits.

The DirecTV installation includes an access card, which provides security and encryption information and allows customers to control the use of the system. The access card also enables DirecTV to capture billing information. A standard telephone connection is also used to download billing information from the decoder box to the DirecTV billing center. This telephone line link enables DirecTV subscribers to order pay-per-view transmission as desired.

DirecTV allows users to integrate local broadcast channels with satellite-based transmissions. In markets where broadcast or cable systems are in place, users can maintain a basic cable subscription or connect a broadcast antenna to the DirecTV digital receiver to receive local and network broadcasts. A switch built into the remote control enables consumers to instantly switch between DirecTV and local stations.

HDTV programming from DirecTV is delivered from its 119° west longitude orbital slot location. To receive HDTV programming, consumers must have an HDTV set with a built-in DirecTV receiver or a DirecTV-enabled HDTV set-top converter box. A small elliptical satellite dish is needed to receive HDTV programming from the 119° orbital slot location, as well as core DirecTV programming from the 101° orbital location.

Internet Access

Internet access is provided via two services. The older service is DirecPC, a product that uses DirecTV technology in conjunction with a PC to deliver high-bandwidth, satellite-based access to the Internet. The DirecPC package includes a satellite dish and an expansion card designed for a PC's I/O bus. This receiver card transmits data from the Internet to the computer at 400 kbps, a rate 14 times faster than that of a 28.8-kbps modem connection.

Users connect to the ISP through a modem connection, but the ISP is responsible for routing data through the satellite uplink and transmitting the data to the receiver card and into the computer (Figure D-10). The service also pro-

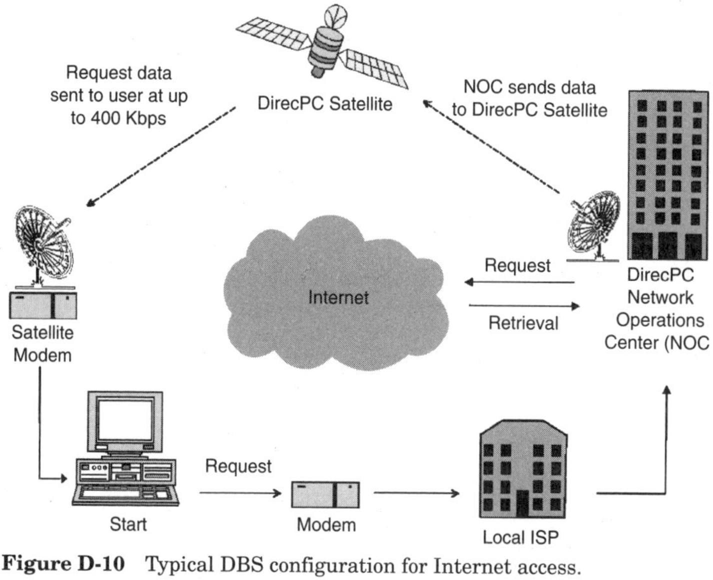

Figure D-10 Typical DBS configuration for Internet access.

vides users with the option to "narrowcast" software from the head end of a network to branch users during off-peak hours. Additionally, DirecPC transmits television broadcasts from major networks, such as CNN and ESPN, to the user's computer system.

The company's newer service, DirecWAY, offers a two-way broadband connection that offers 400 kbps on the downlink and about 150 kbps on the uplink, which eliminates the need for a modem and separate phone line. A new dish antenna provides access to the Internet and cable programming. A business-class DirecWAY service is also available. Multiple-seat account options (2 seats is the entry-level service; 5-, 10-, and 20-seat options are available), LAN software routing, and firewall security are offered as part of the business class service.

Summary

While DBS competes well against cable television in terms of television programming, it may not be able to compete with cable on the data front. In contrast with the finite bandwidth available to wireless and satellite systems, the terrestrial broadband pipe technologies available to cable systems offer bandwidth that is virtually limitless for almost all current practical purposes. Duplication of this pipe requires an investment of tens of billions of dollars and therefore would be impractical. Realizing this, DBS services limit downlink throughput per subscriber at about 400 kbps and reserve the right to limit bandwidth-hogging activities, such as audio and video streaming, and automatic file exchange applications. These restrictions are justified as being necessary to preserve an adequate level of service for all subscribers.

See also

Local Multipoint Distribution Service
Microwave Communications

DOMAIN NAME SYSTEM

The Domain Name System (DNS) is a collection of servers on the Internet that translate the plain text names assigned to Web pages, FTP sites, e-mail servers, and other resources into numerical IP addresses. The DNS acts like a phone book, translating widely known Web site names into numbers for a browser to contact.

For example, instead of having to remember an IP address such as 204.177.193.22, a Web server can be assigned a simple name such as "Amazon," which is much easier to remember. The DNS performs a lookup procedure to translate the text name into the assigned numerical IP address.

The DNS is, in fact, its own network of databases. If one DNS server does not know how to translate a particular domain name, it asks another one until the correct IP address is returned. A set of root servers in 13 locations around the world, mostly in the United States, keep the master records that map plain language domain names to IP addresses.

The organization charged with managing the worldwide DNS is the International Corporation for Assigned Names and Numbers (ICANN). This nonprofit corporation was formed in 1998 to assume responsibility for IP address space allocation, protocol parameter assignment, DNS management, and root server system management functions previously performed under U.S. government contract by the Internet Assigned Numbers Authority (IANA) and other entities.

Operation

DNS servers are arranged as a hierarchical database. At the top of the DNS database tree are root name servers, which contain pointers to master name servers for each of the top-level domains. Currently, there are seven top-level domains: com, edu, gov, int, mil, net, and org. To find out the numeric address of www.amazon.com, for example, a DNS server would ask the root name server for the address of the master name server that has responsibility for the .com domain.

In turn, the master name servers for each of the top-level domains contain a record and the name-server address of each domain name. In trying to find out the numeric IP address of www.amazon.com, for example, the DNS server asks the .com server for the name of the server that handles the amazon.com domain.

The individual name servers for each domain name, such as amazon.com, contain detailed address information for the hosts in that domain. Thus, in this example, the DNS server would then ask the amazon.com server for the name of the server that handles the amazon.com domain.

Finally, this most specific name server supplies the DNS server with the IP address of the machine called www.amazon.com. The entire process (Figure D-11) takes only a few seconds.

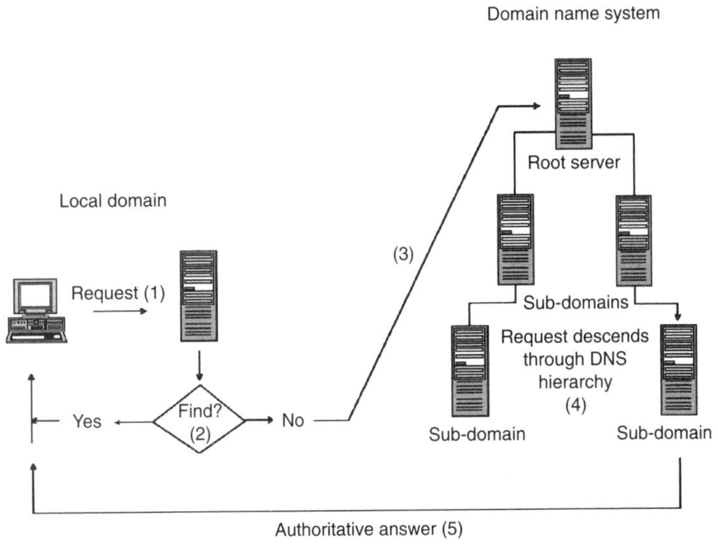

Figure D-11 When an Internet name is submitted to a DNS server (stage 1), the server checks its information and attempts to respond with the appropriate numerical IP address (stage 2). If the server cannot respond, it directs the request to a top-level DNS server (stage 3), which then sends the request down the DNS hierarchy (stage 4). Once an authoritative DNS server for the domain and machine is found, the response is sent to the request originator (stage 5). With this information, the client can then access the resource having that name.

DNS Configuration

There are two ways to implement DNS. One way is to use an ISP's DNS server. Many ISPs provide this as part of their service to subscribers; if not, they link to a specific DNS server on the Internet that does. The ISP will provide its subscribers with the numeric IP addresses of the primary and secondary DNS servers. Two DNS servers are required to avoid cutting off users from the Internet if the one DNS server goes down. Subscribers enter this address information when configuring their computers for dedicated or dial-up Internet access (Figure D-12).

Another way to implement the DNS is to set up the primary and secondary servers on a private intranet. This gives an organization more control over the administration of IP addresses. Having a DNS server inside a private network lets staff make changes, additions, and deletions on their own schedule instead of having to wait for ISP staff to do it. And if the server sits behind a firewall, security is enhanced because internal IP addresses can be hidden from public view. This helps prevent hackers on the Internet from gaining access to a company's network resources. In addition, name resolution will be faster for internal users because the organization's DNS server usually will not be as heavily loaded as an ISP's server.

To aid in the administration of a large number of IP addresses and conserve scarce IP addresses, the Dynamic Host Configuration Protocol (DHCP) is usually implemented. This is a server program that automatically assigns IP addresses to users as they log on the network. Each client is configured to automatically retrieve an IP address from a pool of unused IP addresses that are assigned to the organization. This capability is referred to as "dynamic IP addressing." Many ISPs also use DHCP to manage the IP addresses of large numbers of subscribers rather than assign them permanent addresses, known as "static IP addresses."

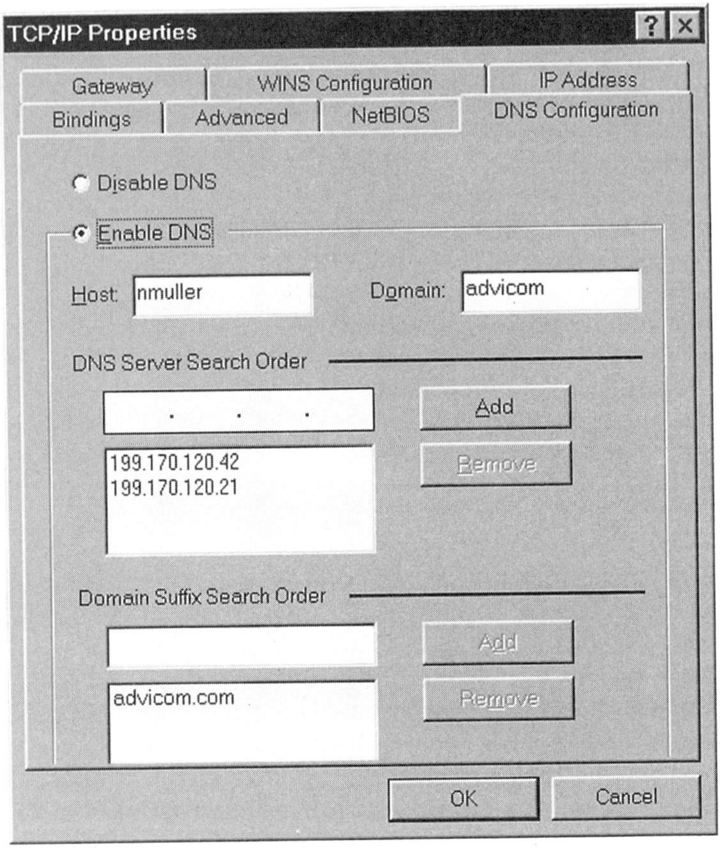

Figure D-12 An ISP provides its subscribers with the IP addresses of its primary and secondary DNS servers. As part of the dial-up networking configuration procedure, a Windows 98 user, for example, would enter these IP addresses into the TCP/IP "Properties" dialog box, along with the host name and domain name. When configured with this and other information, the TCP/IP dialer that comes with Windows 98 can be used to access the Internet.

Summary

The DNS is the foundation for message delivery and navigation on the Internet as well as private intranets. It enables plain text names to be assigned to various network resources, which are easier to remember than numerical IP addresses. Whether operated by an ISP or a company, two DNS servers are required—primary and secondary—to prevent users from being cut off from the network if one DNS server becomes disabled.

See also

Internet

Internet Service Providers (ISPs)

Proxy Servers

E

ELECTRONIC COMMERCE

Electronic commerce is simply the use of the Internet to conduct business. The increasing popularity of the Internet has awakened companies to the numerous commercial opportunities that entail the sale of goods and services to a vast global marketplace. Better security, simplified payment systems, and easy-to-use e-commerce development tools have contributed to the success of e-commerce in recent years. E-commerce can take many forms, including

- Business to business (B2B), such as companies seeking supplier bids over the Internet or companies buying office supplies online from Office Depot.
- Business to consumer (B2C), such as sites operated by traditional catalogue companies and sites operated by online banking services or brokerages that allow customers to view account information online.
- Consumer to consumer (C2C), such as sites that facilitate barter transactions or act as online auctions between buyers and sellers, as popularized by sites such as eBay.

Payment Systems

An essential function of any e-commerce site is the processing of payments. There are a number of tools available for creating and managing secure electronic payment systems. Such tools typically offer a shopping cart that can be accessed by a Web browser (Figure E-1). With these tools, retail merchants of any size can set up and manage virtual storefronts on the Internet with templates included in the software. With the storefront in place, companies can then

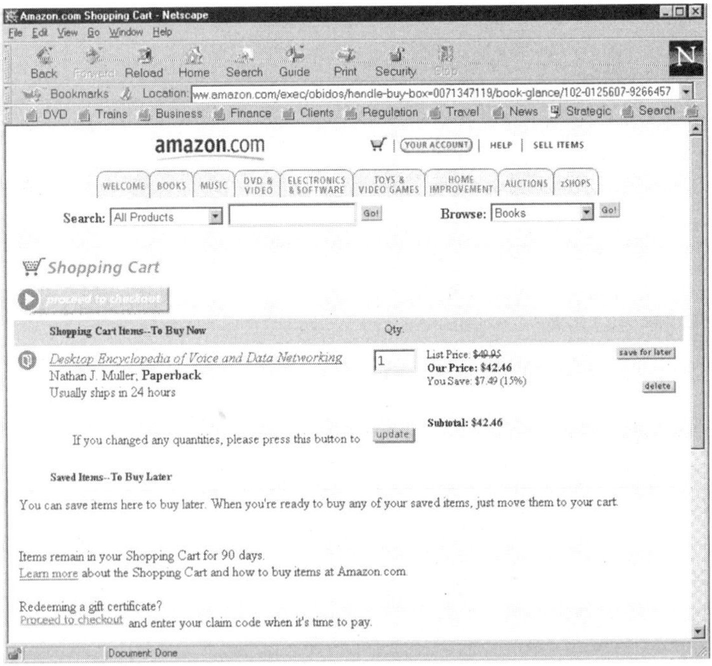

Figure E-1 Once users find an item of interest, they simply click the "Add to Shopping Cart" button. In this example from Amazon.com, the shopping cart page lists the item, quantity, and shipping schedule. When users have finished adding items to their shopping cart, they click the "Proceed to Checkout" icon, which takes them through the rest of the ordering process.

process credit card payments over the Internet. The application takes payment information from Internet clients protected with 128-bit Secure Sockets Layer (SSL) encryption. It also enables retailers to add product and related searching, sales tax and audit reports, discounts, and coupons.

There are other types of payment systems that make online shopping faster and more convenient. One of the newest payment systems is the virtual credit card, a virtual version of a consumer's plastic credit or debit card that resides on the cardholder's PC and autopopulates online merchant payment forms.

The virtual credit card is issued by the cardholder's issuing bank and contains the cardholder's basic card information. The virtual credit card application sits on the cardholder's PC desktop or toolbar until it is invoked for online payment. Once activated, by dragging and dropping it onto the merchant payment form, the virtual credit card automatically fills in the payment fields on the form. The cardholder information resides on the issuer's secure server and is transmitted from the secure server only at the time an online purchase is made. The user types in a personal password for verification and then hits the "Submit" button on the Web page to complete the purchase. Safe and secure shopping is ensured by SSL encryption of payment card and address information.

Online shopping is becoming easier with the introduction of computer keyboards that have a built-in credit card scanner. With the Compaq Smart Credit Card Internet Keyboard (Figure E-2), for example, the buyer simply inserts a credit card into the reader, enters a PIN, and the purchase is made. The keyboard scanner accepts a smart credit card, such as the Fleet Fusion Visa Smart Card promoted by Compaq. A customer can apply for the card by pressing a dedicated "Smart Credit Card Zone" button on the keyboard to complete an online application.

Other smart credit cards can be used with the keyboard as long as they support the EMV (Europay, MasterCard, Visa)

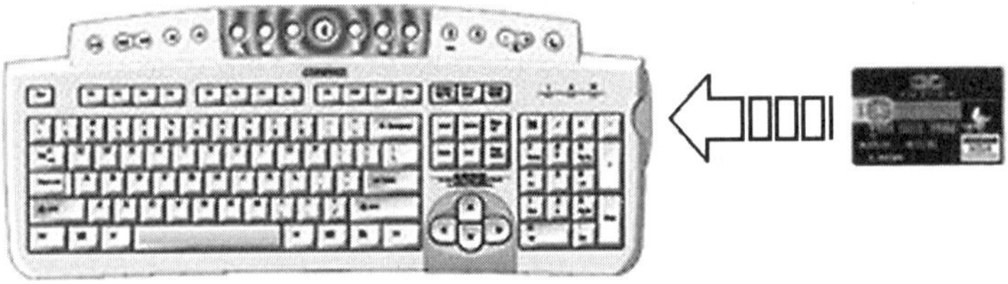

Figure E-2 Credit card scanners for home and small business use, such as the Compaq Smart Credit Card Internet Keyboard, facilitate safe, convenient, and value-added online transaction payments. Buyers will be able to shop with multiple online merchants enabled to accept the smart card technology.

standard. However, the end user will need to install the necessary application software originally included with the smart credit card in order to enable card functionality, which may differ according to the issuer.

Online Banking

Not only is the Internet suited for buying and selling, but it is suitable for online banking as well. Via the Internet, bank customers can monitor account balances, transfer money from their savings account to checking account, pay bills electronically, apply for car loans, and prequalify for mortgages. Hundreds of financial service organizations currently offer, or are implementing, remote-access banking and/or bill payment services.

Among the growing number of virtual banks on the Internet is Atlanta Internet Bank, a service of Carolina First Bank. The FDIC-insured virtual bank offers a comprehensive portfolio of services that can be accessed online via its Web page. Customers can choose any service a traditional bank has (Figure E-3). Clicking on the “Transfer Funds” icon, for example, allows the user to open a secure connection for transferring money from a money market fund to his or her checking account. Some virtual banks, such as

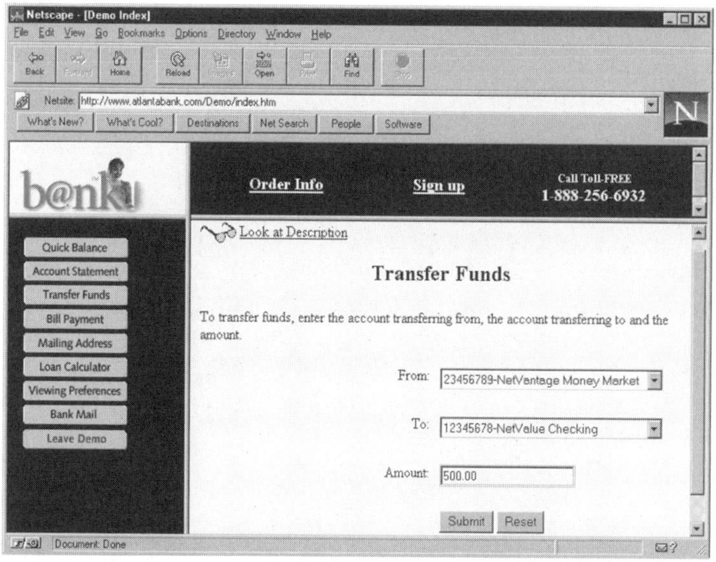

Figure E-3 The Web page of Atlanta Internet Bank, an FDIC-protected bank that operates on the Internet.

Security First Network Bank (SFNB), even offer check imaging, a feature that lets users view and print out a copy of each check that has cleared in the last 90 days. Customers also can transfer funds back and forth between accounts through point and click commands.

Internet checking accounts at SFNB have no monthly fees or minimum balance requirements. Customers get 20 free electronic payments each month, unlimited check-writing privileges, and a free ATM or debit card. SFNB is even insured by the FDIC, just like bricks-and-mortar banks. SFNB expects to offer customers online discount stock brokerage, insurance purchasing, and consumer loans. A single balance sheet will display all assets and liabilities, brokerage activity, and account information.

To extend these and other financial services over the Internet requires that retail companies, credit card issuers, and financial institutions implement secure transaction systems that guard against tampering by hackers and protect sensitive financial information from getting into the wrong hands.

Summary

From the consumer's perspective, certain online transactions finally make economic sense. On a transaction-by-transaction basis, online banking, for example, actually can be cheaper than postage stamps. With point and click ease, users now can launch Quicken, Microsoft Money, or a proprietary banking software package and pay bills, open accounts, close accounts, transfer money between accounts, monitor stock portfolios, make adjustments to investments, and even create monthly ledgers for tax purposes. These online services offer automation, immediacy, and flexibility—key selling points, especially among harried urban consumers. And even the smallest businesses can now automatically execute monthly financial transactions and save money on bookkeeping—all over the Internet through participating financial institutions.

See also

Electronic Data Interchange

ELECTRONIC DATA INTERCHANGE

Since the 1970s, electronic data interchange (EDI) has been promoted as the way to enhance business transactions with speed, accuracy, and cost savings. Essentially, EDI is the electronic transfer of business documents between companies in a structured, computer-processed data format. When the information is transmitted over a value-added network (VAN), it does not have to be rekeyed at the other end, which greatly improves the speed and accuracy of business transactions. Since business documents are in electronic form, cost savings can range from $3 to $10 per transaction over manual paper-based procedures.

The idea behind EDI is simple: Instead of processing a purchase order, for example, with multiple paper forms, and mailing it to a supplier, the data are passed through an application link, where software maps the data into a standard machine-readable data format. Those data are then transmitted to the supplier's computer, where they are passed to an application link that maps the data into the internal format expected by the supplier's order-entry system.

While electronic mail also transfers business data electronically, it generally uses a free format rather than a structured format. Since the sender may choose any format, it would be difficult to design an application program that would directly accept electronic mail input from many different sources without significant manual editing. This is the situation EDI standards are intended to overcome.

Benefits of EDI

EDI offers many benefits to participants. Data flow within an organization is streamlined, making it easier to develop and maintain complete audit trails for all transactions. Having all this information online also provides the means to track vendor performance, conduct cost-benefits analyses, improve project management, and enhance overall corporate financial control.

A reduced order-to-pay cycle is the natural result of eliminating the use of regular mail and cutting the time required to process paper at both ends of the transaction. This means that buyers can wait longer before replenishing stock ordered from suppliers, thereby reducing inventory and associated costs. The inherent efficiency of EDI means that buyers can pay for goods sooner, perhaps qualifying for further discounts. At the same time, sellers can improve their cash position through timely payments from buyers.

With more accurate and timely data, planning and forecasting can be greatly improved. This means that companies can better plan the receipt of materials to coincide with assembly-line schedules or continuous-control processing operations, thus eliminating unnecessary downtime caused by material shortages.

Finally, dramatic cost savings associated with daily business transactions can accrue to users of EDI, since the manual tasks of sorting, matching, filing, reconciling, and mailing documents are eliminated.

For many small to midsized companies, the pressure for an EDI implementation is increasingly coming from larger customers who indicate that EDI is the preferred method of transacting business with them. In some cases, EDI implementation is becoming a prerequisite for continuation of the business relationship. Some larger companies are even subsidizing EDI implementation for their smaller suppliers, recovering the cost by realizing even greater efficiencies and economies in transaction processing.

Role of VANs

The main service a value-added network (VAN) offers is a reliable, secure clearinghouse for EDI transmissions. It operates similar to an electronic mail system. The user sends the VAN a single transmission containing several interchanges destined for various trading partners. The VAN provides assurances that the data were received intact and distributes them to each recipient's inbox. In turn, the service provider sends the contents of each inbox to the appropriate subscriber on request. The VAN also provides security to guarantee that no one but the addressee can access the data. The VAN performs other services, including compliance checking and translation of data from one protocol standard to another. VANs also provide software installation and troubleshooting assistance to their subscribers.

The VAN supplies EDI translation software to its subscribers, which maps the data from an application (order entry) and translates them into the EDI format. This software runs on a modem-equipped PC with communications software to provide dial-up access to the VAN. If the volume of transactions justifies, a dedicated connection to the VAN may be used instead.

Another approach, which represents a more thorough adoption of EDI, entails a company integrating several business applications into the EDI system and using its own communication facilities and services to carry the transaction messages to an EDI gateway that connects to a VAN (Figure E-4). Each application is reengineered so that the output is already in the proscribed EDI format, making further translations unnecessary.

Interactive versus Batch Processing

The problem with traditional batch processing for EDI transactions is that as a company adds trading partners and

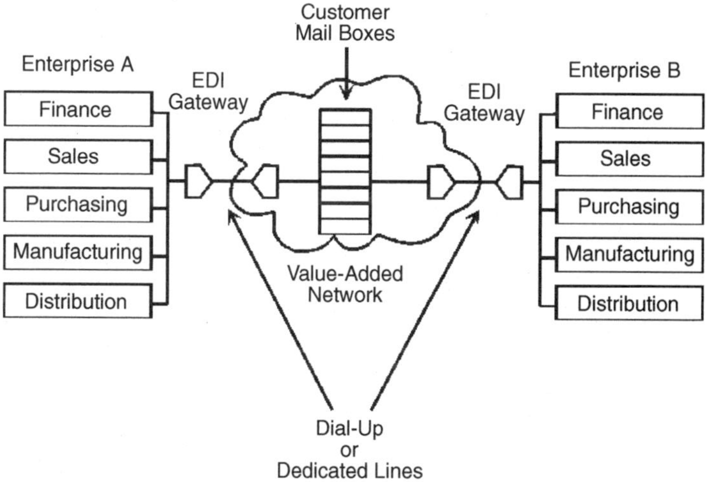

Figure E-4 Typical network configuration for EDI.

increases the volume of messages, batch jobs begin to contend with one another for access to computer and communications resources. EDI messages, such as shipping notices, may be delayed because the EDI translator is busy processing a large batch job or all available communications channels are being used. Aside from rescheduling jobs, these problems can be solved through the purchase of additional equipment or the leasing of more lines.

To improve this situation, fast-batch EDI and interactive EDI are available. Fast-batch EDI essentially speeds up the store-and-forward process by sending messages through electronic mail boxes on VANs directly into a recipient's computer system. Interactive EDI typically involves establishing a two-way link that enables trading partners to rapidly exchange records or fields within an EDI message rather than the entire message.

Web-Based EDI

A newer innovation in EDI entails use of the Web for sending business documents. This allows smaller firms that traditionally have not used EDI software and subscribed to the services of a VAN to participate in the process at a very low cost of entry. The common format for Web-based document exchanges is the HyperText Markup Language (HTML). This can work in several ways. In one scenario, when a company fills out HTML-based forms, they are converted behind a VAN's Web site (i.e., gateway) into EDI-formatted messages before being passed on to the recipient.

In another scenario, two trading partners might exchange documents over the Web directly, without going through a VAN. A back-end program puts the information in the right format so that it can be stored in a database. Security is maintained during transmission over the Web through the Secure HyperText Transfer Protocol (SHTTP) or Secure Sockets Layer (SSL) protocol.

Increasingly, the eXtended Markup Language (XML) is being used for EDI transactions over the Web.[1] XML is a technology that allows the creation of an unlimited number of different markup languages for different purposes, including electronic business. The attribute of XML that makes it so popular is that the various special-purpose languages that it can be used to define all can be parsed by a single standardized processor small enough to be built into every Web browser.

XML can be integrated with existing EDI systems by providing application-specific forms that users can complete to generate EDI messages for transmission between computers over the Internet or through existing VANs. XML allows data received in EDI format to be interpreted according to sets of predefined rules for display by the receiver on standardized browsers using a user-defined template rather than having to rely on customized display packages.

XML can enhance existing EDI applications by allowing message creators to add application-specific data to standardized message sets. This allows message creators or receivers to display the contents of each field in conjunction with explanatory material that is specific to the application and the language preferences of the user. System developers not only can customize the help information associated with the data for each field but also can embed field value checking and integrate it with checks on the validity of the data with respect to information stored on local databases.

Summary

In the future, businesses that have not or will not implement EDI will be at a competitive disadvantage. Companies that derive considerable benefit from EDI will be reluctant to deal with those who force them to get bogged down in

[1]XML is a simplified form of the Standard Generalized Markup Language (SGML), not an extended form of the HyperText Markup Language (HTML). The difference between XML and SGML is that the designers of XML took out a number of advanced SGML features that make a full SGML parser difficult to implement in a Web browser.

manually processing paper transactions. This goes against the trend toward flatter, decentralized organizational infrastructures, which many companies believe is necessary to compete more effectively on a global scale. The traditionally high cost of EDI is plummeting as more vendors offer inexpensive software that enables EDI to be run over the Web with XML. Even traditional VAN providers see the writing on the wall and are starting to offer Web-based EDI service. This will widen the appeal of EDI to even the smallest companies and further stimulate the growth of electronic commerce.

See also

Electronic Commerce
World Wide Web

ELECTRONIC MAIL

The popularity of electronic mail (e-mail) in recent years has paralleled the growth of the Internet. More mail is now delivered in electronic form over the Internet than is delivered by the U.S. Postal Service. This amounts to tens of billions of messages per month. Message delivery usually takes only minutes over the Internet, instead of days or weeks, as is typical with postal services in many countries.

Ray Tomlinson (Figure E-5) invented e-mail in 1971 while working at Bolt, Beranek and Newman (BBN), a company that had a government contract to work on the Advanced Research Projects Agency Network (ARPANET), the precursor of the Internet. An MIT graduate, Tomlinson was part of a team building an operating system when he came up with a "Send Message" program. At first it worked only on a local system, but he developed it further into cross-ARPANET mail. Tomlinson needed a character to separate a name from a place so that computers sending messages would not

Figure E-5 Ray Tomlinson invented e-mail in 1971.

confuse the two. He immediately came up with the @ sign, the only prepositional character on the keyboard.

Advantages of E-Mail

Automating information delivery and processing with electronic mail can reduce the cost of doing business dramatically, since the manual tasks of sorting, matching, filing, reconciling, and mailing paper files are virtually eliminated. There are also attendant cost savings on overnight delivery services, supplies, file storage, and clerical personnel.

Users can even send e-mail from within any application that supports Microsoft's Messaging Applications Programming Interface (MAPI) specification. Thus a file done in a MAPI-compliant word processor or spreadsheet application, for example, can be e-mailed as an attachment without having to leave that application.

Through mail gateways, e-mail can be sent to people who may subscribe to some other type of service, such as America Online or Microsoft Network (MSN). Some gateway services even transport messages from the Internet to wireless devices such as personal digital assistants (PDAs) and portable computers equipped with radio modems. And with more paging services supporting short-text messaging, e-mail can even be sent over the Internet to alphanumeric pagers.

The servers and gateways on the Internet take care of message routing and delivery. The e-mail address contains all the information necessary to route the message. If a message cannot be delivered because of some problem on the Internet, an error message is returned that explains the reason and estimates the time of delivery.

Internet Protocols

Currently, the dominant mail protocols used on the Internet are the Simple Mail Transfer Protocol (SMTP) and Post Office Protocol 3 (POP3). When installed on a server, SMTP gives it the capability to send and route messages over the Internet. POP3 is also installed on a server, giving it the capability to hold incoming e-mail until the recipient is ready to download it to his or her own computer. Once downloaded, the e-mail message can be opened, edited for reply, cut and pasted into another application, saved as a document, filed for future reference, forwarded, or deleted (Figure E-6).

POP was designed to support offline message access. Once downloaded, messages can be opened at any time and then marked for deletion from the mail server on next logon. This mode of access is not compatible with access from multiple computers because it tends to distribute messages across all the computers used for mail access.

A newer protocol, Internet Mail Access Protocol (IMAP4), allows users to access messages on the server as if they were local. For example, e-mail stored on an IMAP server can be

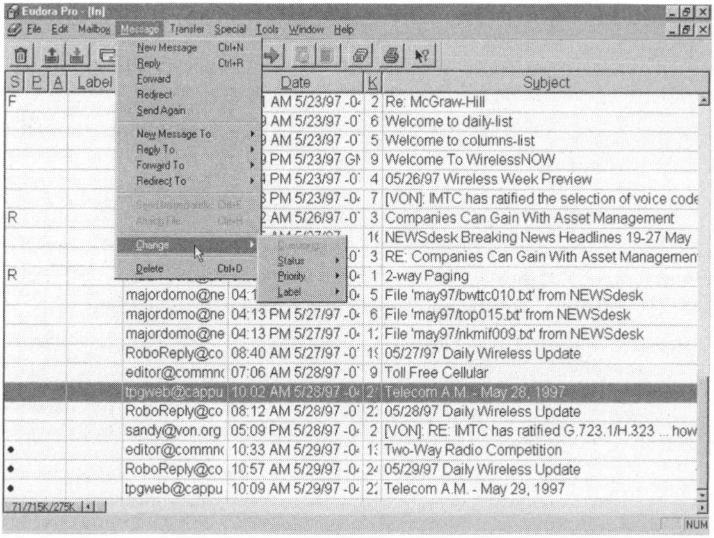

Figure E-6 A typical e-mail interface—in this case, QualComm's Eudora Pro.

manipulated from a desktop computer at home, a workstation at the office, and a notebook computer while traveling—without the need to transfer messages or files back and forth between them.

IMAP's ability to access messages (both new and saved) from more than one computer has become extremely important as reliance on electronic messaging and the use of multiple computers among mobile professionals increases. The key advantages of IMAP include

- Full compatibility with other Internet messaging standards, such as MIME[2]
- Message access and management from more than one computer
- Message access without reliance on less efficient file access protocols

[2]MIME stands for Multipurpose Internet Mail Extensions. It is a technique for encoding text, graphics, audio, and video files as attachments to SMTP-compatible Internet mail messages.

- Support for "online," "offline," and "disconnected" access modes
- Support for concurrent access to shared mailboxes
- Client software needs no knowledge about the server's file store format

IMAP supports operations for creating, deleting, and renaming mailboxes; checking for new messages; permanently removing messages; setting and clearing flags; server-based MIME parsing (relieving clients of this burden) and searching; and selective fetching of message attributes and text. In certain circumstances, IMAP also allows sent messages to be recalled.

When a user has second thoughts about sending a message, it can be recalled with another message to save embarrassment (Figure E-7). Alternatively, the original message can be replaced with a new message. The status of the recall attempt, whether it succeeded or failed, is reported back to the originator as a message in the Inbox. However, the recall/replace feature works only if the recipient has not yet opened the message.

Outsourcing Arrangements

Many businesses find it difficult and time-consuming to run their own e-mail systems. Businesses with up to 1000 employees may find outsourcing their e-mail a more attractive option. By subscribing to a carrier-provided e-mail service, these companies can save from 50 to 75 percent on the cost of buying and maintaining an in-house system.

Carriers even offer guaranteed levels of service that will minimize downtime and maintain 24 × 7 support for this mission-critical application. An interface allows the subscribing company to partition and administer accounts. Companies can even create multiple domains for divisions or contractors. Administration is all done through the service provider's Web site. Changes are easy and intuitive—no programming experience is necessary—and the changes

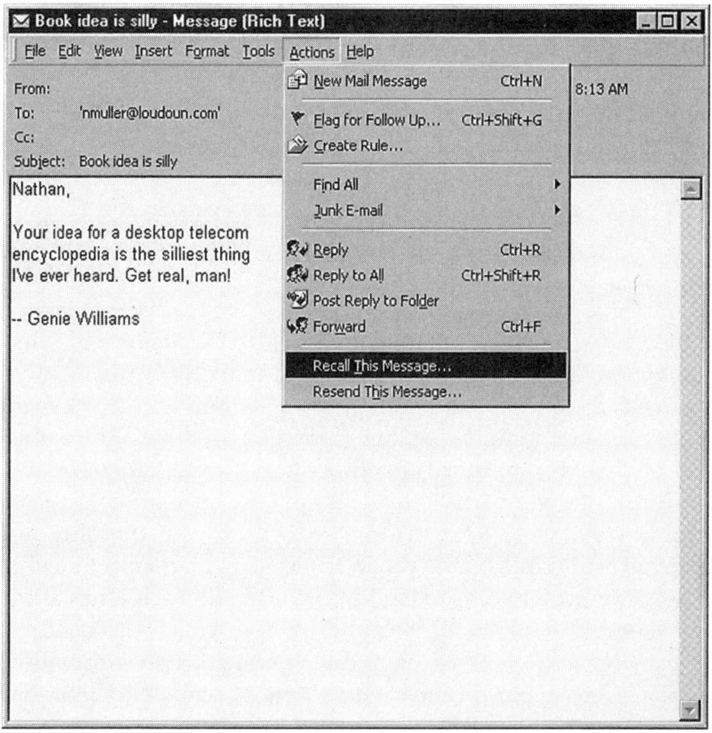

Figure E-7 When Microsoft Outlook is configured as an IMAP client, it can recall or replace a sent message, provided that the recipient has not yet opened it.

take place immediately. Many services offer spam controls to virtually eliminate junk mail.

Outsourcing can level the playing field for companies competing in the Internet economy. Businesses no longer need to budget for endless rounds of software upgrades. A carrier-provided service also can include value-added applications such as fax, collaboration, calendaring, and unified messaging services.

For companies that are not ready to entrust their entire e-mail operation to a third-party service provider, there are customized solutions that allow them to selectively outsource certain aspects and functions. For example, a company can choose to have its e-mail system hosted on the service provider's server. Such "midsourcing" can reduce the costs of administration and support and provide improved performance. In addition, the company can add new functionality easily, without incurring a costly upgrade.

Summary

Once considered by most companies to be just a fad, electronic mail is valued for its role in supporting daily business operations. In fact, the popularity of e-mail is now so great that many companies are being forced to upgrade the capacity of their communications links and systems to accommodate the growing traffic load. Steps are also being taken to minimize unnecessary traffic, such as by storing only one copy of e-mail attachments on a server rather than allowing the same attachment to be duplicated to all recipients of the message. Other companies are considering outsourcing their e-mail operations to carriers and Internet service providers (ISPs). This is part of the relatively new trend of applications outsourcing.

See also

Electronic Commerce
Internet Service Providers
Unified Messaging

ELECTRONIC SOFTWARE DISTRIBUTION

With a growing population of PC and workstation users deployed across widely dispersed geographic locations—each potentially using different combinations of operating systems, applications, databases, and network protocols—

software has become more complex and difficult to install, maintain, and meter. The ability to perform these tasks over a network from a central administration point can leverage investments in software, enforce vendor license agreements, qualify the organization for discounts on network licenses, and greatly reduce network support costs.

Industry experts estimate that the average 5-year cost of managing a single desktop PC exceeds $45,000 and that the 5-year cost of deployment and managing changes to new client-server applications averages an additional $45,000 to $55,000 per user. Automating the distribution and maintenance of software can cut support costs in half. Electronic software distribution (ESD) tools also can provide useful reports that can aid in problem resolution and determining the need for license upgrades.

Automating File Distribution

The complexity of managing the distribution and implementation of software at the desktop requires that network administrators make use of automated file distribution tools. By assisting a network administrator with tasks like packaging applications, checking for dependencies, and offering links to event and fault management platforms, these tools reduce installation time, lower costs, and speed problem resolution.

One of these tools is a programmable file distribution agent. It is used to automate the process of distributing files to particular groups or workstations. A file distribution job can be defined as software installations and upgrades, startup file updates, or file deletions. Using a file distribution agent, these types of changes can be applied to each workstation or group automatically.

The agent can be set up to collect file distribution status information. The network administrator can view this information at the console to determine if files were distributed successfully. The console allows the administrator to review status data, such as which workstations are set up for file distributions, the stations to which files have been distributed, and the number of stations waiting for distributions.

Because users can be authorized to log on to the network at one or more workstations, the file distribution agent determines where to distribute files based on the primary user (owner) of the workstation. The owner is established the first time that a hardware/software inventory is taken of the workstation. Before automated file distributions are run, the hardware inventory agent is usually run to check for resource availability, including memory and hard disk space. Distributions are made only if the required resources are available to run the software.

To help network administrators prepare for a major software distribution, some products offer routines called "wizards" that walk administrators through the steps required to assemble a "package." A package is a complete set of scripts, files, and recipients necessary to successfully complete a distribution and install the software. To reduce network traffic associated with software distributions, some products automatically compress packages before they are sent to another server or workstation. At the destination, the package is automatically decompressed when accessed.

When a file distribution job is about to run, target users receive a message indicating that files are about to be sent and requesting that they choose to either continue or cancel the job or postpone it for a more convenient time.

Managing Installed Software

Maintaining a software inventory allows the network administrator to quickly determine what operating systems and applications are installed on various servers and clients. In addition to knowing what software components are installed and how many are in use on the network (Figure

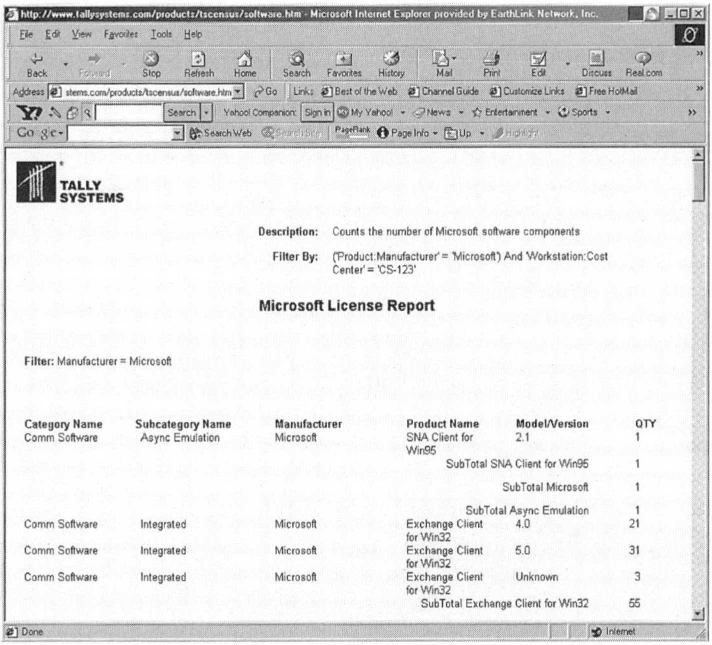

Figure E-8 Software license compliance summary report from Tally Systems' TS.Census. With the Microsoft filter applied, the report shows what Microsoft software components are installed on the network.

E-8), the network administrator can track application usage to ensure compliance with vendor license agreements.

The ESD tool creates and maintains a software inventory by scanning all the disk drives on the network. Usually software management tools come with preset lists of software packages they can identify during a scan of all disk drives on the network. Some tools can recognize several thousand software packages. Software that cannot be identified during a scan is tagged for further inquiry and manual data entry. The next time a scan is done, the added software packages will be identified properly.

Metering Software Usage

The ability to track software usage helps the network administrator ensure that the organization complies with software license agreements while making sure that users have access to required applications. Tracking software usage also helps reduce software acquisition costs, since accurate usage information can be used to determine which applications are run most before deciding on upgrades and how many copies to buy.

Metering allows the network administrator to control the number of concurrent users of each application. The network administrator also can choose to be notified of the times when users are denied access to particular applications because all available copies are in use. This may identify the need to purchase additional copies of the software or pay an additional license charge to the vendor so that more users can access the application.

Before users are granted access to metered software, the software inventory is checked to determine whether there are copies available. If no copies are available, a status message is issued indicating that there are no copies available. The user waits in a queue until a copy becomes available.

Metering software can save money on software purchases and ensure compliance with software copyright laws. For example, if there are 100 users of Microsoft Word on an enterprise network and only half that number use it concurrently, the software metering tool's load-balancing feature automatically handles the transfer of software licenses from one server to another on a temporary or permanent basis. Load balancing helps network administrators purchase licenses based on need rather than on the number of potential users at a given location.

License management capabilities are important to have because it is a felony under U.S. federal law to copy and use (or sell) software. Companies found guilty of copyright infringement face civil penalties of up to $150,000 for each work infringed on. In criminal cases, the maximum penalty for copyright infringement is up to $250,000 and a jail term of up to 5 years.

The Software & Information Industry Association (formerly the Software Publishers Association) runs a toll-free hotline and receives about 40 calls a day from whistle-blowers.[3] It sponsors an average of 250 lawsuits a year against companies suspected of software copyright violations. Having a license management capability can help the network administrator track down illegal copies of software and eliminate a company's exposure to litigation and financial risk.

Distribution over the Internet

Since 1996, corporations have had the option of subscribing to Internet-based software distribution services offered by application service providers (ASPs). Such services are usually hosted on a Web server and provide managed Internet-based delivery and tracking of business-critical software applications and documents. Available via a link from any Web or extranet site, such services offer all the functionality of an enterprise-class software management system, with the convenience and flexibility of a hosted Web service.

Such services are designed for large and midsized companies that are using the Web to extend their businesses and to lower the cost of customer support and product distribution. As these companies make broader use of their extranets, there is an increasing need to distribute proprietary digital files to support the online business service. On an outsource basis, service providers offer confirmed delivery and tracking of these software and document packages securely and cost effectively.

The entire service is available from a standard browser, offering users the flexibility and ease of use of an intuitive Web interface. Businesses seeking to outsource their software distribution operations simply create authorized user lists and packages, software, documents, and digital files that will be available for download. The service provider's

[3] Cases of software piracy can be reported via the SIIA's antipiracy hotline at 800-388-7478.

management server authenticates users and maintains a database to track distribution transactions, package versions, and customer accounts, providing comprehensive reports to support business operations.

Once the system is set up, users are assigned an authorization code and are directed to a secure, fully customized download portal site hosted by the service provider, where they can browse through available software, check their download histories, and select packages by name, version, or category. The download manager can be a Java applet that loads dynamically when a customer requests an update package. The applet executes a managed download, confirming receipt of the files on the customer's computer. As additional packages are made available, the service supplies automatic e-mail notification of the updates, providing a way to instantly notify authorized users and groups.

The service also may include a block-level restart feature, which automatically resumes the file transfer if a download is interrupted and synchronizes from the point of failure. All download activities are secure, using multilevel encryption and authentication features.

In addition to a completely managed service, some service providers also offer a lower-cost "self-drive" option, enabling IT and Web site managers to run the service from a browser, with all distribution and tracking activity operated from the service provider's distribution center.

Summary

ESD tools are essential for managing software assets. They can help trim support costs by permitting software to be distributed and installed from a central administration point; ensure compliance with vendor license agreements, thereby eliminating exposure to lawsuits for copyright infringement; and help companies manage software to minimize their investments while meeting the needs of all users.

See also

Application Service Providers
Asset Management

ETHERNET

Ethernet is a type of local area network (LAN) that uses a contention-based method of access to allow computers to share resources, send files, print documents, and transfer messages. Like other types of networks, Ethernet also carries Internet Protocol (IP) traffic, allowing multiple users on the LAN to conveniently access the Internet through a common router.

The Ethernet LAN originated as a result of the experimental work done by Xerox Corporation at its Palo Alto Research Center (PARC) in the mid-1970s. However, Robert Metcalfe (Figure E-9) is the individual generally credited with the development work that led to Ethernet. Once developed, Ethernet quickly became a de facto standard with the

Figure E-9 Robert M. Metcalfe began working for Xerox Corporation at its Palo Alto Research Center (PARC) in 1972 while working on his Ph.D. at Harvard. It was at PARC in 1973 that Dr. Metcalfe and D. R. Boggs invented Ethernet. In 1979, Metcalfe founded 3Com Corp., a computer networking company in Santa Clara, California.

backing of DEC and Intel. Xerox licensed Ethernet to other companies that developed products based on the specification issued by Xerox, Intel, and DEC. Much of the original Ethernet design was incorporated into the 802.3 Standard adopted in 1980 by the Institute of Electrical and Electronics Engineers (IEEE).

Ethernet is contention-based, meaning that stations compete with each other for access to the network, a process that is controlled by a statistical arbitration scheme. Each station "listens" to the network to determine if it is idle. On sensing that no traffic is currently on the line, the station is free to transmit. If the network is already in use, the station backs off and tries again. If multiple stations sense that the network is idle and transmit at the same time, a "collision" occurs, and each station backs off to try again at staggered intervals. This media access control scheme is known as Carrier Sense Multiple Access with Collision Detection (CSMA/CD).

Frame Format

The IEEE 802.3 Standard defines a multifield frame format, which differs only slightly from that of the original version of Ethernet, known as "pure" Ethernet (Figure E-10).

Preamble The frame begins with an 8-byte field called a "preamble," which consists of 56 bits having alternating 1 and 0 values. These are used for synchronization and to mark the start of the frame. The same bit pattern used in the pure Ethernet preamble is used in the IEEE 802.3 preamble, which includes the 1-byte start-frame delimiter field.

Start-Frame Delimiter The IEEE 802.3 Standard specifies a start-frame delimiter field, which is really a part of the preamble. This is used to indicate the start of a frame.

Address Fields The destination address field identifies the station(s) that are to receive the frame. The source address

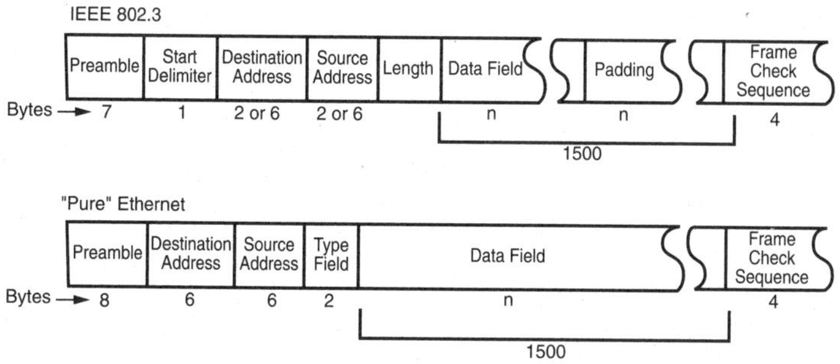

Figure E-10 Comparison of Ethernet frame formats: IEEE 802.3 and "pure" Ethernet.

field identifies the station that sent the frame. If addresses are assigned locally, the address field can be either 2 bytes (16 bits) or 6 bytes (48 bits) in length. A destination address can refer to one station, a group of stations, or all stations. The original Ethernet specifies the use of 48-bit addresses, while IEEE 802.3 permits either 16- or 48-bit addresses.

Length Count The length of the data field is indicated by the 2-byte count field. This IEEE 802.3–specified field is used to determine the length of the information field when a pad field is included in the frame.

Pad Field To detect collisions properly, the frame that is transmitted must contain a certain number of bytes. The IEEE 802.3 Standard specifies that if a frame being assembled for transmission does not meet this minimum length, a pad field must be added to bring it up to that length.

Type Field Pure Ethernet does not support length and pad fields, as does IEEE 802.3. Instead, 2 bytes are used for a type field. The value specified in the type field is only meaningful to the higher network layers and was not defined in the original Ethernet specification.

Data Field The data field of a frame is passed by the client layer to the data link layer in the form of 8-bit bytes. The minimum frame size is 72 bytes, while the maximum frame size is 1526 bytes, including the preamble. If the data to be sent use a frame that is smaller than 72 bytes, the pad field is used to stuff the frame with extra bytes. In defining a minimum frame size, there are fewer problems to contend with in handling collisions. If the data to be sent use a frame that is larger than 1526 bytes, it is the responsibility of the higher layers to break it into individual packets in a procedure called "fragmentation." The maximum frame size reflects practical considerations related to adapter card buffer sizes and the need to limit the length of time the medium is tied up in transmitting a single frame.

Frame Check Sequence A properly formatted frame ends with a frame check sequence, which provides the means to check for errors. When the sending station assembles a frame, it performs a cyclic redundancy check (CRC) calculation on the bits in the frame. The sending station stores the results of the calculation in the 4-byte frame check sequence field before sending the frame. At the receiving station, an identical CRC calculation is performed, and a comparison is made with the original value in the frame check sequence field. If the two values do not match, the receiving station assumes that a transmission error has occurred and requests that the frame be retransmitted. In pure Ethernet, there is no provision for error correction; if the two values do not match, notification that an error has occurred is simply passed to the client layer.

Media Access Control

Several key processes are involved in transmitting data across the network, among them data encapsulation/decapsulation and media access management, which are performed by the Media Access Control (MAC) sublayer of OSI's Data Link Layer.

Data Encapsulation/Decapsulation Data encapsulation is performed at the sending station. This process entails adding information to the beginning and end of the data unit to be transmitted. The data unit is received by the MAC sublayer from the Logical Link Control (LLC) sublayer. The added information is used to perform the following tasks:

- Synchronize the receiving station with the signal
- Indicate the start and end of the frame
- Identify the addresses of sending and receiving stations
- Detect transmission errors

The data encapsulation function is responsible for constructing a transmission frame in the proper format. The destination address, source address, type, and information fields are passed to the Data Link Layer by the client layer in the form of a packet. Control information necessary for transmission is encapsulated into the offered packet. The CRC value for the frame check sequence field is calculated, and the frame is constructed.

When a frame is received, the data decapsulation function performed at the receiving station is responsible for recognizing the destination address, performing error checking, and then removing the control information that was added by the data encapsulation function at the sending station. If no errors are detected, the frame is passed up to the LLC sublayer.

Specific types of errors are checked in the decapsulation process, including whether the frame is a multiple of 8 bits or exceeds the maximum packet length. The address is also checked to determine whether the frame should be accepted and processed further. If it is, a CRC value is calculated and checked against the value in the frame check sequence field. If the values match, the destination address, source address, type, and data fields are passed to the client layer. What is passed to the station is the packet in its original form.

Media Access Management The method used to control access to the transmission medium is known as "media access management" in IEEE terms but is called "link management" in Ethernet parlance. Link management is responsible for several functions, starting with collision avoidance and collision handling, which are defined by the IEEE 802.3 Standard for contention networks.

Collision Avoidance Collision avoidance entails monitoring the line for the presence or absence of a signal (carrier). This is the "carrier sense" portion of CSMA/CD. The absence of a signal indicates that the channel is not being used and that it is safe to begin transmission. Detection of a signal indicates that the channel is already in use and that transmission must be withheld. If no collision is detected during the period of time known as the "collision window," the station acquires the channel and can complete the transmission without risking a collision.

Collision Handling When two or more frames are offered for transmission at the same time, a collision occurs, which triggers the transmission of a sequence of bits called a "jam." This is the means whereby all stations on the network recognize that a collision has occurred. At that point, all transmissions in progress are terminated. Retransmissions are attempted at calculated intervals. If there are repeated collisions, link management uses a process called "backing off," which involves increasing the retransmission wait time following each successive collision.

On the receiving side, link management is responsible for recognizing and filtering out fragments of frames that resulted from a transmission that was interrupted by a collision. Any frame that is less than the minimum size is assumed to be a collision fragment and is not reported to the client layer as an error.

Methods have been developed to improve the performance of Ethernet by reducing or totally eliminating the chance for collisions without having to segment the LAN into smaller subnetworks. Special algorithms sense when frames are on a collision course and will temporarily block one frame while allowing the other to pass. This is called "collision avoidance."

Summary

Ethernet is the most popular type of LAN. Its success has spawned continued innovations leading to higher speeds and overcoming distance limitations. 10BaseT Ethernet, for example, enables the LAN to operate over ubiquitous unshielded twisted-pair (UTP) wiring instead of thick or thin coaxial cable. For those who find 10 Mbps inadequate for supporting large file transfers and graphic-intensive applications, there are higher-speed versions of Ethernet, including Fast Ethernet at 100 Mbps, Gigabit Ethernet, and 10× Gigabit Ethernet—all of which can operate over various grades of UTP wiring within a building. There are also versions of Ethernet that run over optical fiber for metropolitan area or wide area connectivity.

See also

Token Ring

EXTRANETS

An extranet is a network that is shared among multiple organizations, usually strategic business partners, which are given controlled access to selected information and applications. The extranet is based on the TCP/IP suite of protocols and provides the same capabilities as the public Internet, but with security features and access privileges that guard against unauthorized entry.

Businesses build extranets to improve communication among key constituents, facilitate information distribution, broaden access to each other's resources, enable group scheduling, and provide a browser front end to various corporate databases to expedite inventory tracking, supply-side management, and invoicing. Extranets have become the means through which companies engage in business-to-business (B2B) e-commerce on a global basis while reaping the added benefits of reduced operational costs, improved productivity, and timely response to changing market conditions.

Extranets also can be made accessible to the general public, providing customers with secure access to certain types of data. Banks, brokerage houses, and other financial institutions, for example, provide customers with secure access to their extranets to check on the status of their accounts. Delivery services, such as Federal Express and UPS, give customers access to their extranets to check on the delivery status of their packages. Airlines, hotels, and resorts provide extranet access to allow customers to place reservations online.

Benefits of Extranets

Among the advantages of establishing an extranet is that it provides self-service opportunities for constituents, who can order products directly, get immediate answers to their inquiries, and solve product-related problems by looking through a database. In some cases, an extranet can be used to allow customers to configure and price the products they want before placing the order electronically.

In turn, companies can serve constituents at a very low cost—24 hours a day, 7 days a week. Cost savings come from reductions in sales and support staff and the elimination of dedicated lines and services. In fact, an extranet allows an organization to create the equivalent of a corporate wide area network (WAN) with global reach but without the cost of leased lines. Like corporate intranets, an extranet uses the same hardware, software, protocols, and development tools as the public Internet.

In many cases, the same staff that maintains the corporate intranet also maintains the extranet, since the knowledge and skills to do both are not significantly different. The real challenges are in securing and managing the extranet, since the cooperation of several companies is necessary. Ensuring interoperability between the different systems, databases and applications of the participants also merits technical expertise.

Planning

In the design of an extranet, a document needs to be developed that discusses the network architecture, the goals of the architecture, and the system specifications that will support the goals. This document should be shared with partners for their input, since they will be the primary "customers" making the decision to join the extranet.

Before committing any resources to the extranet project, the hub company should determine what resources it already has that can be leveraged or redeployed. This could save on capital equipment costs and shorten the time to service cutover. A more realistic baseline budget can then be established for the project.

When building an extranet, personnel decisions should be made before resources are committed. There will be staff turnover to contend with, and as the extranet evolves, each person's responsibilities may increase or change. It must be determined early on how extra resources will become available when needed, in what time frame, and in a way that will not erode confidence in the project among partners.

It is recommended that performance criteria be drawn up for the extranet and acceptance criteria developed that will satisfy management and partners that the job has been completed within specifications. In some cases, it might be worthwhile to offer a service level agreement (SLA) to partners governing such key areas as application performance, network availability, and response time to repair. An SLA could become a further inducement for partners to join the extranet.

If the resources of the hub company are constrained, consideration should be given to outsourcing extranet setup and management to an integrated communications provider (ICP). Since 24 × 7 monitoring is the responsibility of the ICP, there is no need for the hub company to have a support facility of its own, which is a critical requirement for the smooth functioning of an extranet. The ICP handles security with a combination of authentication and filtering techniques and the use of security protocols and firewalls—all of which can be difficult and time-consuming for organizations to set up and manage by themselves. By outsourcing extranet management, the hub company can save a substantial amount of money in startup costs and free up IT staff to meet other business needs. Other reasons to outsource the extranet include

- Faster extranet development
- Easier integration of new technologies and capabilities
- The availability of best-of-breed equipment and higher-speed lines than a company cannot otherwise afford
- The availability of a wider range of expertise
- Quality-of-service (QoS) guarantees
- Continuous network management and faster response to problems
- One-stop service and support

Management

Extranet management is a natural extension of systems management, except users are allowed to pass the firewall. Extranets present special management problems because they allow access to information systems that are normally considered private, including inventory databases, order-entry and accounting systems, and product configuration and pricing tools. As with any network-based information system, there are management requirements to contend with, including

- Configuring new users and adding new nodes
- Setting up and changing access controls in response to changing needs
- Ensuring that system security works properly
- Adding new software and features to satisfy changing business requirements
- Ensuring that information systems, applications, and links continue to work properly

These duties usually fall within the traditional domain of system administration but are made slightly more difficult when they involve multiple organizations. For the extranet to yield the anticipated benefits to all participants, managers must ensure that the right information is available to the right constituents, that frequently changing information—such as product prices, catalogs, and inventory—is kept up to date, and that all supporting systems function properly.

Management of a multicompany extranet usually is the responsibility of the largest organization, which acts as the hub. This company encourages its suppliers, sales channels, and other partners to join the extranet for mutual benefit and assists the others in configuring their information systems, routers, and firewalls. The hub company also may arrange for dedicated access lines and dial-up service providers for access to the extranet. To get inexperienced companies up to speed on using the extranet to optimal advantage, the hub company also may offer consulting services and online training.

Security

Although extranets are extensions of corporate intranets and usually allow access via the global Internet, sensitive business data can be kept private via the use of a firewall, which may use a number of strategies—including packet filtering and intrusion detection—to keep private data and resources off-limits. If implemented properly, extranets provide access to appropriate information while effectively securing other data from the general public, as well as from strategic partners, on a selective basis.

Security is problematic when applied to extranets for reasons that have nothing to do with the technologies for controlling access. The real challenge comes from the fact that the extranet involves partners whose relationships are dynamically changing and complex. Today's partners may become tomorrow's competitors, and a partner may be both a competitor and a partner simultaneously. If an extranet system does not permit dynamic changes in access control, there is the possibility of sensitive information getting into the wrong hands that cannot simply be called back.

There are several key requirements to securing information on an extranet. First, the identity of an individual wishing to access the extranet should be authenticated. This process is complicated when employees or business partners access information from multiple computers and, often, from remote locations over the Internet. Users should be able to authenticate from a Web browser, with no client software requirements. In addition, there are often hundreds of Web servers in a large enterprise, and users need access privileges for each server they access. This can lead to many problems: Users must remember passwords for many servers, administrators need to manage the access controls for each individual server, and many separate entries must be added or removed when a user's access privileges change or when employees join or leave the company. A security solution that lets the organization manage access controls for all these servers centrally and presents users with a single sign-on to the Web space can greatly simplify security management as well as enhance the user's experience.

Once a user's identity has been authenticated, the access privileges should be determined. An authenticated user does not necessarily have permission to access resources. Security policies should explicitly grant access rights to Web resources. An access control decision function must establish whether requests for specific information should be granted or denied. Administration is complicated if access controls must be configured at each Web server, and it is difficult to construct a comprehensive picture of a user's privileges in the Web space if an administrator must consult each Web server's configuration information. A centralized authorization framework greatly simplifies administration.

Large extranets will require that certain management tasks be delegated. It is often necessary to delegate the management of security and privileges for certain information resources to either the individual or group responsible for maintaining them. An effective security system should facilitate secure delegation of permissions to ease the management burden.

Another important concern with any security management solution is how easy it is to implement and administer. For any security solution to be effective, it should integrate easily with the organization's existing infrastructure and be easy to administer. Any complexities in security management increase the possibility of human errors, make the extranet difficult to navigate, and expose the extranet to attack or misuse.

Content Management

Within the context of the extranet, security and content management are closely interrelated; after all, security is about protecting information. Content management is a critical issue with extranets because the owner has virtually no control over the information once it is downloaded and has no idea of its ultimate destination and how the information is really being used.

For example, extranets lend themselves to the aggregation of tactical information that can have strategic value.

This can become a threat if content control and the time value of information are not considered as part of the security equation. While production figures for a specific month are considered tactical information and may help partners who are supplying raw materials to the assembly line, 3 months of production information may be of strategic value to another partner who is also a competitor. Therefore, the same information should not be made available in the same form to all partners.

An extranet is the only situation where a firm exposes its most proprietary information to a semiopen audience. Sound content management entails never assuming that the target user at the partner organization will be the final user of that information. While the success of the partnership is certainly important, and ease of communication is inherent to achieving that success, it must be recognized that failure can come from revealing too much information. It is therefore necessary for content owners to implement information security systems that permit content control and dynamic changes to minimize exposure to risk.

Summary

Extranets are simply intranets that are opened to external constituents. The underlying facilities, applications, services, protocols, and development tools are pretty much the same. An application developed for use on the corporate intranet can be modified to include an extranet component; that is, it can be modified to allow controlled access by external parties. Both the intranet and extranet are ways to leverage existing Internet technology to achieve business goals efficiently and economically.

See also

Internet

Intranets

F

FIBER IN THE LOOP

Fiber in the loop (FITL) is a network architecture in which services to contiguous groupings of residential and business customers are delivered using fiberoptic media. A number of broadband services can be offered over the fiber network, including entertainment programming, video on demand, and Internet access. FITL is an umbrella term that encompasses various systems that depend to a greater or lesser degree on optical fiber, including

- *Fiber to the neighborhood (FTTN)* A technology that involves bringing fiber into the neighborhood. From there, signals would be carried to businesses and residences via the existing copper wiring.
- *Fiber to the curb (FTTC)* A technology that involves bringing fiber into the neighborhood and up to the curb. From there, signals would be carried to businesses and residences via the existing copper wiring.
- *Fiber to the home (FTTH)* A technology that involves bringing fiber all the way to the home. Inside the home, signals would be carried over the existing copper wire or coaxial cable.

- *Fiber to the building (FTTB)* A technology that involves bringing fiber all the way to the building. Inside the building, signals would be carried over the existing copper wire or coaxial cable or optical fiber.

Very High Speed Digital Subscriber Line (VDSL) service is based on the FITL architecture because it overcomes the normal distance limitation of high-speed digital signals over local telephone lines. VDSL systems are able to offer a dedicated rate of 52 Mbps on the downstream path and a dedicated rate of 1.6 Mbps on the upstream path. But with copper lines, the subscriber would have to be within 4000 feet of a central office equipped with an access multiplexer.

With the access multiplexer installed in the neighborhood, however, most subscribers would be within that distance (Figure F-1). The access multiplexer combines all the

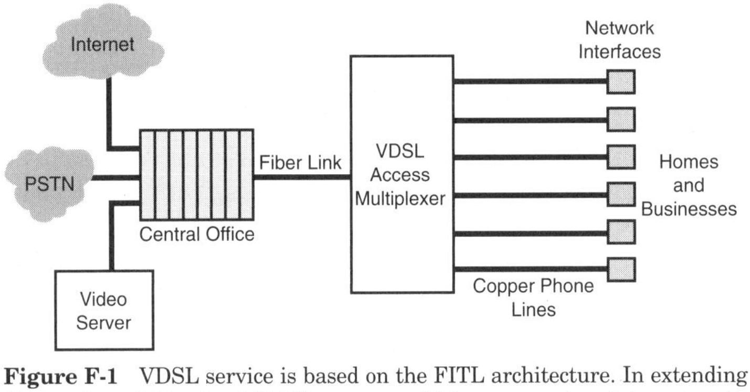

Figure F-1 VDSL service is based on the FITL architecture. In extending fiber to the neighborhood, VDSL access multiplexers are placed closer to subscribers, allowing ordinary telephone lines to deliver broadband services.

traffic offered to it from the copper lines and sends it via fiber to the data switching equipment in the telephone company's central office. With this much bandwidth available, the service provider would be able to support multiple services over VDSL, including broadband Internet access and video on demand. This would allow service providers to compete directly with cable TV operators.

Summary

Many service providers have concluded that the cost of offering FTTC on a mass-market basis is currently out of reach because of the high cost of installing extra fiber and electronics in the local loop. Instead, hybrid fiber/coax (HFC) systems and broadband wireless technologies such as Local Multipoint Distribution Service (LMDS) offer a more economical approach. Nevertheless, FTTC may yet emerge as the ultimate architecture, at least in high-density metropolitan areas. Until then, FITL architectures that bring fiber to the neighborhood and use systems that interconnect with existing copper lines to homes and businesses offer the optimal blend of economy, capacity, and reliability, enabling VDSL to become a viable service for broadband applications.

See also

Digital Subscriber Line Technologies

Hybrid Fiber/Coax

Local Multipoint Distribution Service

FIREWALLS

Firewalls occupy a strategic position between a "trusted" corporate network and an "untrusted" network, such as the Internet (Figure F-2). They implement perimeter security by monitoring all traffic to and from the enterprise network to determine which packets can pass and which cannot pass. A firewall can identify suspected break-in events, issue appropriate alarms to a management station, and then invoke a predefined action to head off the attack. Firewalls also can be used to trace attempted intrusions from the Internet through logging and auditing functions.

Firewalls can be stand-alone devices that are dedicated to safeguarding the enterprise network. Similar functionality can be added to routers, in which case the security features

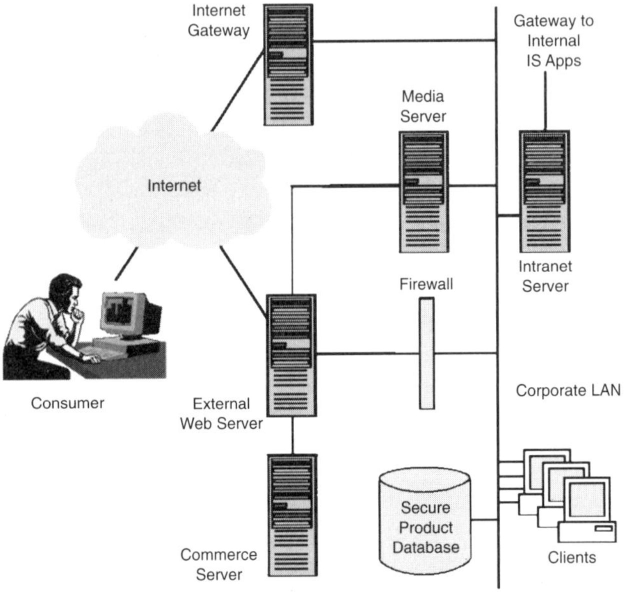

Figure F-2 Firewalls guard resources on the corporate network from access by unauthorized persons on the greater Internet.

are programmed through the router's operating system. Internet appliances can have firewall capabilities as well, such a DSL routers and cable modems. In addition, there is firewall software that can be loaded into desktop computers that gives users personal control over security (Figure F-3).

Operation

A packet-filtering firewall examines all the packets passed to it and then forwards them or drops them according to predefined rules. The network administrator can control how packet filtering is performed, permitting or denying connections using criteria based on the source and destination host or network and the type of network service.

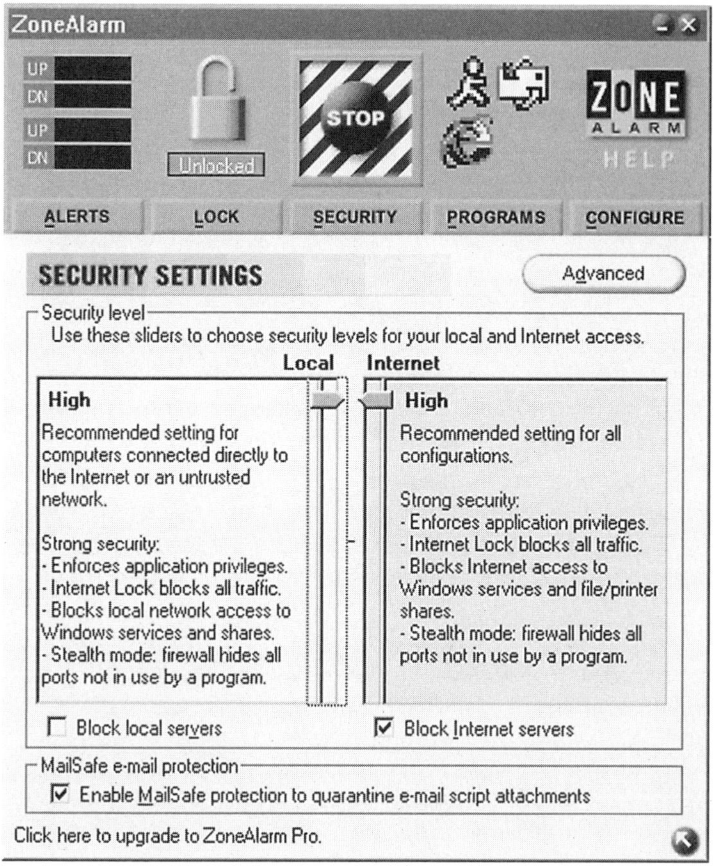

Figure F-3 Firewall software for individual computers allows users to control their own level of security. Shown is ZoneAlarm from Zone Labs, which is available free at the company's Web site for personal use.

In addition to packet filtering, a firewall offers other useful security features, such as

- *Stateful packet inspection* State information is derived from past communications and other applications to make the control decision for new communication attempts. With this method of security, the packet is intercepted by an inspection engine, which extracts state-related information. It maintains this information in dynamic state tables for evaluating subsequent connection attempts. Packets are allowed to pass only when the inspection engine examines the list and verifies that the attempt is in response to a valid request. The list of connections is maintained dynamically so that only the required ports are opened. As soon as the session is closed, the ports are locked, ensuring maximum security.

- *Network Address Translation (NAT)* This hides internal Internet Protocol (IP) addresses from public view, preventing them from being used for "spoofing"—a technique for impersonating authorized users by using a valid IP address to gain access to an internal network.
- *Denial-of-service detection* This defends the network against SYN flooding,[1] port scans, and packet injection. This is accomplished by inspecting packet sequence numbers in TCP connections. If they are not within expected ranges, the firewall drops them as suspicious packets. When the firewall detects unusually high rates of new connections, it issues an alert message so appropriate action can be taken.
- *Virus scanning* In addition to hiding viruses, Java applets and ActiveX controls can be used to hide intelligent agents that can give intruders access to corporate

[1]SYN flooding involves sending a continuous stream of bogus messages to a targeted computer, keeping it busy and locking out legitimate users. This method of attack exploits the synchronization (SYN) feature of the Transmission Control Protocol (TCP). When users connect to a Web site, they are actually asking the computer to send back requested information. That request initiates an interaction, called a "handshake," between the computer looking for data and the computer sending data. When the first computer begins to talk to the second, it sends a message that essentially means "Hello." The second computer answers with the equivalent of, "Hello, how are you?" on answering back, both computers have established that they are listening, or synchronized. They exchange confirmation of the connection so that the transfer of information can begin. In a SYN-flood attack, hackers send a series of forged messages that do not contain real return addresses to which the Web site's computer can send its response. As a result, the computer under attack waits a long time—as long as a minute—for the second computer to respond. Soon its storage buffers fill up as it struggles to complete the connections, preventing new requests from legitimate users from being answered while it attempts to deal with the congestion problem.

resources once they get inside the enterprise network. With the increasing use of Java applets and ActiveX controls on Web sites, more firewalls offer the means to either deny corporate users access to Web pages that contain these elements or filter such content from the Web pages when they are downloaded.

- *Probe detection* Firewalls offer alarms that are activated when port probing is detected. The alarm system can be configured to watch for TCP or UDP probes from either external or internal networks. Alarms can be configured to trigger e-mail, pop-up windows, or output messages to a local printer.
- *Event logging* This automatically logs system error messages to a console terminal or system log server, allowing administrators to track potential security breaches or other nonstandard activities on a real-time basis.

Automated Intrusion Detection

Many vendors now offer automated intrusion detection tools for their firewalls. These real-time tools monitor the audit trails of distributed systems for "footprints" that signal suspicious or unauthorized activity on all major operating systems, Web servers, firewalls, routers, applications, databases, and simple network management protocol (SNMP) traps from other network devices. Unlike other intrusion detection tools, which typically report suspicious activity hours or even days after it occurs, the new breed of real-time tools instantly takes action to alert network administrators, shut systems down, terminate offending sessions, execute commands, and take other actions to stop intrusions before they damage critical systems.

As new security threats emerge, network administrators can quickly protect their systems by loading new drop-and-detect scenarios into their firewalls. These are available from the firewall vendors, whose computer security staff focus on hacking techniques and the latest computer security threats. These new scenarios, which can be downloaded from the vendor's Web site and installed enterprise-wide, make it easy for network administrators to keep systems safe from evolving threats.

From a single management workstation, network administrators can quickly drag new security policies and attack scenarios to different enterprise domains, implementing additional protection for hundreds or thousands of systems in a matter of minutes. The management console also provides a correlated, graphical view of security trends, letting network administrators view graphs that illustrate real-time security trends and drill-down to additional details on activity.

Security Appliances

For small remote offices or remote users who cannot justify the expense of an enterprise firewall, there are appliances that combine firewall and virtual private network (VPN) security capabilities. Security is particularly important when Digital Subscriber Line (DSL) or cable is used for Internet access. One drawback of these "always on" services is that they are vulnerable to security attacks.

NetScreen Technologies, for example, offers its NetScreen-5 security appliance for use in small offices and telecommuter sites with DSL or cable modem access. Its stateful-inspection firewall prevents hacker attacks, and the VPN delivers secure remote access to a corporate network through encrypted tunnels. NetScreen-5 unit is installed between the PC and cable/DSL modem. Since small remote offices typically do not have staff with technical expertise, the box can be configured by a network administrator and shipped out to a user for plug-and-play installation. The network administrator can then centrally manage and reconfigure dispersed units through the NetScreen Global Manager or a Web browser.

Administrators can check the status of multiple NetScreen appliances, monitor performance, troubleshoot existing configurations, or add remote sites to the network from one location. All such activities are conducted via VPN tunnels for the highest level of security.

Risk Assessment

Even after a firewall solution is implemented, it is recommended that a comprehensive security risk assessment be conducted periodically. This helps network administrators identify and resolve security breaches before they are discovered and exploited by hackers and cause serious problems later. There are a number of risk-assessment tools available.

Among these tools is bv-Control for Internet Security (formerly known as Hacker-Shield) from BindView, a provider of network scanning and response software. This tool scans and detects networks for potential security holes and offers the user patches or corrective actions to fix the breaches before they become a threat. The tool identifies and resolves security vulnerabilities at both the operating system level and the network level, protecting against both internal and external threats. It also monitors key system files for unauthorized changes and, by referencing a 1-million-word dictionary, identifies vulnerable user passwords through a variety of password-cracking techniques.

A detailed report provides network administrators with a description of each vulnerability and corrective action as well as a ranking of vulnerabilities by the risk they pose to a site's security. Network administrators are also presented with a high-level overview of the vulnerability and its solution with an option to link to a more detailed explanation and reference materials.

Employing an implementation model similar to antivirus products, BindView provides ongoing security updates via the Internet to ensure that users are protected from the latest threats. BindView uses secure push technology to broadcast the vulnerability updates. Users are not required to reinstall the software in order to integrate the updates.

Load Balancing

While the value of firewalls is undisputed, they can degrade performance and create a single point of network failure. This is so because such tasks as stateful packet inspection, encryption, and virus scanning require significant amounts of processing. As traffic load increases, the firewall can become bogged down. And since the firewall sits directly in the data path, it constitutes a single point of failure. If the firewall cannot keep up with filtering all the packets coming through, it will go down and isolate the whole network behind it.

A solution to this problem is to add a firewall for redundancy and put dynamic load-balancing switches on each side of them. That way, the switches can distribute incoming requests to the firewalls according to their availability. This configuration eliminates the firewalls as single points of failure, dramatically simplifies configuration, and increases end-user performance by balancing traffic among multiple firewalls.

The switches monitor the health of attached firewalls through automatic, periodic health checks. They also monitor physical link status of the switch ports connected to each firewall. Since the firewalls are no longer directly in line and traffic is evenly distributed among them, the end-user experience is improved. The switches automatically recognize failed firewalls and redirect entire sessions through other available firewalls while maintaining the state of each session.

Managed Firewall Services

For small and medium-sized companies, maintaining a full-time staff of network security professionals at a reasonable cost is virtually impossible. Knowing this, hackers have been known to use the unprotected networks of smaller companies to launch denial-of-service attacks against the e-commerce servers of larger companies. Now smaller companies can implement effective security and mitigate their exposure to risk by subscribing to a managed firewall service.

The service provider performs an initial security assessment, which includes reviewing the customer's security requirements, configuring any required equipment, and managing the remote turn-up of management services. Continuous remote monitoring is performed via encrypted channels over the Internet. In scenarios where the firewall resides in a router, that device is managed as well. Performance reports may be accessed from the service provider's secure Web site.

As part of the managed firewall service, periodic reports are furnished that focus on "hot spots" or anomalies in the firewall. Such reports include a performance analysis and recommendations for modifications that will improve throughput and close potential breach points. These recommendations might include software changes, hardware upgrades or changes, or topologic or transport changes.

Summary

Although firewalls can provide a formidable defense against many kinds of attacks, they are not a panacea for all network security problems, particularly those that originate from the corporate side. For example, even if virus scanning is provided at the firewall, it will protect only against viruses that come from the Internet. It does nothing to guard against more likely sources, such as floppy disks brought into the company by employees who upload the contents to a desktop computer and inadvertently (or deliberately) spread the virus throughout the network. Companies must take appropriate internal security measures to safeguard mission-critical resources.

See also

Network Security

FIXED WIRELESS ACCESS

Fixed wireless access technology provides a wireless link to the public switched telephone network (PSTN) as an alternative to traditional wire-based local telephone service. Since calls and other information (e.g., data, images) are transmitted through the air rather than through conventional cables and wires, the cost of providing and maintaining telephone poles and cables is avoided. Unlike cellular technologies, which provide services to mobile users, fixed wireless services require a rooftop antenna to an office building or home that is lined up with a service provider's hub antenna.

Fixed wireless access systems come in two varieties: narrowband and broadband. A narrowband fixed wireless access service can provide bandwidth up to 128 kbps, which can support one voice conversation and a data session such as Internet access or fax transmission. A broadband fixed wireless access service can provide bandwidth in the multi-megabit-per-second range, which is enough to support telephone calls, television programming, and broadband Internet access.

A narrowband fixed wireless service requires a wireless access unit that is installed on the exterior of a home or business (Figure F-4) to allow customers to originate and receive calls with no change to their existing analog telephones. This transceiver is positioned to provide an unobstructed view to the nearest base station receiver. Voice and data calls are transmitted from the transceiver at the customer's location to the base station equipment, which relays the call through carrier's existing network facilities to the appropriate destination. No investment in special phones or facsimile machines is required; customers use all their existing equipment.

Narrowband fixed wireless systems use the licensed 3.5-GHz radio band with 100-MHz spacing between uplink and downlink frequencies. Subscribers receive network access over a radio link within a range of 200 meters (600 feet) to

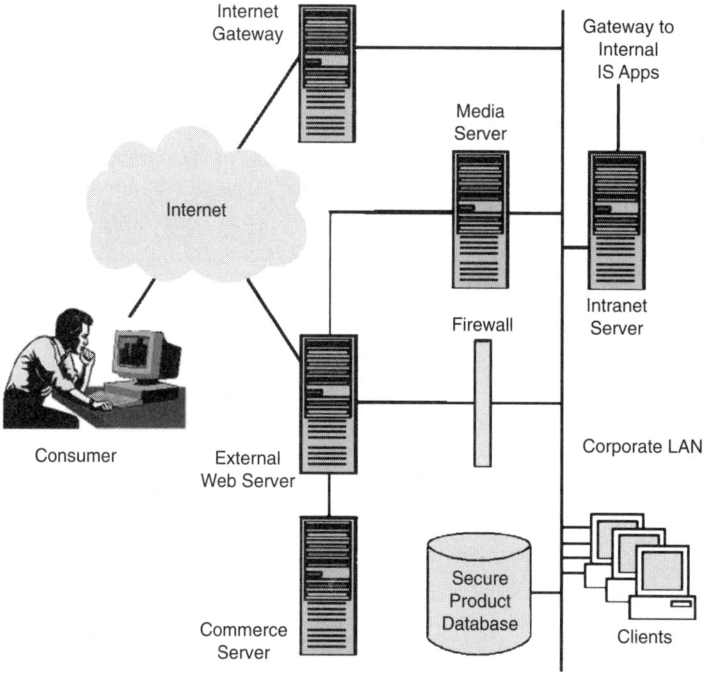

Figure F-4 Fixed wireless access configuration.

40 kilometers (25 miles) of the carrier's hub antenna. About 2000 subscribers can be supported per cell site.

Broadband fixed wireless access systems are based on microwave technology. Multichannel Multipoint Distribution Service (MMDS) operates in the licensed 2- to 3-GHz frequency range, while Local Multipoint Distribution Service (LMDS) operates in the licensed 28- to 31-GHz frequency range. Both services are used by Competitive Local Exchange Carriers (CLECs) primarily to offer broadband Internet access. These technologies are used to bring data traffic to the fiberoptic networks of Interexchange Carriers (IXCs) and nationwide CLECs, bypassing the local loops of the Incumbent Local Exchange Carriers (ILECs).

Summary

Fixed wireless access technology originated out of the need to contain carriers' operating costs in rural areas, where pole and cable installation and maintenance are more expensive than in urban and suburban areas. However, wireless access technology also can be used in urban areas to bypass the local exchange carrier for long-distance calls. Since the IXC or CLEC avoids having to pay the ILEC's local loop interconnection charges, the savings can be passed back to the customer. This arrangement is also referred to as a "wireless local loop."

See also

Local Multipoint Distribution Service
Multichannel Multipoint Distribution Service

FRAME RELAY

Frame relay is a stripped-down version of the X.25 protocol for packet data networking, which was designed in the 1970s to transport data very reliably over noisy analog-line networks. In running over cleaner digital-line networks, however, frame relay does not need many of the overhead functions of X.25, including node-to-node error correction. In stripping away unnecessary functions and relegating error correction to the frame relay access devices (FRADs) or routers at the edge of the network, frame relay could transport data at much higher speeds than X.25, making it suited for interconnecting LANs over a WAN. Frame relay also can be used to carry IP traffic through a carrier's data network to a network access point (NAP) on the Internet backbone.

About X.25

In the 1970s, WANs were built using low-speed analog facilities that were used primarily for voice traffic. For reliable data transmission, however, private and public packet-switched networks had to employ the X.25 suite of protocols to overcome noise and other impairments that made the transmission of data difficult. X.25 was endowed with substantial error-correction capabilities so that any node on the network could request a retransmission of errored data from the node that sent it. Errors had to be detected and corrected within the network, since the user's equipment typically did not have the intelligence and spare processing power to devote to this task.

However, the error-correction and flow-control capabilities of the X.25 protocol, plus its many other functions, entail an overhead burden that limits network throughput. This, in turn, limited X.25 to niche applications, such as point-of-sale transaction processing, where the reliable transmission of credit card numbers and other financial information—not speed—was the overriding concern. At the same time, LANs were becoming popular in the 1980s, and there was a growing need to interconnect them over the WAN. Point-to-point T-carrier lines were becoming commercially available, but they were cost prohibitive for all but the largest companies. Frame relay was developed specifically to provide LAN interconnectivity as a service, eliminating costly leased-line charges based on distance.

With the increasing use of digital facilities, there is less need for error protection. At the same time, the end devices increased in intelligence, processing power, and storage capacity, making them better at handling error control and diverse protocols. Consequently, the communications protocols used over the network may be scaled down to its bare essentials to greatly increase throughput. This is the idea behind frame relay, which can support voice traffic, as well as data, packaged in variable-sized frames of up to 4000 bytes in length.

Frame relay was introduced commercially in May 1992. It initially gained acceptance as a method for providing end users with a solution for data connectivity requirements, such as LAN-to-LAN connections. Frame relay provided both an efficient and flexible data transport mechanism and also allowed for a cheaper bandwidth cost associated with connecting legacy systems network architecture (SNA) networks.

Advantages of Frame Relay

The most compelling advantages of a carrier-provided frame relay service include

- *Improved throughput / low delay* Frame relay service uses high-quality digital circuits end to end, making it possible to eliminate the multiple levels of error checking and error control. The result is higher throughput and less delay compared with legacy packet-switched networks like X.25.
- *Any-to-any connectivity* Any node connected to the frame relay service can communicate with any other node via predefined permanent virtual circuits (PVCs) or dynamically via switched virtual circuits (SVCs).
- *No long-distance charges* Since frame relay is offered as a service over a shared network, the need for a highly meshed private line network is eliminated, for substantial cost savings. There are no distance-sensitive charges with frame relay, as there is with private lines.
- *Oversubscription* Multiple PVCs can share one access link, even exceeding the port speed of the frame relay switch. In oversubscribing the port, multiple users can access the frame relay network—but not all at the same time—eliminating the cost of multiple private-line circuits and their associated customer premises equipment (CPE), for further cost savings.
- *Higher speeds* Whereas X.25 tops out at 56 kbps, frame relay service supports transmission speeds up to 44.736 Mbps. If the frame relay switches in the network support Frame Relay Forum Implementation Agreement 14 (FRF

14), speeds at the OC-3 rate of 155 Mbps and the OC-12 rate of 622 Mbps over fiber backbones are possible.

- *Simplified network management* Customers have fewer circuits and less equipment to monitor. In addition, the carrier provides proactive monitoring and network maintenance 24 hours a day.
- *Intercarrier connectivity* Frame relay service is compatible between the networks of various carriers, through network-to-network interfaces (NNIs), enabling data to reach locations not served by the primary service provider.
- *Customer-controlled network management* Frame relay allows customers to obtain network management information via in-band SNMP queries and pings launched from their own network management stations.
- *Performance reports* These reports enable customers to manage their frame relay service to maximum advantage. Available network reports, accessible on the carrier's secure Web site, include those for utilization, errors, health, trending, and exceptions.
- *Service level guarantees* Frame relay service providers offer customers service level agreements (SLAs) that specify availability as a percentage of uptime, round-trip delay expressed in milliseconds, and throughput in terms of the committed information rate (CIR). If the carrier cannot meet the SLA, it credits the customer's invoice accordingly.

Types of Circuits

Packet networks make use of virtual circuits, sometimes referred to as "logical channels." The two primary types of virtual circuits supported by frame relay are switched virtual circuits (SVCs) and permanent virtual circuits (PVCs). SVCs are analogous to dial-up connections, which require path setup and teardown. A key advantage of SVCs is that they permit any-to-any connectivity between devices connected to the frame relay network. PVCs are more like dedicated private lines; once set up, the predefined logical connections between sites attached to the frame relay network stay in place. This allows logical channels to be dedicated to specific terminals. The SVC requires fewer logical channels at the host because the terminals contend for a lesser number of logical channels. Of course, it is assumed that not everyone will require access to the host at the same time.

Another type of virtual circuit is the multicast virtual circuit (MVC), which is used to broadcast the same data to a group of users over a reserved data link connection in the frame relay network. This type of virtual circuit might be useful for expediting communications among members of a single workgroup dispersed over multiple locations or to facilitate interdepartmental collaboration on a major project. It also can be used for broadcast faxing, news feeds, and "push" applications.

The same frame relay interface can be used to set up SVCs, PVCs, and MVCs. All three may share the same digital facility. In supporting multiple types of virtual circuits, frame relay networks provide a high degree of configuration flexibility, as well as more efficient utilization of the available bandwidth.

The virtual circuits have a committed information rate (CIR), which is the minimum amount of bandwidth the carrier agrees to provide for each virtual circuit. If some users are not accessing the frame relay network at any given time, extra bandwidth becomes available to users who are online. The CIR of their virtual circuit can burst up to the full port speed. As other users come online, however, the virtual circuits that were bursting beyond their CIR must back down to the assigned CIRs.

Congestion Control

Real-time congestion control must accomplish the following critical objectives in a frame relay network:

- Maintain high throughput by minimizing timeouts and out-of-sequence frame deliveries.
- Prevent session disconnects, unless required for congestion control.
- Protect against unfair users who attempt to hog the available network resources by exceeding their CIR or established burst size.
- Prevent the spread of congestion to other parts of the network.
- Provide delays consistent with application requirements and service objectives.

In the frame relay network, congestion can be avoided through control mechanisms that provide backward explicit congestion notification (BECN) and forward explicit congestion notification (FECN), which are depicted in Figure F-5.

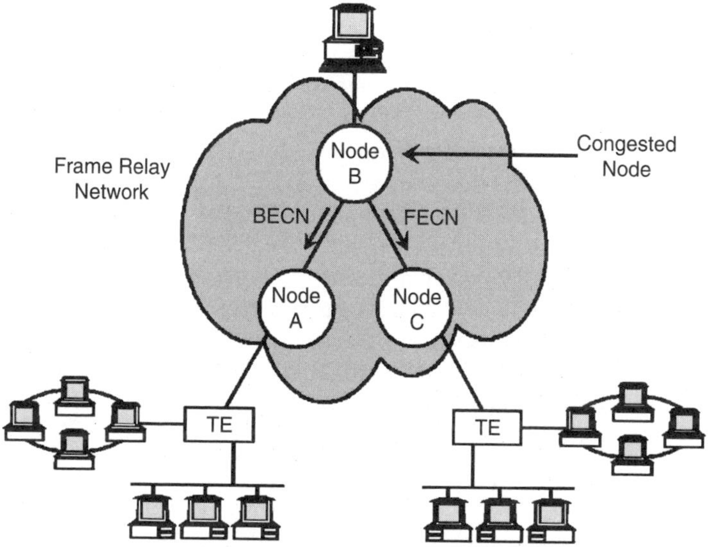

Figure F-5 Congestion notification on the frame relay network.

BECN is indicated by a bit set in the data frame by the network to notify the user's equipment that congestion-avoidance procedures should be initiated for traffic in the opposite direction of the received frame. FECN is indicated by a bit set in the data frame by the network to notify the user that congestion-avoidance procedures should be initiated for traffic in the direction of the received frame. On receiving either indication, the end point (i.e., bridge, router, or other internetworking device) takes appropriate action to ease congestion.

The response to congestion notification depends on the protocols and flow-control mechanism employed by the end point. The BECN bit typically would be used by protocols capable of controlling traffic flow at the source. The FECN bit typically would be used by protocols implementing flow control at the destination.

On receipt of a frame with the BECN bit set, the end point must reduce its offered rate to the CIR for that frame relay connection. If consecutive data frames are received with the BECN bit set, the end point must reduce its rate to the next "step" rate below the current offered rate. The step rates are 0.675, 0.50, and 0.25 of the current rate. After the end point has reduced its offered rate in response to receipt of BECN, it may increase its rate by a factor of 0.125 times the current rate after receiving two consecutive frames with the BECN bit clear.

If the end point does not respond to the congestion notification, or the user's data flow into the network is not significantly reduced as a result of the response to the congestion notification, or an end point is experiencing a problem that exacerbates the congestion problem, the network switches collaborate in implementing congestion-recovery procedures. These procedures include discarding frames, in which case the end-to-end protocols employed by the end points are responsible for detecting and requesting the retransmission of missing frames.

Frame discard can be done on a priority basis; that is, a decision is made on whether certain frames should be discarded in preference to other frames in a congestion situation based on predetermined criteria. Frames are discarded on the basis of their "discard eligibility" setting of 1 or 0, as specified in the data frame. A setting of 1 indicates that the frame should be discarded during congestion, while a setting of 0 indicates that the frame should not be discarded unless there are no alternatives.

The discard eligibility may be determined in several ways. The user can declare whether the frames are eligible for discard by setting the discard eligibility bit in the data frame to 1. Or the network access interface may be configured to set the discard eligibility bit to 1 when the user's data has exceeded the CIR; in which case, the data are considered excess and subject to discard. For users who subscribe to CIR = 0, which moves data through the frame relay network on a best-effort basis subject to bandwidth availability, all traffic is discard eligible.

Frame Relay Charges

Frame relay charges differ by carrier and may differ further by configuration. Accordingly, frame relay service charges may include

- Port charge for access to the nearest frame relay switch, which is applied to every user location attached to the frame relay network.
- Local loop charge, which is the monthly cost of the facility providing access to the frame relay network. This charge may not apply if the customer's building is directly connected to the carrier's metro fiber ring, in which case the customer is charged only a one-time setup fee.
- Charges for the PVCs and SVCs, which are determined according to the CIR assigned to each virtual circuit.
- Burst capability, usually determined by the burst excess size. Most carriers do not specifically charge customers for bursting beyond the CIR.
- Customer premises equipment, which includes the frame relay or internetworking access equipment optionally

leased from the service provider and bundled into the cost of the service.

- IntraLATA/interLATA service. Usually, there is one price for "local" frame relay service and another price for "national" frame relay service. Neither is distance-sensitive, however.

Voice over Frame Relay

Voice over frame relay (VoFR) is receiving growing attention. Most data-oriented frame relay access devices (FRADs) and routers use the first-in, first-out (FIFO) method of handling traffic. In order to achieve the best voice quality, however, voice frames cannot be allowed to accumulate behind a long queue of data frames. Voice FRADs and routers, therefore, employ traffic prioritization schemes to minimize delay for voice traffic.

Traffic prioritization schemes ensure that voice packets have preference over data. During times of network congestion, one of the easiest prioritization methods is to simply discard frames. In such cases, data rather than voice frames will be discarded first, giving voice a better chance of making it through the network.

Some service providers offer prioritization of PVCs within the frame relay network. Prioritization features on both the CPE and the frame relay network can result in better voice application performance. The CPE ensures that higher-priority traffic is sent to the network first, while PVC prioritization within the network ensures that higher-priority traffic is delivered to its destination first.

VoFR equipment compresses the voice signal from 64 kbps to at least 32 kbps. In most cases, compression to 16 kbps or even 8 kbps is possible. Some equipment vendors support dynamic compression options. When bandwidth is available, a higher voice quality is achieved using 32 kbps, but as other calls are placed or other traffic requires bandwidth, the 16- or 8-kbps compression algorithm is implemented. Most voice FRADs also support fax traffic. A fax can take up as little as 9.6 kbps of bandwidth for each active line.

VoFR usually allows a company to use its existing phones and numbering plan. In most cases, an internal dialing plan can be set up that allows users to dial fewer digits to connect to internal locations.

A persistent myth about VoFR is that voice calls can be carried free on an existing frame relay network. In fact, VoFR requires special customer premises equipment and entails an increase in the port speed and possibly an increase of the CIR—all of which have a cost.

Summary

The need for frame relay arose largely out of the need to interconnect LAN at different locations. Since frame relay was designed to operate over digital networks, which are faster and less prone to transmission errors than older analog lines, there was less need for the network to perform error correction. This could be effectively handled by the CPE at the edges of the network. The X.25 protocol overcame the limitations of analog lines but did so with a significant performance penalty, due mainly to its extensive error-checking and flow-control capabilities. In being able to do without these and other functions, frame relay offers higher throughput, less delay, and more efficient utilization of the available bandwidth. In addition to LAN traffic, frame relay carries Internet, voice, and legacy SNA traffic.

See also

Asynchronous Transfer Mode
Internet

G

GATEWAYS

Gateways are used to interconnect dissimilar networks or applications. Gateways operate at the highest layer of the Open Systems Interconnection (OSI) reference model—the Application Layer (Figure G-1). A gateway consists of protocol-conversion software that usually resides in a server, minicomputer, mainframe, or front-end device. One application of gateways is to interconnect disparate networks or media by processing the various protocols used by each so that information from the sender is intelligible to the receiver despite differences in network protocols or computing platforms.

For example, when a systems network architecture (SNA) gateway is used to connect an asynchronous PC to a synchronous IBM mainframe, the gateway acts as both a conduit through which the computers communicate and a translator between the various protocol layers. The translation process consumes considerable processing power, resulting in relatively slow transmission rates when compared with other interconnection methods—hundreds of packets per second for a gateway versus tens of thousands of packets per second for a bridge.

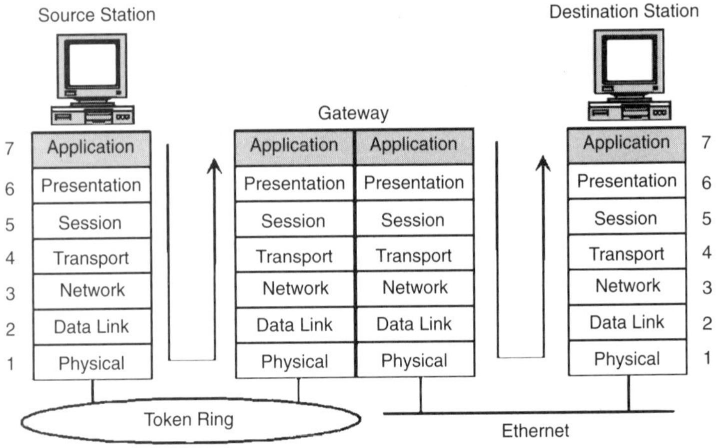

Figure G-1 Gateway functionality in reference to the OSI model.

In addition to its translation capabilities, a gateway can check on the various protocols being used, ensuring that there is enough protocol processing power available for any given application. It also can ensure that the network links maintain a level of reliability for handling applications in conformance with predefined error rate thresholds.

IP-PSTN Gateways

Gateways have a variety of other applications. A relatively new type of gateway provides connections between the Internet and the Public Switched Telephone Network (PSTN), enabling users to place phone calls from their multimedia PCs or conventional telephones over the Internet or a carrier's managed Internet Protocol (IP) data network and vice versa. This arrangement allows users to save on long-distance and international call charges.

The IP-PSTN gateways perform the translations between the two types of networks. When a standard voice call is received at a near-end gateway, the analog voice signal is digitized, compressed, and packetized for transmission over an

IP network. At the far-end gateway the process is reversed, with the packets decompressed and returned to their original digital form for delivery to the nearest Class 5 central office.

The gateways support one or more of the internationally recognized G.7xx voice codec specifications for toll-quality voice. The most commonly supported codec specifications are as follows:

- G.711 describes the requirements for a codec using Pulse Code Modulation (PCM) of voice frequencies to achieve 64 kbps, providing toll-quality voice on managed IP networks with sufficient available bandwidth.
- G.723.1 describes the requirements for a dual-rate speech codec for multimedia communications (e.g., videoconferencing) transmitting at 5.3 and 6.3 kbps. This codec provides near-toll-quality voice on managed IP networks.[1]
- G.729A describes the requirements for a low-complexity codec that transmits digitized and compressed voice at 8 kbps. This codec provides toll-quality voice on managed IP networks.

The specific codec to be used is negotiated on a call-by-call basis between the gateways using the H.245 control protocol. Among other things, the H.245 protocol provides for capability exchange, enabling the gateways to implement the same codec at the time the call is placed. The gateways may be configured to implement a specific codec at the time the call is established, based on predefined criteria, such as

- Use G.711 only, in which case the G.711 codec will be used for all calls.
- Use G.729 (A) only, in which case the G.729 (A) codec will be used for all calls.
- Use highest common bit rate codec, in which case the codec that will provide the best voice quality is selected.

[1]The mean opinion score (MOS) used to rate the quality of speech codecs measures toll-quality voice as having a top score of 4.0. With G.723.1, voice quality is rated at 3.98, which is only 2 percent less than that of analog telephone.

- Use lowest common bit rate codec, in which case the codec that will provide the lowest packet bandwidth requirement is selected.

This capability exchange feature provides carriers and ISPs with the flexibility to offer different quality voice services at different price points. It also allows corporate customers to specify a preferred proprietary codec to support voice or a voice-enabled application through an intranet or IP-based virtual private network (VPN).

Summary

Gateways are available as software or may be dedicated hardware systems equipped with appropriate software to make the translations between different applications, networked devices, or different types of networks. Gateways can even be used to reconcile the differences between network management systems or operations support systems (OSSs), enabling them to interoperate with the systems of other vendors.

See also

Bridges
Repeaters
Routers
Voice over IP

H

HOME PHONE-LINE NETWORKING

Home phone-line networking refers to the ability of consumers to implement a local area network (LAN) in the home over ordinary telephone wire, enabling them to connect computers and peripherals at speeds of up to 10 Mbps and share the same Internet access connection. A home network permits file and application sharing, messaging, and multiuser gaming among family members or with others via the Internet.

The number of multiple-PC homes is growing faster than the number of single-PC homes. Currently, there are about 35 million homes in the United States that have at least two PCs. The need for a network in the home to share printers, modems for Internet access, CD-ROMs, hard drives, and other equipment becomes essential if only to save money on expensive resources, including multiple phone lines and Internet access accounts. Beyond cost savings, a home network offers convenience. Users no longer have to waste time looking all over the house to check each computer for the documents, spreadsheets, images, or software they need for office work, school work, tax preparation, or to play their favorite games. Further, a network in the home provides the opportunity to implement special features such as a family message board and voice intercom.

The cost of installing and configuring a basic two-node network in the home is less than $100 (U.S.), and several vendors offer kits that include all the necessary components. Third-party software is usually required for features such as internal messaging and voice intercom. More computers can be added to the network by purchasing extra adapter cards to make the phone-line connections. Such networks can be built and configured at a leisurely pace, with professional results achievable within a Saturday afternoon.

The first home phone-line products were introduced in 1998 but offered a top transmission speed of only 1 Mbps. Today's phone-line networking products offer a top transmission speed of 10 Mbps. In either case, the existing phone wiring within the home provides the connectivity among computers, including shared Internet access, without the need for a hub device. In addition, the newer 10-Mbps products are backward-compatible with the older 1-Mbps products.

Data run over the home's existing phone wiring without disrupting normal phone service. This is accomplished through the use of Frequency Division Multiplexing (FDM), which essentially divides the data traveling over the phone lines into separate frequencies—one for voice, one for high-bandwidth net access such as Digital Subscriber Line (DSL), and one for the network data. These frequencies can coexist on the same telephone line without interfering with each other. The technology is designed to operate with computers and devices up to 500 feet apart, making it suitable for homes of up to 10,000 square feet.

Components

For each computer that will be networked in the home, a network interface card (NIC) is required, just as at the office. This is an adapter that plugs into a vacant slot of the computer. For phone-line networking, the NIC will have two

RJ11 ports into which a segment of phone wire will be connected—one to the phone and the other to the wall jack. Many NICs are optimized to work in Windows environments, which provide the software drivers for most brands of NICs, making installation and configuration essentially a plug-and-play affair.

With the computer turned off, the user opens the case and inserts the NIC into the appropriate type of card slot. The short white slots are for PCI (peripheral component interconnect) cards, and the longer black slots are for the older ISA (industry standard architecture) cards. Because of the limited number of vacant card slots in each computer—usually only three to five—it is best to inventory each computer before actually buying the NICs or a network kit. Whenever possible, PCI-type cards should be used for their higher performance.

Once the card is installed, the computer can be restarted. Windows will recognize the new hardware and configure it automatically. If the driver or a driver component is not found, however, it may be necessary to get out the Windows CD-ROM or the NIC vendor's 3.5-inch installation disk so that Windows can find the components it needs.

After Windows has recognized and configured the NICs, it is time to connect each computer. One segment of phone line is inserted into the wall jack and the other end into an RJ11 port of the computer's NIC. Another phone line segment plugs into the phone and terminates at the other RJ11 port on the NIC. This allows the phone to be used even while a file transfer is in progress between computers.

Instead of using one of the two RJ11 ports for connection to a telephone, the extra RJ11 port can be used for daisy-chaining computers together over the same phone line. Some vendors also include an RJ45 port on the NIC, which lets users migrate to 100 Mbps using Category 5 LAN cabling.

To actually share files and peripherals over the phone-line connections, each computer must be configured with the right protocols and be set up for file and printer sharing.

Once this is done, a modem also can be shared. Up to 25 users can share that modem for Internet access, eliminating the need for separate telephone lines, modems, and accounts with an Internet service provider (ISP). If Internet access is provided over a broadband service such as cable or DSL, the computers connected over the phone-line network can share these resources as well.

Configuring the Network

As noted, after installing a NIC and powering up the computer, Windows will recognize the new hardware and automatically install the appropriate network-card drivers. If the drivers are not already available on the system, Windows will prompt the user to insert the manufacturer's disk containing the drivers, and they will be installed automatically.

Next, the user must select the client type. Since this is a Microsoft peer-to-peer network that is being created, the user must add "Client for Microsoft Networks" as the primary network logon (Figure H-1). Since the main advantage of networking is resource sharing, it is important to enable the sharing of both printers and files, which is done by clicking on the "File and Print Sharing" button and choosing one or both of these capabilities. (refer again to Figure H-1.)

Identification and security are the next steps in the configuration process. From the "Identification" tab of the dialog box, the user must select a unique name for the computer and the workgroup to which it belongs, as well as a brief description of the computer (Figure H-2). This information will be visible to others when they use Network Neighborhood to browse the network.

From the "Access Control" tab of the dialog box, the user selects the security type. For a small peer-to-peer network, share-level access is adequate (Figure H-3). It allows printers, drives, directories, and CD-ROM drives to be shared and enables the user to establish password access for each of

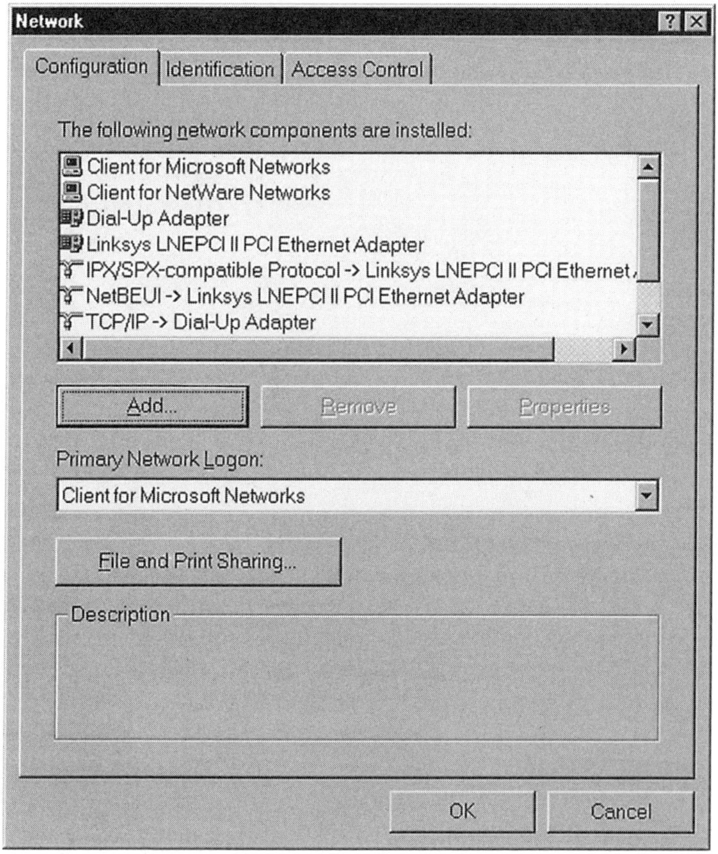

Figure H-1 By accessing the Windows control panel and double clicking on the network icon, this dialog box opens, where the user can configure the computer for the primary network logon plus file and printer sharing.

these resources. In addition, read-only access allows users to view (not modify) a file or directory.

To allow (or disallow) disk drives to be shared, the user double-clicks on the "My Computer" icon on the desktop and then right-clicks on the drive to be shared. Next, the user selects "Sharing" from the pick list (Figure H-4). The type of

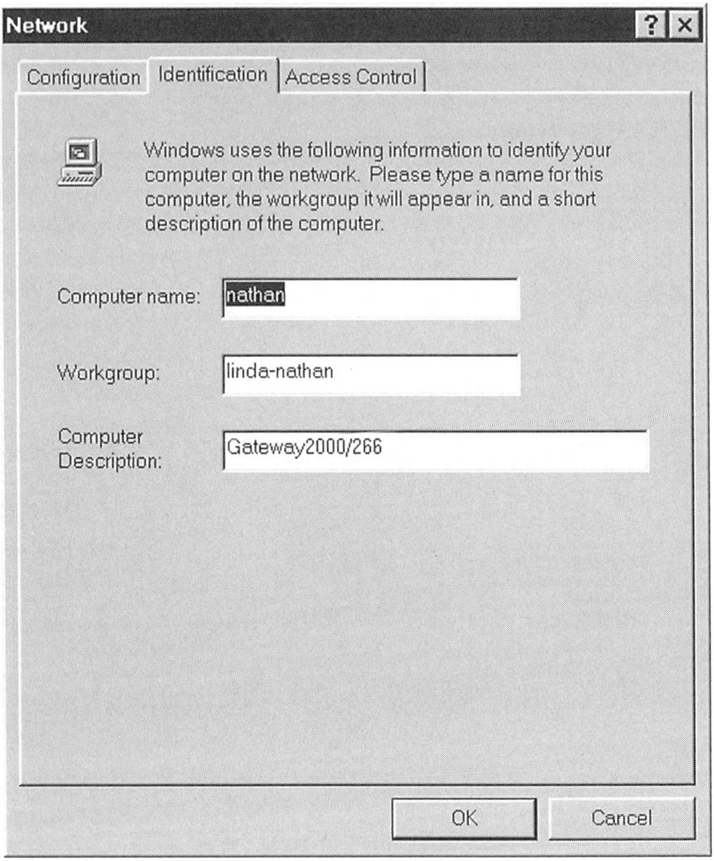

Figure H-2 A unique name for the computer and the workgroup to which it belongs and a brief description of the computer identify it to other users when they access Network Neighborhood to browse the network.

access can be set as read-only, full, or depends on password. This is done for each drive the user wants to share, including CD-ROM, CD-RW, CD-R, and DVD optical disk drives. Individual directories can be set for no sharing in the same way.

To allow a printer to be shared, the user right-clicks on the "Printer" icon in the Control Panel and selects "Sharing"

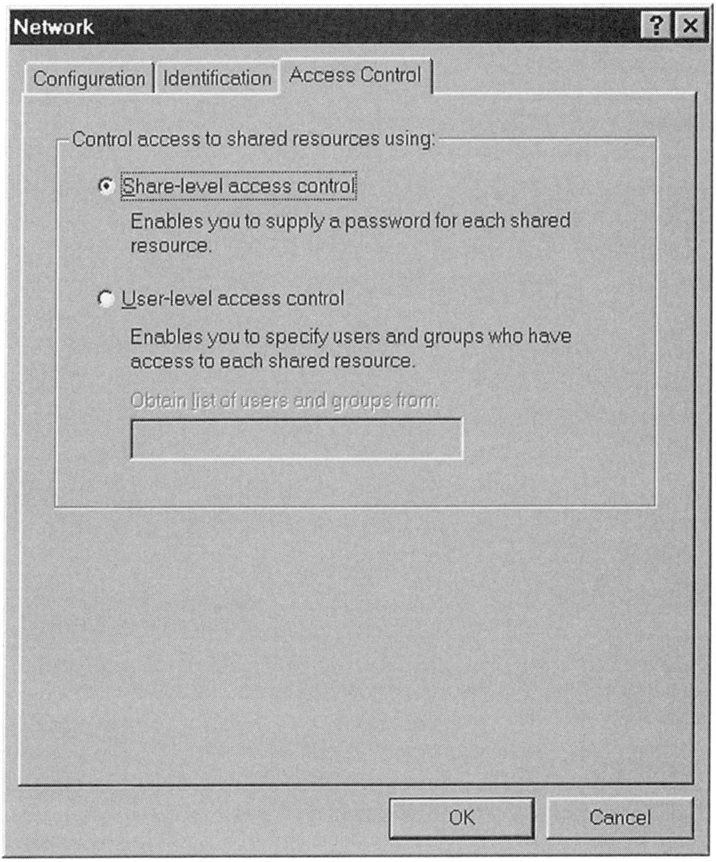

Figure H-3 Choosing share-level access allows the user to password-protect each shared resource.

from the pick list (Figure H-5). Next, the user clicks on the "Shared As" radio button and enters a unique name for the printer (Figure H-6). If desired, this resource can be given a password as well. When another computer tries to access the printer, the user will be prompted to enter a password. If a password is not necessary, the password field is left blank.

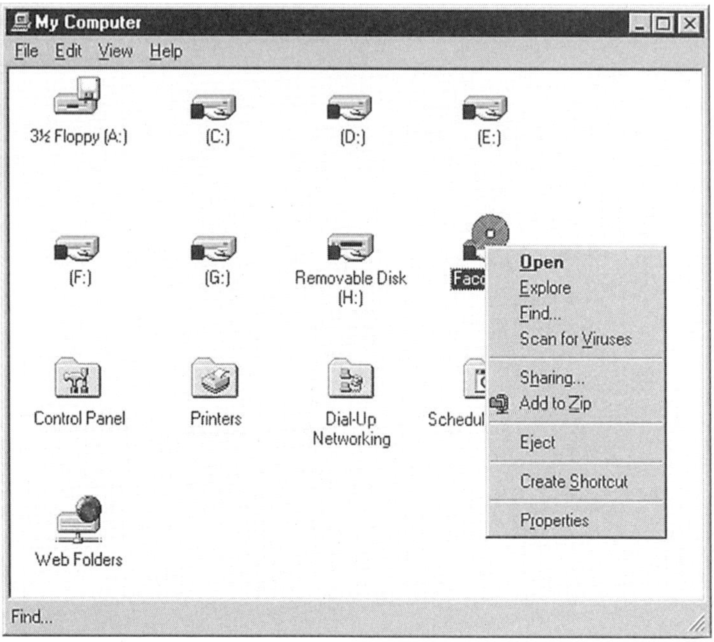

Figure H-4 Any type of hard drive or optical drive on any computer can be shared. Access privileges can be associated with each drive, such as read-only, full, or depends on password.

Another security option in the "Access Control" tab is user-level access, which is used to limit resource access by user name. This function eliminates the need to remember passwords for each shared resource. Each user simply logs onto the network with a unique name and password; the network administrator governs who can do what on the network. However, this requires the computers to be part of a larger network with a central server—one running Windows NT server, for example—which maintains the access-control list for the whole network. Since Windows 95/98/XP and Windows NT/2000 workstations support the same protocols, Windows 95/98/XP computers can participate in a Windows NT/2000 server domain.

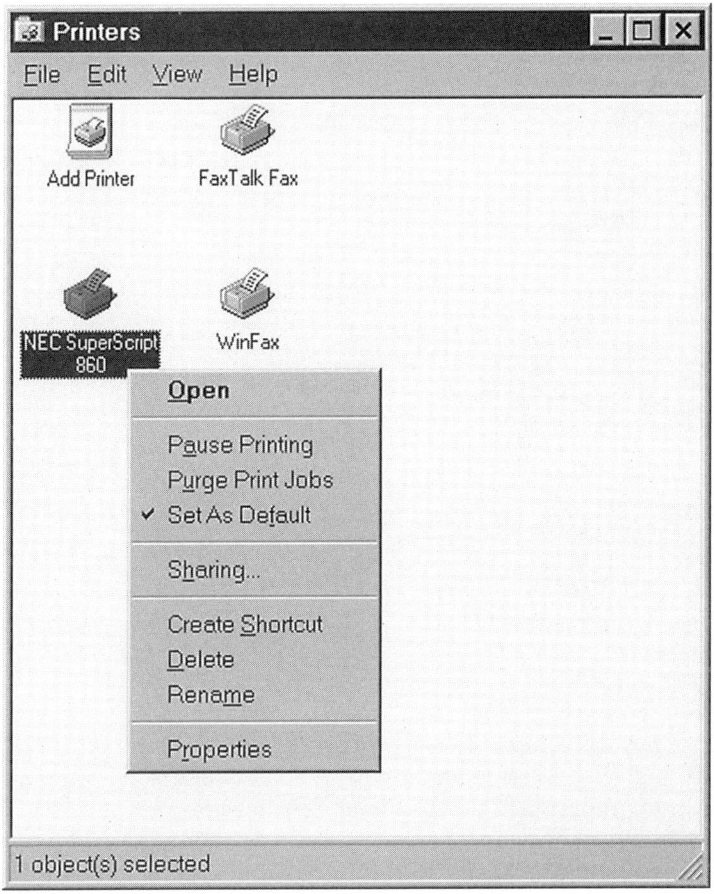

Figure H-5 Right-clicking on a printer sets it up for sharing.

Peer services can be combined with standard client-server networking. For example, if a Windows 98 computer is a member of a Windows NT network and has a color printer to share, the resource "owner" can share that printer with other computers on the network. The server's access-control list determines who is eligible to share resources.

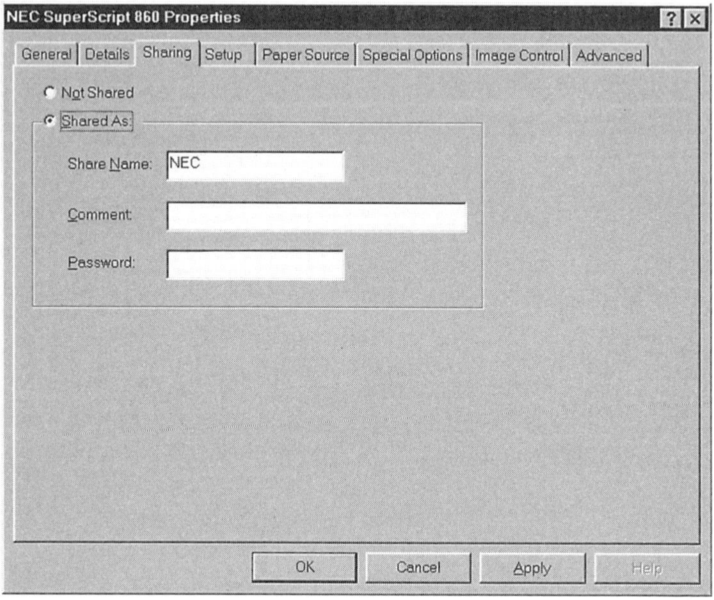

Figure H-6 A shared printer is named and may be password protected if necessary.

Once the computers are properly configured and connected through the existing phone lines, the network is operational. Although designed for the home, this peer-to-peer network is an inexpensive way for small companies within a building to share resources among a small group of computers. This type of network provides many of the same functions as the traditional client-server network, including the ability to run network versions of popular software packages.

Standards

To standardize the products that interconnect computers over standard telephone wiring in the home, the vendor-oriented Home Phoneline Networking Alliance was established in

1998. Products that adhere to the HomePNA standard permit the creation of simple, cost-effective home networks using existing phone wiring. The use of phone wiring for this purpose eliminates the need for cable installation and a hub yet allows shared Internet access with one ISP account without interfering with regular phone service. HomePNA Specification 2.0 delivers a 10-Mbps data rate for home phone-line networking while maintaining full backward-compatibility and interoperability with the previous specification, which offered a data rate of 1 Mbps.

Adding data to voice over existing phone wiring does not pose interference problems because different frequencies are used for each. Standard voice occupies the range from 20 Hz to 3.4 kHz in the United States (slightly higher internationally), while phone-line networking operates in a frequency range above 2 MHz. By comparison, DSL services like ADSL occupy the frequency range from 25 kHz to 1.1 MHz. The frequencies used are far enough apart that the same wiring can support all three simultaneously.

Summary

As the number of PCs and peripherals in the home continues to increase, the need for a network to leverage these assets and provide shared access to the Internet becomes more apparent. PCs on home phone networks must run Windows 95, 98, or NT/2000 or some other software that supports file sharing. Once networked, the PCs can share a printer, as well as a modem dial-up connection for Internet or corporate network access. Users also can work high-speed DSL and cable modems into the mix. In fact, HomePNA lets the consumer choose the method of WAN access, which also can include ISDN and wireless services.

See also

Cable Television Networks

Digital Subscriber Line Technologies
Ethernet
Hubs
Internet
Modems

HUBS

With today's networks becoming increasingly more complex, the conventional bus and ring LAN topologies have exhibited shortcomings, especially with regard to cable installation and maintenance. Furthermore, a fault anywhere in the cabling often brought down the entire network or a significant portion of it. This weakness was compounded by the inability of technicians to readily identify the point of failure from a central administration point, which tended to prolong network downtime. This situation led to the development of the wiring hub in the mid-1980s.

Hubs provide a central point at which all wires meet. They are at the center of the star configuration, with the wires (i.e., segments) radiating outward to connect the various network devices, which may include bridge/routers that connect to remote LANs via the wide area network (WAN). Wiring hubs physically convert the networks from a bus or ring topology to a star topology while logically maintaining their Ethernet or token ring characteristics. The advantage of this configuration is that the failure of one segment—which may be shared among several devices or dedicated to just one device—does not necessarily impact the performance of other segments.

Not only do hubs limit the impact of cabling faults to a particular segment, they also provide a centralized administration point for the entire network. If the wiring hub also employs some CPUs and management software to automate fault isolation, network reconfiguration, and statistics gathering operations, it is no longer just a hub but an "intelligent hub," capable of solving a wide range of connectivity problems efficiently and economically.

Types of Hubs

High-end hubs are modular in design, allowing the addition of ports, network interfaces, and special features as they are needed by the organization. These enterprise-level hubs can support networks that combine different LAN topologies and media types in a single chassis. Ethernet, token ring, and FDDI networks can coexist in a single hub. LAN segments using twisted-pair wiring, coaxial cable, and optical fiber also can be interconnected through the hub. In such cases, the hub is equipped with media conversion modules.

Other hubs are available in fixed configurations for departments or workgroups that do not anticipate future growth. A variation of the fixed-configuration hub is the stackable hub. A unique feature of stackable hubs is that they can be interconnected through a modular backplane. This offers managers the ability to economically expand workgroup and departmental networks as needed. Whereas the high-end modular systems are used to build large-scale enterprise networks, "stackables" are designed for small to medium-sized networks.

There are also workgroup hubs, which typically have eight ports or less. This type of hub not only provides connectivity among connected workstations but also connects to a DSL or cable router to allow all the workstations to share the same Internet access connection. This type of hub is becoming popular for small businesses, small office–home office (SOHO) environments, and consumers who have multiple PCs in the home.

A relatively new category of hub is the "superhub." These are modular units that provide at least an uplink to a standalone Asynchronous Transfer Mode (ATM) switch, if not some level of integral ATM switching, in addition to 100-

Mbps LAN support and integrated LAN switching and routing. Fully populated superhubs support in excess of 500 ports of mixed-media, shared and switched connectivity over a gigabit-per-second backplane in a software-manageable, fault-tolerant, hot-swappable modular chassis that can cost well over $100,000.

Hub Components

Enterprise-level intelligent hubs contain four basic components—chassis, backplane, plug-in modules, and a network management system.

Chassis The chassis is the hub's most visible component. It contains an integral power supply and/or primary controller unit and varies in the number of available module slots. The modules insert into the chassis and are connected by a series of buses, each of which may constitute a separate network or integrate with one or more backbone networks. The chassis holds the individual modules. In fitting into the chassis, each module is instantly connected to other modules via the hub's high-speed backplane.

Backplane The main artery of the hub is its backplane, a board that contains one or more buses that carry all communications between LAN segments. The hub's backplane is analogous to a PC bus through which various interface cards may be interconnected. The data path that carries traffic from card to card is often called a channel; unlike the PC, though, the hub's backplane typically consists of multiple physical or logical channels. Minimally, the hub accommodates one LAN segment for each channel on the backplane.

Segmenting the backplane in this way allows multiple independent LANs or LAN segments to coexist within the same chassis. There is usually a separate backplane channel to carry management information. The segmented backplane typically has dedicated channels for Ethernet, token ring, and FDDI. Some hubs employ a multiplexing technique across the backplane to divide the available bandwidth into multiple logical channels. Other hubs support load sharing that allows network modules to select the backplane channel that will transport the traffic. Still other hubs are designed to allow backplanes to be added or upgraded to accommodate network expansion and new technologies.

The potential bandwidth capacity of newer backplane designs supporting ATM switching is quite impressive, reaching well into the gigabit-per-second range—more than enough to accommodate several Ethernet, token ring, and FDDI networks simultaneously.

Modules The functionality of hubs is provided by individual modules, the types of which depend on the hub vendor. Typically, the vendor will provide multiuser Ethernet and token ring cards, LAN management, and LAN bridge and router cards. The use of bridge and router modules in hubs overcomes the distance limitations imposed by the LAN cabling and facilitates communication between LANs and WANs.

There are even plug-in modules for terminal servers, communications servers, file and application servers, and systems network architecture (SNA) gateways. Hub vendors also offer a variety of WAN interfaces, including those for X.25, frame relay, ISDN, T-carrier, SMDS, and ATM. As many as 60 different types of modules may be available from a single hub vendor, many of them provided under third-party OEM, technology-swap, and other vendor-partnering arrangements.

Modules plug into vacant chassis slots. Depending on the vendor, the modules can plug into any vacant slot or slots specifically devoted to their function. Hubs supporting any-slot insertion automatically detect the type of module that is inserted into the chassis and establish the connections to other compatible modules. In addition, many vendors offer a "hot swap" capability that permits modules to be removed or inserted without powering down the hub.

Management Hubs occupy a strategic position on the network, providing the central point of connection for workstations, servers, hosts, bridges, and routers on the LAN and over the WAN. The hub's management system is used to view and control all devices connected to it, providing information that can greatly aid troubleshooting, fault isolation, and administration. Most of these management systems support the Simple Network Management Protocol (SNMP), enabling them to be controlled and managed through an existing management platform such as Hewlett-Packard's OpenView. Some hubs have remote monitoring (RMON) embedded in the hub, making possible more advanced network monitoring and analysis up to OSI Layer 7, the Application Layer.

Summary

Hubs are now the central point of control and management for the elements that make up departmental and enterprise networks. Hubs, which were developed to simplify the management of structured wiring as networks became bigger and more complex, allow the wiring infrastructure to expand in an orderly and cost-effective manner as the organization's computer systems grow and move and as interconnectivity requirements become more sophisticated.

See also

Bridges
Gateways
Repeaters
Routers

HYBRID FIBER/COAX

As its name implies, hybrid fiber/coax (HFC) is the combination of optical fiber and coaxial cable on the same network.

Such networks are used to provide high-speed digital services to homes and businesses. An HFC system could deliver to each customer

- Multiple telephone lines
- 25 to 40 broadcast analog TV channels
- 200+ broadcast digital TV channels
- 275 to 475 digital pointcast channels that deliver programs at a time selected by the customer
- High-speed two-way digital link for Internet and corporate LAN access

HFC divides the total bandwidth into a downstream band (to the home or business) and an upstream band (to the network). The downstream band typically occupies the 50- to 750-MHz frequency range, while the upstream band typically occupies the 5- to 40-MHz frequency range. The higher-bandwidth downstream band is necessary for delivering cable television broadcasts and multimedia Internet content to the user.

Originally, cable operators envisioned a coaxial tree-and-branch architecture to bring advanced services to the home. However, the capacity of fiberoptic transmission technology led many cable operators to shift to an approach that combined fiber and coax networks for optimal advantage. Transmission over fiber has two key advantages over coaxial cable:

- A wider range of frequencies can be sent over the fiber, increasing the bandwidth available for transmission.
- Signals can be transmitted over greater distances without amplification.

Fiber to the Neighborhood

A key disadvantage of fiber is that the optical components required to send and receive data are still too expensive to deploy to each subscriber location. Therefore, cable operators have adopted an intermediate approach known as "fiber to the neighborhood" (FTTN). In this approach, fiber reaches into the neighborhood, and coaxial cable branches out to each subscriber location. This arrangement increases the bandwidth that the plant is capable of carrying while reducing the number of amplifiers needed and the number of amplifiers in cascade between the cable operator's head-end office and each subscriber.

The total number of amplifiers is an important economic consideration because each amplifier must be upgraded or, more typically, replaced to pass the larger bandwidth that the fiber and shorter coaxial cable runs allow. The number of amplifiers in cascade is important for ensuring signal quality. Since each amplifier is an active component that can fail, the fewer amplifiers in cascade, the lower is the chance of failure. Fewer amplifiers and shorter trees also introduce less noise into the cable signal. These improvements translate into higher bandwidth, better-quality service, and reduced maintenance and operating expense for the cable operator.

HFC Advantages

In HFC networks, fiber is run from a service distribution hub (i.e., head-end office) to an optical feeder node in the neighborhood, with tree-and-branch coax distribution in the local loop (Figure H-7). Two overriding goals of the HFC architecture are to minimize the fiber investment by distributing it over the maximum number of subscribers and to use the upstream bandwidth efficiently for the highest subscriber fan-out.

There are over 200 million TV sets and 100 million VCRs deployed in the United States—and few of them are digital. This casts doubt about the immediate viability of fiber-to-the-curb (FTTC) networks. Any serious attempt to provide video services must deal with this embedded base. Services

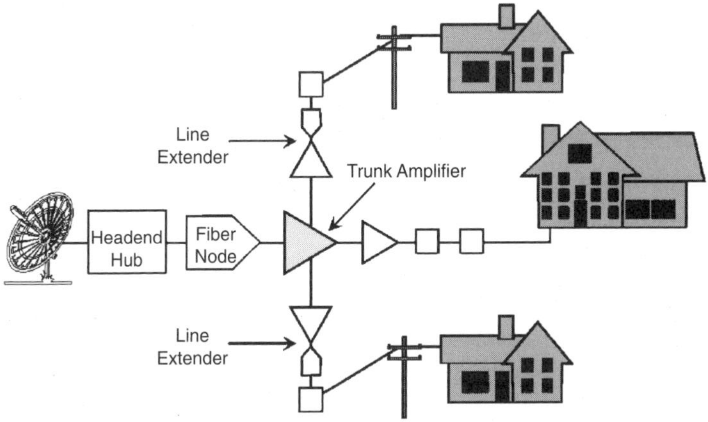

Figure H-7 Typical topology of an HFC network.

that provide digital television signals require a separate digital decoder for each TV tuner, including the VCR, making FTTC an expensive proposition for subscribers, especially since the digital set-top converter would be required even to receive basic television channels. The advantage of HFC networks, however, is that they carry radiofrequency (RF) signals, delivering video signals directly to the home in the exact format that most television sets and VCRs are able to receive.

HFC networks also have the ability to evolve over time from a basic broadcast plant to a two-way network with interactive bandwidth. This evolution is achieved as needed by activating unused dark fiber and subdividing existing nodes to serve fewer homes per node.

During the initial stages of interactive service development in each market, it is anticipated that consumer demand will be low, with early subscribers scattered throughout the service area. To provide services, FTTC network operators must adhere to standard network design concepts, which entail bringing fiber to every curb whether or not a particular household or business wants to subscribe to the service. With HFC networks, scattered demand can be met by providing cable modems as needed. Using cable modems, digital bandwidth can be allocated flexibly on demand, with virtually arbitrary bit rates. This allows the service provider to inexpensively add bandwidth capacity as demand builds, spreading out capital expenditures to meet subscriber growth in any given service area.

Modulation Technologies

For example, if the HFC network operator wants to supply high-definition television (HDTV) service, today's HFC network designs can already accommodate individual users of digital HDTV at 20 Mbps—even if the demand is scattered. With a technique called Quadrature Amplitude Modulation (QAM), a single HDTV channel could be supplied in a standard 6-MHz channel slot within the digital band between 550 and 750 MHz. HFC network equipment operates equally well with National Television Standards Committee (NTSC) signals, which occupy the 6-MHz channels, or Phase Alternating by Line (PAL) standard broadcast analog signals, all on the same network. Channel assignments are completely arbitrary within the 750-MHz spectrum.

One of the major difficulties with HFC is that the cable system was never intended for reliable high-speed data transmission. Cable was installed with the aim of providing a low-cost conventional (analog) broadcast TV service. As a result of the original limited performance requirements, CATV networks are very noisy, and communications channels are subject to degradation. In particular, they can suffer badly from very strong narrowband interference called "ingress noise." Ingress noise plays havoc with conventional single-carrier modulation schemes such as QAM, which cannot avoid the noisy region.

A newer modulation technique—Discrete Wavelet Multitone (DWMT)—is a wavelet transform–based multicarrier modulation scheme that provides better channel isolation, thereby increasing bandwidth efficiency and noise immunity. DWMT divides the channel bandwidth into a large number of narrowband subchannels and adaptively optimizes the number of bits per second that can be transmitted over each subchannel. DWMT provides throughput of 32 DS0s (64 kbps each) over each megahertz of bandwidth on the coaxial cable. When a subchannel is too noisy, DWMT does not use it, thereby avoiding channel impairments and maintaining a reliable high-bit-rate throughput.

Using DWMT, higher data rates can be achieved—which translates into more channels—over longer distances. It uses the cable bandwidth and communications infrastructure more efficiently, and this allows cable operators to offer more services at a reasonable cost for implementation.

Summary

The growth of video and interactive communications services, coupled with developments in digital compression, has driven both CATV and telephone operators to seek effective ways to integrate interactive video and data services with traditional communications networks. With two-way capability, HFC enables cable operators to offer telephone service over the cable as well as hundreds of TV programs, digital services, broadband Internet access, and more. By using fiber links from the central site to a neighborhood hub and coax cable from there to a few hundred homes and businesses, HFC provides an efficient and economical way to deliver the next generation of communications services while supporting current services.

See also

Cable Telephony

Fiber in the Loop

Digital Subscriber Line Technologies

HYPERTEXT MARKUP LANGUAGE

The HyperText Markup Language (HTML) is a set of tags that enables documents to be published on the Web. HTML is an application of ISO 8879: the Standard Generalized Markup Language (SGML), published in 1986. HTML started as a subset of SGML and has been in general use since 1990. Since then, HTML has branched off into new directions—some of them proprietary—and is currently in its fourth revision.

The purpose of the HTML tags is to give Web browsers the information they need to properly render the document so that it appears as the author intended. The tags encapsulate or surround various elements of a document such as headings, paragraphs, lists, forms, tables, and frames. Tags also specify such things as hypertext links, fonts, colors, and backgrounds. In combination with various scripts, tags also can point to Java applets, JavaScript functions, and Common Gateway Interfaces (CGIs) for forms processing and database access.

HTML is not a programming language in the normal sense. It is more like the simple notations a magazine editor uses to get an article ready for publication, which is why it is referred to as "markup language." HTML is essentially ASCII or plain text embedded with special tags that are used to specify how documents are to be rendered by Web browsers for viewing by clients.

HTML documents can be created using a simple text editor, a word processor, or a graphical development tool that makes it unnecessary for authors to learn the growing number of HTML tags and their proper usage. With graphical development tools, the user simply drags and drops various elements into a workspace or highlights portions of the document for such attributes as font size, font style, paragraph, or list as if using a word processor.

Use of Tags

Tags consist of one or more characters and are enclosed within the smaller-than (<) and greater-than (>) brackets. In most cases, there are start tags and end tags: The start tag turns on or opens certain features such italics and bold, while an end tag turns them off or closes them. A feature's end tag is indicated by a slash (/). Together the start and end tags have the following format:

<TAG> . . . </TAG>

The dots between the tags represent the specific text areas of the document that will be rendered by a browser according to the tags' instructions.

The World Wide Web makes extensive use of hypertext links. A hypertext link is usually identified by an underlined word or phrase, or a graphic symbol that points the way to other information, which may be found virtually anywhere: the same document, a different document on the same server, or another document on a different server that may be located anywhere in the world. A hypertext link does not necessarily point to text documents; it can point to maps, forms, images, sound and video clips, or e-mail programs. Hypertext links can even point to other Internet services such as Gopher or FTP sites. Table H-1 lists some of the most frequently used HTML tags, including those used to build hypertext links.[1]

Development Aids

Most Web browsers have a "view source" or similarly named feature that is accessible from the menu bar. This lets users view the HTML coding for any document currently displayed on their screen. If a user likes a particular form, for example, and wants to see how it is written, the source code can be viewed and even downloaded into a word processor, where it can be modified to suit individual needs.

The task of manually encoding documents with HTML tags can be quite tedious. This job is made easier with various HTML editors that can be fully integrated with popular

[1]There are many more HTML tags, but a complete description of all of them is beyond the scope of this book.

Table H-1 A Selection of Some Common HyperText Markup Language (HTML) Tags Used for Rendering Web Pages

Tag	Description
<HTML> . . . </HTML>	Delimits the start and end of the HTML document. (This is now optional.)
<HEAD> . . . </HEAD>	Delimits the start and end of the header portion of the HTML document.
<BODY> . . . </BODY>	Delimits the start and end of the body portion of an HTML document.
<TITLE> . . . </TITLE>	Delimits the title string of the HTML document, which appears above the menu bar of the Web browser's GUI.
<P>	Indicates the end of a paragraph and separates two paragraphs with one line of white space.
 	Equivalent to a hard return, and does not add a line of white space.
<ADDRESS> . . . </ADDRESS>	Delimits the address text in an HTML document, which is used to frame such information as the name of the document author, an email address, and document modification date.
<TEXTAREA> . . . </TEXTAREA>	Delimits more than one line of text in a scrollable area, the dimensions of which are defined by the ROWS and COLS attributes.
<PRE> . . . </PRE>	Delimits text to be displayed in a nonproportional font with all spacing intact and without automatic line wrap.
<H1> . . . </H1>	Delimits a level one heading, providing the largest font size. H6 provides the lowest font size.
<B> . . . </B>	Delimits the start and end of boldface text.
<I> . . . </I>	Delimits the start and end of italic text.
<U> . . . </U>	Delimits underlined text. (Not often used because it may be confused with a hypertext link, which also is indicated by underlined text.)
<UL> . . . </UL>	Delimits an unordered (bulleted) list.
<OL> . . . </OL>	Delimits an ordered (numbered) list.

<LI>	Indicates a bulleted item in an unordered list or a numbered item in an ordered list.
<DL> . . . </DL>	Delimits a definition list, consisting of a variable number of alternating terms and
definitions.	
<DT>	Indicates the start of a term within a definition list.
<DD>	Indicates the start of a definition within a definition list.
<SELECT> . . . </SELECT>	Delimits a multiple-choice list, which is typically rendered as a drop-down or pop-up menu. Each item in the menu list starts with an OPTION tag.
<A HREF="goto.html">. . . </A>	Anchor tag that indicates a hypertext link. The text within quotation marks refers to the name of a target document or program.
<A NAME="# . . .">. . . </A>	Indicates a label within the same or target document that is used as the target of a hypertext link. The text within quotation marks is the label and is preceded by a pound sign (#).
<IMG SRC="graphic.gif">	Indicates the location of an image within the document. Images are usually in GIF or JPG format.
<TABLE> . . . </TABLE>	Delimits a table, the rows of which are defined by table row <TR> . . . </TR> tags and cell contents by table data <TD> . . . </TD> tags.
<FRAMESET> . . . </FRAMESET>	Delimits and specifies two or more HTML files that will be rendered adjacent to each other as separate display areas.
<FORM> . . . </FORM>	Delimits the area of the HTML document that provides input fields of a form.
<INPUT>	Describes the input field using such attributes as type, name, and size.
<SCRIPT> . . . </SCRIPT>	Delimits the area of the HTML document that contains the coding for a scripting language such as JavaScript, as in <SCRIPT language="JavaScript"> . . . </SCRIPT>.
<APPLET> . . . </APPLET>	Delimits the area of the HTML document that identifies a Java applet.

word processors, which allow documents to be saved in the HTML format. In addition, many third-party HTML editors are available that facilitate the creation of HTML documents with point-and-click ease.

Summary

The latest release of HTML is version 4.0, approved by the W3C HTML Working Group in December 1997. HTML 4.0 improves the look and functionality of Web pages, offering several key improvements over the previous version, which essentially collected and standardized the most popular Netscape and Microsoft extensions that were in use in 1996. Features of HTML 4.0 include advanced forms, which allow publishers to display "rich" HTML on any button and build keyboard shortcuts into page controls.

Other features include in-line frames, enhanced tables, and support for objects and scripts. Additionally, HTML 4.0 provides the markup needed for any language, including multilingual documents, allowing authors to manage differences in language, text direction, and character encoding schemes. HTML 4.0 is also more accessible to users with disabilities, allowing table and form text to be rendered into Braille or speech for access by specially equipped clients. HTML version 4.1, released in December 1999, added new style sheets for the document based on W3C technical report styles and fixed document scripts to remove markup-caused crashes on some browsers.

See also

Java
World Wide Web

I

INCUMBENT LOCAL EXCHANGE CARRIERS

Incumbent local exchange carriers (ILECs) is a term that refers to the 22 former Bell Operating Companies (BOCs) divested from AT&T in 1984, as well as Cincinnati Bell, Southern New England Telephone (SNET), and the larger independent telephone companies of GTE and United Telecommunications. In addition, some 1300 smaller telephone companies are also in operation, serving mostly rural areas. These, too, are considered incumbents, but the small markets they serve do not attract much competition.

After being spun off by AT&T in 1984, the BOCs were assigned to seven regional holding companies: Ameritech, Bell Atlantic, BellSouth, Nynex, Pacific Telesis, Southwestern Bell Communications (SBC), and US West. Over the years, some of these regional companies merged to the point that today only four are left. Bell Atlantic and Nynex were the first to merge in 1994. Bell Atlantic also completed a $53 billion merger with GTE in mid-1999 and changed its name to Verizon. SBC Communications merged with Pacific Telesis in 1997 and then with Ameritech in 1999. It also acquired Southern New England Telephone (SNET).

Regulatory Approval

All the mergers passed regulatory approval at the state and national levels. The Federal Communications Commission (FCC) approves mergers with input from the Department of Justice (DoJ). In the case of the SBC-Ameritech merger, the FCC imposed 28 conditions on SBC in exchange for approving the transaction.

The approval package contained a sweeping array of conditions designed to make SBC-Ameritech's markets the most open in the nation, boosting local competition by providing competitors with the nation's steepest discounts for resold local service and full access to operating support systems (OSSs).

It also required SBC to accelerate by 6 months its entry into new markets, forcing the company to compete in 30 new markets within 30 months after completion of the merger. The FCC's rationale was that increased competition in out-of-region territories would help offset reduced competition in the SBC-Ameritech service areas.

The conditions also required stringent performance monitoring, reporting, and enforcement provisions that could trigger more than $2 billion in fines if these goals were not met. Fortunately for SBC, the agreement required it to serve only three customers in each out-of-region market. According to SBC, it will not begin to seriously market its out-of-region services until it has obtained approval to offer long-distance services in its 13 home states.

Summary

The monopoly status of the ILECs officially ended with passage of the Telecommunications Act of 1996. Not only can other types of carriers enter the market for local services in competition with them, but also their regional parent companies can compete in each other's territories. Through mergers, the reasoning went, the combined companies can enter out-of-region markets on a broad scale quickly and efficiently enough to become effective national competitors.

Unfortunately, this has not occurred on a significant scale. In fact, the lack of out-of-region competition among the "Baby Bells" means that consumers and businesses do not have as much choice in service providers, especially now that many competitive local exchange carriers (CLECs) are being hit hard by financial problems and the lack of venture capital. The ILECs are more concerned with being able to qualify for long-distance services in their own markets so that they can bundle local and long-distance services and Internet access—a package few, if any, competitors would be able to match.

See also

Competitive Local Exchange Carriers

Interexchange Carriers

Local Exchange Carriers

INTEGRATED ACCESS DEVICES

Integrated access devices (IADs) support voice, data, Internet, and video services over the same $N \times$ T1/E1 access lines. They typically support Asynchronous Transfer Mode (ATM) technology (Figure I-1), which turns the different traffic types into fixed-length 53-byte cells for transmission through the carrier's ATM network. The consolidation of multiple traffic types over the same aggregate access facility eliminates the need for separate lines for each type of application and having to subscribe to separate services. Even though the IAD resides at the customer location, management of the device and the access links into the network is usually the responsibility of the service provider.

Carriers benefit from this arrangement as well. Today's IADs are feature-rich. They offer end-to-end support for

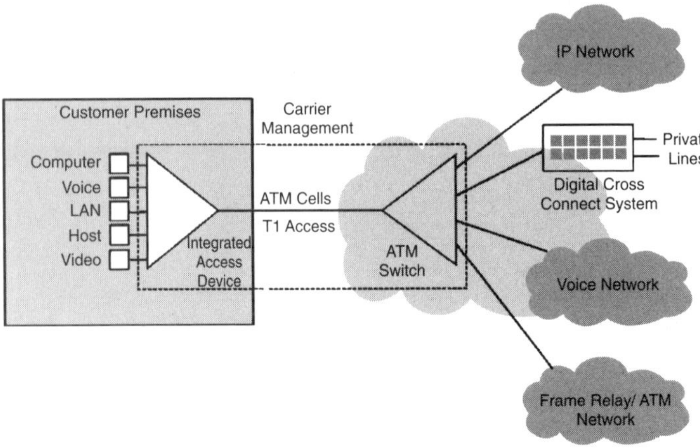

Figure I-1 An integrated access device (IAD) supports voice, data, and video services over the same access lines, which can be bonded together and used as a single higher-speed channel shared by multiple applications. Traffic is split out on the carrier side of the network and routed to the appropriate destinations.

ATM Adaptation Layer 2 (AAL2), the industry standard for transporting compressed voice over ATM. This enables service providers to deliver eight times the call traffic of traditional T1/E1 lines without degradation in the quality of voice. Not only do IADs allow carriers to offer bandwidth, but they also provide them with a way to deliver integrated services to their customers with quality, reliability, and flexibility over that bandwidth. IADs also provide carriers with the means to support any service—ATM, frame relay, or Internet Protocol (IP)—over any channel within the same T1/E1 line simultaneously.

In addition, such products provide queuing algorithms, which provide traffic prioritization capabilities that allow time-sensitive traffic like voice to use the access bandwidth ahead of routine data traffic, which can tolerate delay. With multiple classes of service available, carriers can guarantee differentiated levels of service for both data and voice traffic, enabling them to offer customers virtual private networks (VPNs) that support voice as well as data.

Quality of voice features such as silence suppression and echo cancellation, as well as support for fax/modem call detection and international code conversion, are also available. The common platform—customer premises and carrier point of presence (POP)—ensures interoperability and simplifies operations and management for carriers while providing cost savings to customers.

Modular routers can be turned into IADs very easily, starting with multiple T1/E1 ports and voice compression options, and be incrementally upgraded to support Inverse Multiplexing over ATM (IMA), DS3/E3, and OC-3 wide area network (WAN) connections. This scalability enables service providers to meet the demands of a growing enterprise customer with a single customer premises equipment (CPE) solution.

Summary

IADs allow users to aggregate a variety of lower-speed services onto a high-speed ATM infrastructure. A CLEC, for example, can deploy IADs to provide integrated voice, data, Internet, and video services to business customers, letting them take advantage of applications necessary to succeed in a highly competitive marketplace. At the same time, IADs allow CLECs and other types of carriers to meet their business needs for cost containment in service provisioning.

See also

Asynchronous Transfer Mode

Multiservice Networking

Quality of Service

INTEGRATED SERVICES DIGITAL NETWORK

The Integrated Services Digital Network (ISDN) made its debut in 1980 with the promise of providing a high-quality, ubiquitous switched digital service for multimedia applications. Although ISDN was intended to become a worldwide standard to facilitate global communications, this was not to be the case for many years. In the United States, nonstandard carrier implementations, incompatibilities between customer and carrier equipment, the initial high cost of special adapters and telephones, spotty coverage, and configuration complexity hampered user acceptance of ISDN through the mid-1990s. This changed as the Internet started coming into mainstream use in 1996 with the advent of Web browsers that made navigation easy through a graphical user interface (GUI).

Applications

By 1996, many of the problems with ISDN had been resolved. The increasing popularity of the Internet sparked consumer demand for ISDN as a means of accessing the Web and improving the response time for navigating, viewing, and downloading multimedia content. For other applications, ISDN offers the benefits of

- Faster call setup and teardown
- Increased network management and control facilities
- Improved configuration flexibility
- The capability to streamline networks through integration, reducing the complexity and cost of cabling and equipment

Other applications of ISDN include videoconferencing, the delivery of multimedia training sessions, and temporarily rerouting of traffic around failed leased lines or frame relay networks and handling peak traffic loads (Figure I-2). ISDN also

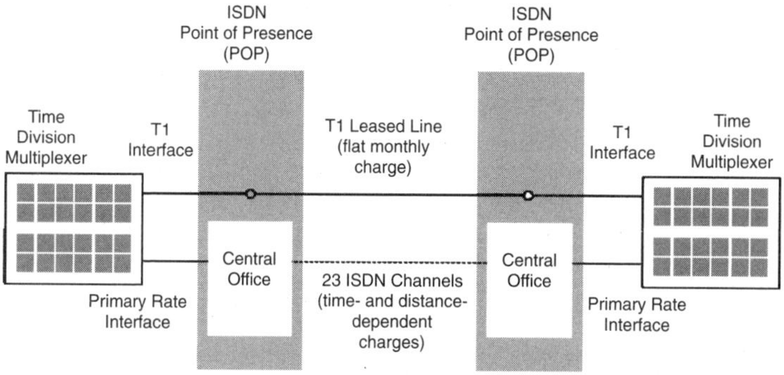

Figure I-2 ISDN has many applications. In addition to carrying IP traffic into the Internet, ISDN PRI can be used to back up T1 leased lines in case of failure, provide an additional source of temporary bandwidth to handle peak traffic loads, or support special applications such as videoconferencing on an as-needed basis. In addition, ISDN BRI can be used to back up frame relay permanent virtual circuits (PVCs).

can play a key role in various call center, computer-telephony integration (CTI), telecommuting, and remote access (i.e., remote control and remote node) applications.

ISDN Channels

ISDN is a circuit-switched digital service that comes in two varieties. The basic rate interface (BRI) provides two bearer channels of 64 kbps each, plus a 16-kbps signaling channel. The primary rate interface (PRI) provides 23 bearer channels of 64 kbps each, plus a 64-kbps signaling channel. Any combination of voice and data can be carried over the B channels.

ISDN PRI was designed to be compatible with existing digital transmission infrastructures, specifically T1 in North America and E1 in Europe (2.048 Mbps). In fact, ISDN PRI is a digital service that rides over a T1/E1 facility. Because ISDN can evenly reduce both T1 and E1 into 64-kbps increments, the 64-kbps channels became the worldwide standard. The use of 64-kbps channels also allows users to migrate more easily from private T1/E1 networks to ISDN and build hybrid networks consisting of both public and private facilities. Table I-1 compares ISDN PRI with T1/E1.

Table I-1 A Comparison of ISDN PRI with T1/E1

ISDN PRI	T1/E1
Digital service	Digital facility
Circuit-switched	Dedicated
Any-to-any connectivity	Point-to-point connectivity
Shared bandwidth	Assigned bandwidth
Dial-up	Always on
Out-of-band signaling (D channel)	In-band signaling (bit-robbed)
Call handling features	No call handling features
Call-by-call channel assignment	Fixed channel assignments
Efficient bandwidth usage	Inefficient bandwidth usage
Optimized for voice, computer-telephony integration (CTI) in call centers, videoconferencing	Optimized for data, compressed voice, PBX trunks

In both BRI and PRI, ISDN's separate D channel is used for signaling. As such, it has access to the control functions of the various digital switches on the network. It interfaces with Signaling System 7 (SS7) to provide message exchange between the user's equipment and the network to set up, modify, and clear the B channels. Via SS7, the D channel also gathers information about other devices on the network, such as whether they are idle, busy, or off. In being able to check ahead to see if calls can be completed, network bandwidth can be conserved. If the called party is busy, for example, the network can be notified before resources are committed.

The D channel's task is carried out very quickly, so it remains unused most of the time. For this reason, PRI users can assign the 64-kbps D channel to perform the signaling function for as many as eight PRI lines.[1] For BRI users, whenever the D channel is not being used for signaling, it can be used as a bearer channel (if the carrier offers this capability as a service) for point-of-sale applications such as automatic teller machines (ATMs), lottery terminals, and cash registers. Some carriers offer BRI users the option of using idle D channel bandwidth for e-mail.

With regard to ISDN PRI, there are two higher-speed transport channels called "H channels." The H0 channel operates at 384 kbps, while the H11 operates at 1.536 Mbps. These channels are used to carry multiplexed data, data and voice, or video at higher rates than that provided by the 64-kbps B channel. The H channels also are ideally suited for backing up FT1 and T1 leased lines.

Multirate ISDN lets users select appropriate increments of switched digital bandwidth on a per-call basis. Speeds, in increments of 64 kbps, are available up to 1.536 Mbps. Multirate ISDN is used mostly for multimedia applications such as collaborative computing and videoconferencing, where the number of channels may vary with each session.

ISDN Architectural Elements

The architectural elements of ISDN include several reference points that define network demarcations between the telephone company and the customer premises:

- *R* The reference point separating non-ISDN (TE2) equipment and the terminal adapter (TA), which provides TE2 with ISDN compatibility.
- *S* The reference point separating terminals (TE1 or TA) from the network terminal (NT2).

[1]This is rarely done because if the PRI line with the D channel goes out of service, the other PRI lines that depend on it for signaling also go out of service.

- *T* The reference point separating NT2 from NT1 (not required if NT2 and NT1 functionality is provided by the same device).
- *U* The reference point separating the subscriber's portion of the network (NT1) from the carrier's portion of the network (LT).

There are also interfaces between various types of CPE. Network terminators provide network control and management functions, while terminal equipment devices implement user functions. Figure I-3 illustrates the reference points and architectural elements of ISDN.

Although ISDN BRI and PRI services consist of different configurations of communications channels, both services require the use of distinct functional elements to provide network connectivity. One of these elements is the Network Terminator 1 (NT1), which resides at the user's premises and performs the four-wire to two-wire conversion required by the local loop. Aside from terminating the transmission line from the central office, the NT1 device is used by the telephone company for line maintenance and performance monitoring.

Network Terminator 2 (NT2) devices include all NT1 functions in addition to protocol handling, multiplexing, and switching. These devices are usually integrated with PBX and key systems.

ISDN terminal equipment (TE) provides user-to-network digital connectivity. TE1 provides protocol handling, maintenance, and interfacing functions and supports such devices as digital telephones, data terminal equipment, and integrated workstations—all of which comply with the ISDN user-network interface. The large installed base of non-ISDN TE2 devices (e.g., telephones and PCs) can communicate with ISDN-compatible devices when users attach or install a terminal adapter (TA) to/in the non-ISDN device. A TA takes the place of a modem. Users can connect a maximum of eight TE/TA devices to a single NT2 in a multidrop configuration.

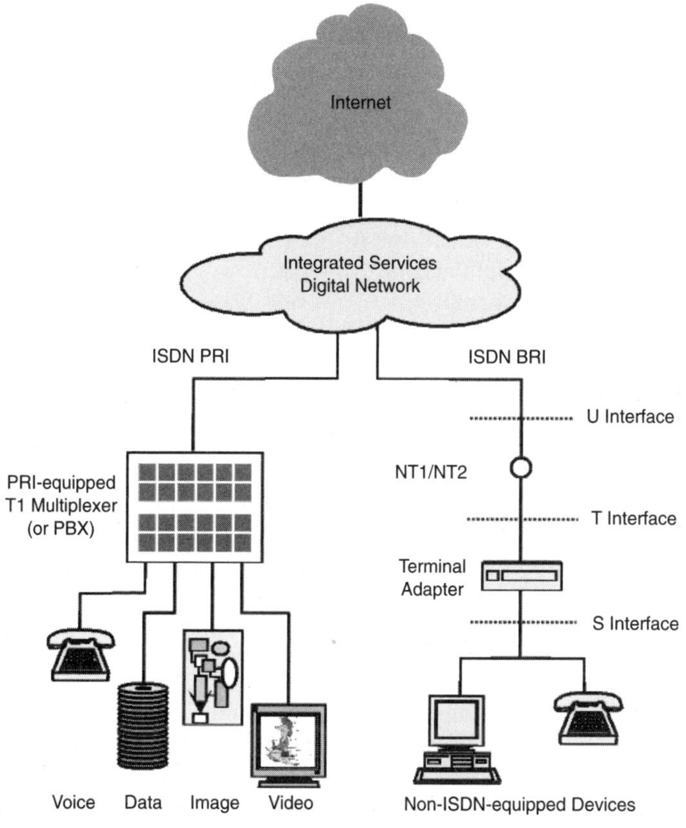

Figure I-3 ISDN architectural elements and reference points.

Summary

Over the years, ISDN has been touted as a breakthrough in the evolution of worldwide telecommunications networks—the single most important technological achievement since the advent of the telephone network itself in the nineteenth century. Others disagree and note that technologies such as IP, frame relay, and ATM have overtaken ISDN to the point of making it virtually obsolete. There are still problems with

ISDN, however, due to inconsistencies in carrier implementation. For example, Global Crossing does not support 64-kbps "clear channel," which is required for video applications over ISDN, whereas WorldCom and other carriers do support clear channel and can pass video over ISDN with no problem. Global Crossing supports only 56 kbps, which is not suited for video applications. The debate over ISDN's relevance to today's telecommunications needs must be put into the context of specific applications. Just like any other service, ISDN will be adequate to serve the needs of some users but not others.

See also

Digital Subscriber Line Technologies

Multiplexers

T1 Lines

INTEREXCHANGE CARRIERS

Interexchange carriers (IXCs), otherwise known as long-distance carriers, include the big three—AT&T, Worldcom, and Sprint. The incumbent local exchange carriers (ILECs) have been limited since 1984 to providing local calling services within their own local access and transport areas (LATAs), except where they have received specific authorization by the Federal Communications Commission (FCC). Generally, long-distance calls between LATAs must be handed off to the IXCs who have established points of presence (POPs) within each LATA for the purpose of receiving and terminating inter-LATA traffic.

In addition to providing long-distance telephone service, the IXCs offer business services like ISDN, frame relay, leased lines, and a variety of other digital services. Many IXCs are also Internet service providers (ISPs), which offer Internet access services, virtual private networks (VPNs), electronic mail, Web hosting, and other Internet-related services.

Bypass

Traditionally limited to providing service between LATAs, IXCs are allowed by the Telecommunications Act of 1996 to offer local exchange services in competition with the ILECs. But because the ILECs charge too much for local loop connections and services and do not deliver them in a consistently timely manner, the larger IXCs have implemented technologies that allow them to bypass the local exchange. Among the methods IXCs use to bypass the local exchange include CATV networks and broadband wireless technologies, such as Local Multipoint Distribution Service (LMDS) and Multichannel Multipoint Distribution Service (MMDS).

With regard to cable, AT&T, for example, has acquired the nation's two largest cable companies, TCI and MediaOne, to bring local telephone services to consumers, in addition to television programming and broadband Internet access. As these bundled services are introduced in each market, they are provided to consumers at an attractive price with the added convenience of a single monthly bill. Sprint uses MMDS to offer Internet access to consumers and businesses that are out of range for Digital Subscriber Line (DSL) services. XO Communications, a nationwide integrated communications provider (ICP), uses LMDS to go beyond its metropolitan fiber loops to reach buildings that are out of the central business districts.

Long-Distance Market

In January 2001, the FCC released the results of a study on the long-distance telecommunications industry. Among the findings from the report:

- In 1999, the long-distance market had more than $108 billion in revenues, compared with $105 billion in 1998. In 1999, long-distance carriers accounted for over $99 billion, and local telephone companies accounted for the remaining $9 billion.

- Interstate long-distance revenues increased by 12.8 percent in 1999 compared with 1.5 percent the year before.
- Since 1984, international revenues have grown more than fivefold from less than $4 billion in 1984 to over $20 billion in 1999. The number of calls has increased from about half a billion in 1984 to almost 8 billion in 1999.
- In 1984, AT&T's market share was about 90 percent of the toll revenues reported by long-distance carriers. By 1999, AT&T's market share had declined to about 40 percent, WorldCom's share was 25 percent, Sprint's was 10 percent, and more than 700 other long-distance carriers had the remaining quarter of the market.
- According to a sampling of residential telephone bills, in 1999 the average household spent $64 monthly on telecommunications. Of this amount, $21 was for services provided by long-distance carriers, $34 for services by local exchange carriers, and the remainder for services by wireless carriers.
- According to the same sampling of residential telephone bills, 38 percent of toll calls in 1999 were interstate and accounted for 50 percent of toll minutes. Also, 33 percent of residential long-distance minutes were on weekdays, 30 percent on weekday evenings, and 37 percent on weekends.

Summary

Growing competition in long-distance services has eroded AT&T's market share from its former monopoly level to about 40 percent. With this competition has come increasing availability of low-cost calling plans for a broad range of consumers. As a result, average revenue per minute earned by carriers has been declining steadily for several years, while long-distance usage has increased substantially to make up for that revenue shortfall. As more ILECs get permission from the FCC to enter the in-region long-distance market, IXCs will come under increasing competitive pressure because the ILECs will be able to bundle local and long-distance service and Internet access into attractively priced service packages.

See also

Building Local Exchange Carriers

Incumbent Local Exchange Carriers

INTERNET

The Internet is a network of networks consisting of millions of interconnected servers worldwide, all of which use the Internet Protocol (IP). The Internet has developed largely without any central plan, and no single entity can control or speak for the entire system.

The architecture of the Internet allows new types of services to be layered on top of existing protocols and for new protocols to be introduced without impacting the rest of the Internet. Numerous users can share the lines and equipment that compose the Internet. The distributed network of thousands of routers moves the traffic along the various paths to their destination.

For purposes of understanding how the Internet works, four basic types of entities can be identified: end users, Internet service providers (ISPs), backbone providers, and network access point operators.

- *End users* These are consumers and businesses that use the Internet primarily to receive information, but they also may be content creators who have set up Web pages to distribute information to anyone who requests it.
- *Internet service providers (ISOs)* These are companies such as America Online and Earthlink that provide subscribers with dial-up or dedicated access to the Internet. They pass customer traffic over their Internet backbone networks, which are comprised of high-speed fiberoptic links.

- *National backbone providers* These are the major carriers—which include AT&T, Sprint, and WorldCom—that provide fiber backbones and interconnect with each other at the major NAPs, where they exchange traffic with each other and national ISPs in an effort to provide the best possible service to their customers.
- *Network access point (NAP) operators* NAP operators maintain public peering points on the Internet through which traffic is routed between the major backbone providers and national ISPs. Among these NAPs are the Chicago NAP managed by SBC Ameritech, the New York NAP managed by Sprint, MAE East managed by WorldCom, and MAE West managed by SBC Pacific Bell. There are dozens of other NAPs where carriers interconnect with each other in private peering arrangements. Traffic is usually brought into the NAP via high-speed links at speeds that vary from the DS3 rate of 45 Mbps to the OC-192 rate of 10 Gbps.

The actual architecture of the Internet is far more complex. Backbone providers typically also serve as ISPs; for example, Earthlink offers dial-up and dedicated Internet access to end users but also connects other ISPs to its nationwide backbone. End users such as large businesses may connect directly to backbone networks or to private peering points where backbone networks exchange traffic. ISPs and backbone providers typically have multiple points of interconnection, and the interrelationships between these providers are changing over time.

End users may access the Internet though several different types of connections. Most residential and many small business users have dial-up connections, which use analog modems to send data over the plain old telephone service (POTS) lines of local exchange carriers (LECs) to ISPs. Consumers also may access the Internet through ISDN BRI, Digital Subscriber Line (DSL), cable, terrestrial wireless, and satellite services. Businesses often have dedicated access connections using T1 lines, frame relay, ATM, and wireless. Many businesses also use DSL, and some use cable. The use of hubs and switches allows all employees to access the Internet from a company's local area network (LAN).

Internet History

The Internet was developed to solve one problem—enable incompatible computer systems to talk to each other. It was not developed as a communications system to withstand nuclear attack, although it turned out that the distributed client-server architecture of the Internet certainly makes this possible.

The roots of the current Internet can be traced to ARPANET, a network developed in the late 1960s with funding from the Advanced Research Projects Administration (ARPA) of the U.S. Department of Defense. ARPANET linked together computers at major universities and defense contractors, allowing researchers at those institutions to exchange data. As ARPANET grew during the 1970s and early 1980s, several similar networks were established, primarily between universities. The TCP/IP protocol suite was accepted as a standard to allow these networks, comprising many different types of computers, to interconnect.

In the mid-1980s, the National Science Foundation (NSF) funded the establishment of NSFNET, a TCP/IP network that initially connected six NSF-funded national supercomputing centers at a data rate of 56 kbps. NSF subsequently awarded a contract to a partnership of Merit (one of the existing research networks), IBM, MCI, and the state of Michigan to upgrade NSFNET to T1 speed (1.544 Mbps) and to interconnect additional research networks. The NSFNET backbone, completed in 1988, initially connected 13 regional networks. Individual sites such as universities could connect to one of these regional networks, which then connected to NSFNET, so that the entire network was linked together in a hierarchical structure. Connections to the federally subsidized

NSFNET generally were free for the regional networks, but the regional networks typically charged smaller networks a flat monthly fee for their connections.

In 1992, the NSF announced its intention to phase out federal support for the Internet backbone and encouraged commercial entities to set up private backbones. Alternative backbones had already begun to develop because NSFNET's acceptable use policy, rooted in its academic and military background, did not permit the transport of commercial traffic. Beginning in the early 1990s, the Internet expanded beyond universities and scientific sites to include businesses and individual users who obtained connections through commercial ISPs and consumer online services. The policies of the Clinton administration accelerated the privatization of the Internet and its development as a vehicle for electronic commerce.

Although federal support for the NSFNET backbone ended in April 1995, the NSF continued to provide funding to facilitate the transition of the Internet to a privately operated and funded network. The NSF also provided transitional funding to the regional research and educational networks because these networks were now required to pay commercial backbone providers rather than receiving free interconnection to NSFNET. Finally, the NSF remains involved in certain Internet research activities. Since termination of federal funding for the NSFNET backbone, Internet development has accelerated and evolved into a must-have communications tool for consumers, businesses, and government.

Operating Characteristics

The fundamental operational characteristics of the Internet are that it is a distributed, interoperable, packet data network. A distributed network has no one central repository of information or control but is composed of an interconnected web of host computers, each of which can be accessed from virtually any point on the network. Routers throughout the network regulate the flow of data at each connection point and reroute data around points of congestion or failure.

The Internet is interoperable in that it uses open protocols so that many different types of networks and facilities can be transparently linked together to allow multiple services to be provided to different users, regardless of what computing platform or operating system they may have. The Internet protocols can run over virtually any type of facility that can transmit data, including copper and fiberoptic circuits of telephone companies, coaxial cable of cable companies, and various types of wireless connections. The Internet protocols can run over any kind of data network or service, including Ethernet and token ring LANs and frame relay and ATM wide area networks (WANs). The Internet also interconnects users of thousands of different local and regional networks, using many different types of computers. The interoperability of the Internet is made possible by the TCP/IP suite, which defines a common structure for Internet data and for the routing of that data through the network.

The data transmitted over the Internet are split up into small chunks, or "packets." Unlike circuit-switched networks, such as the public switched telephone network (PSTN), a packet-switched network is connectionless. In other words, a dedicated end-to-end transmission path does not need to be opened for each transmission. Rather, each router calculates the best routing for a packet at a particular moment in time, given current traffic patterns, and sends the packet to the next router. Thus even two packets from the same message may not travel the same physical path through the network. This mechanism is referred to as "dynamic routing." When packets arrive at the destination point, they must be reassembled, and packets that do not arrive for whatever reason generally must be resent. This system allows network resources to be used more efficiently because many different communications can be routed simultaneously over the same transmission facilities.

Addressing

When an end user sends information over the Internet, the data are first broken up into packets. Each of these packets includes a header, which indicates the point from which the data originates and the point to which it is being sent, as well as other information. TCP/IP defines locations on the Internet through the use of Internet Protocol (IP) numbers. These numbers include four address blocks consisting of numbers between 0 and 256 separated by periods (e.g., 160.130.0.252). Internet users generally do not need to specify the IP number of the destination site because IP numbers can be represented by alphanumeric domain names, such as fcc.gov or ibm.com. domain name servers throughout the network contain tables that cross-reference these domain names with their underlying IP numbers.

Some top-level domains (such as .uk for Britain) are country-specific; others (such as .com) are generic and have no geographic designation. The Domain Name System (DNS) was originally run by the U.S. Department of Defense, through private contractors. In 1993, responsibility for non-governmental registration of generic domains was handed over to the NSF. The NSF established an exclusive agreement with Network Solutions, Inc. (NSI), under which NSI handles domain name registration. NSI currently charges $70 for a 2-year domain name registration. The exclusive agreement ended in 1998. Today, there is competition, and NSI is one of several domain name registries. Country-specific domains outside the United States generally are handled by registration entities within those countries.

Services on the Internet

The actual services provided to end users through the Internet are defined not through the routing mechanisms of TCP/IP but depend instead on higher-level application protocols, such as the Hypertext Transport Protocol (HTTP), the File Transfer Protocol (FTP), the Network News Transport

Protocol (NNTP), and the Simple Mail Transfer Protocol (SMTP). Because these protocols are not embedded in the Internet itself, a new application-layer protocol can be operated over the Internet through as little as one server that transmits the data in the proper format. The utility of a service to users, however, increases as the number of servers that provide that service increases.

By the late 1980s, the primary Internet services included e-mail, Telnet, FTP, and Usenet news. E-mail, which is still the most popular Internet service, allows users to send text-based messages to each other using a common addressing system. Telnet allows Internet users to log into a host and access information and applications from a remote location. FTP allows users to download files from a remote host computer onto their own system. Usenet newsgroups enable users to post and review messages on specific topics. Since 1995, with the advent of graphical browsers, the World Wide Web (WWW) has become one of the most used services on the Internet.

The Web has two primary features that make it a powerful, full-service method of accessing information through the Internet. First, the client software, or Web browsers, can access multimedia information—a combination of text, audio, video, and images embedded in the same file—and provide access to all the other major Internet services such as FTP, e-mail, and news through one standard interface. Second, the Web incorporates a hypertext system that allows individual Web pages to provide direct links to other Web pages, files, and other types of information. Thus complex services such as online shopping, news feeds, and interactive games can be provided through the Internet over a nonproprietary system. The Web is the foundation for virtually all the new Internet-based services currently being developed.

Management

As noted, no single entity or organization governs the Internet. Not even the Federal Communications Commission (FCC) has anything to say about how the Internet is run.

Each facilities-based network provider that is interconnected with the global Internet controls only the operational aspects of its own network. No one can even be sure about the exact amount of traffic that passes across the Internet because each backbone provider can account only for its own traffic, and there is no central mechanism for these providers to aggregate their data.

Despite all this, the Internet does not operate in an environment of pure chaos. Certain functions, such as domain name routing, the issuing of IP addresses, and the definition of the TCP/IP suite, must be coordinated, or traffic would never be able to pass seamlessly between different networks. With tens of thousands of different networks worldwide, it would be impossible to ensure technical and administrative compatibility if each network provider had to separately coordinate implementation issues with all other network providers.

These coordinating functions traditionally have been performed by an array of quasi-governmental, intergovernmental, and nongovernmental bodies. The U.S. government, in many cases, has handed over responsibilities to these bodies through contractual or other arrangements. In other cases, entities have simply emerged to address areas of need.

The broadest of these organizations is the Internet Society (ISOC), a nonprofit professional organization founded in 1992. ISOC organizes working groups and conferences and coordinates some of the efforts of other Internet administrative bodies. Internet standards are developed primarily through the Internet Engineering Task Force (IETF), an open international body mostly composed of volunteers. The work of the IETF is coordinated by the Internet Engineering Steering Group (IESG) and the Internet Architecture Board (IAB), both of which are affiliated with the ISOC. The Internet Assigned Numbers Authority (IANA) manages the root servers of the DNS to promote the stability and robustness of the Internet.

Summary

Limited government intervention is the major reason the Internet has grown so rapidly in the United States. The Telecommunications Act of 1996 adopted such a position. The 1996 act states that it is the policy of the United States "to preserve the vibrant and competitive free market that presently exists for the Internet and other interactive computer services, unfettered by Federal or State regulation." The FCC has a responsibility to implement this statute and has itself refrained from regulating Internet activities, even when it is used for telephone service, which it has the authority to regulate. To date, electronic commerce transactions over the Internet have been exempt from taxation in order to allow enough time for the full potential of the Internet to be realized.

See also

Domain Name System

Extranets

Internet Engineering Task Force

Internet Service Providers

Intranets

Transmission Control Protocol/Internet Protocol (TCP/IP)

World Wide Web

INTERNATIONAL CORPORATION FOR ASSIGNED NAMES AND NUMBERS

The Internet Corporation for Assigned Names and Numbers (ICANN) is a nonprofit corporation that has responsibility for key management functions previously performed under U.S. government contract by the Internet Assigned Numbers Authority (IANA) and other entities. The U.S. Commerce Department oversees ICANN.

Prior to being formed in 1998, many of the essential technical coordination functions of the Internet were handled on an ad hoc basis by U.S. government contractors and grantees and a wide network of volunteers. This informal structure represented the spirit and culture of the research community in which the Internet was developed. However, the growing international and commercial importance of the Internet has necessitated the creation of a technical management and policy development body that is more formalized in structure, more transparent, more accountable, and more fully reflective of the diversity of the world's Internet communities.

Therefore, in a phased, cooperative process, ICANN has been assuming responsibility to coordinate the stable operation of the Internet in four key areas: the Domain Name System (DNS), allocation of IP address space, management of the root server system, and coordination of protocol number assignment.

ICANN does not own the Internet or regulate any aspect of its daily operation or management. Rather, it acts a technical coordinating body. In this consensus-based role, ICANN oversees the management of only those specific technical managerial and policy development tasks that require central coordination: the assignment of the Internet's unique name and number identifiers.

Among the responsibilities of ICANN is to decide who may distribute Internet addresses, how much domain names cost, and what addressing suffixes (.com, .net, .org, .biz, .info, and others) are added to and removed from the system.

Summary

While ICANN continues to make these and other decisions, it has faced criticism from public interest advocates and members of Congress who complain that the group, despite its claimed emphasis on consensus building, has enacted too many key policies by fiat and has failed to include enough ordinary Internet users in its decision-making processes.

See also

American Registry for Internet Numbers

Domain Name System

INTERNET ENGINEERING TASK FORCE

The Internet Engineering Task Force (IETF) is a large, open international community of network designers, operators, vendors, and researchers concerned with the evolution of the Internet architecture and the smooth operation of the Internet. The Internet Activities Board (IAB) officially established the IETF in 1986, although it had existed as an informal organization for some time before that. Much of the work of the IETF occurs through mailing lists. The organization itself meets only three times a year.

Membership in the IETF is open to any interested individual who wants to participate. To become a participant in the IETF, one merely becomes active in one or more working groups by asking to be added to the group's mailing list. Most of the activities of the IETF are conducted on a voluntary basis, including standards work. The Internet Engineering Steering Group (IESG) manages the actual technical work of the IETF, which is organized around 100 working groups pursuing topics in the following areas:

- Applications
- General
- Internet
- Operations and management
- Routing
- Security

- Sub-IP
- Transport
- User services

The IETF working groups are managed by area directors. Area directors sitting as a body, along with the IETF chair, constitute the IESG. The IETF executive director is an ex-officio participant of the IESG, as are the IAB chair and a designated IAB liaison. The IESG approves IETF standards and approves the publication of other IETF documents.

A number of procedural questions and issues will arise over time, and it is the function of the working group chair(s) to manage the group process, keeping in mind that the overall purpose of the group is to make progress toward reaching rough consensus in realizing the working group's goals and objectives.

Internet Architecture Board (IAB) Originally called the Internet Activities Board, this organization was established in 1983 when the Internet was still a research project of the U.S. government. As the technical advisory group of the Internet Society, the responsibilities of the IAB include

- Overseeing the architecture for the protocols and procedures used by the Internet
- Overseeing the standards process used to create Internet standards
- Serving as an appeal board for complaints of improper execution of the standards process
- Editorial management and publication of the request for comments (RFC) document series
- Administration of the various Internet assigned numbers
- Representing the interests of the Internet Society in liaison relationships with other organizations concerned with standards and other technical and organizational issues relevant to the worldwide Internet

- Advising and guiding the board of trustees and officers of the Internet Society concerning technical, architectural, procedural, and policy matters pertaining to the Internet and its enabling technologies
- Appointing a new IETF chair and all other IESG candidates from a list provided by the IETF nominating committee

The IAB consists of 13 voting members, 6 of whom are nominated each year by a committee drawn from the IETF. On approval of the Internet Society's board of trustees, members hold their terms for 2 years. The thirteenth voting member of the IAB is the IETF chair. The IAB elects its own chair from among its 12 IETF-nominated members and has a volunteer executive director.

Internet Society The Internet Society (ISOC) is a professional membership society with more than 150 organizational and 6000 individual members in over 100 countries. It provides leadership in addressing issues that confront the future of the Internet and is the organizational home for the groups responsible for Internet infrastructure standards, including the IETF and the IAB.

The society's individual and organizational members have a common stake in maintaining the viability and global scaling of the Internet. They comprise the companies, governmental agencies, and foundations that have created the Internet and its technologies, as well as innovative new entrepreneurial organizations contributing to maintain that dynamic. The society is governed by its board of trustees, which is elected by its membership around the world.

Standards Development

Each version of an Internet standards-related specification is published as part of the RFC document series. The RFC series of documents on networking began in 1969 as part of the original ARPA wide area networking (ARPANET) project. This archival series is the official publication channel for Internet standards documents and other publications of the IESG, IAB, and Internet community.

An Internet draft that is published as a RFC or that has remained unchanged in the Internet drafts directory for more than 6 months without being recommended by the IESG for publication as an RFC is simply removed from the Internet drafts directory. At any time, an Internet draft may be replaced by a more recent version of the same specification, restarting the 6-month timeout period. An Internet draft is not a means of publishing a specification; these are published through the RFC mechanism. Internet drafts have no formal status and are subject to change or removal at any time.

Specifications that are intended to become Internet standards evolve through a set of maturity levels known as the "standards track." These maturity levels are "proposed standard," "draft standard," and "standard." A specification for which significant implementation and successful operational experience has been obtained may be elevated to the level of standard. An Internet standard is characterized by a high degree of technical maturity and by a generally held belief that the specified protocol or service provides significant benefit to the Internet community. A specification that reaches this status is assigned a number in the STD series while retaining its original RFC number.

Summary

The Internet continues to evolve through the circulation of Internet draft documents. During the development of a specification, draft versions of the document are made available for informal review and comment by their placement in the IETF's Internet drafts directory, which is replicated on a number of Internet hosts. This makes an evolving working document readily available to a wide audience, facilitating the process of review and revision. Through this mechanism, new requirements and technology are continually factored into the design and implementation of the Internet.

See also

Internet

INTERNET FACSIMILE

Internet facsimile refers to the capability of sending imaged documents via the Internet instead of dial-up connections. Companies and individuals seek to leverage their existing Internet connections to fax documents, helping them contain telecommunications costs. There are three easy ways to send faxes over the Internet: subscribe to a commercial service, use an Internet-enabled fax software package, or devise a do-it-yourself method that entails scanning documents and saving them in a graphical format for e-mailing as an attachment.

Commercial Services

With a commercial service such as eFax Messenger Plus, sending a document is as easy as printing it. Once the eFax driver is installed, it appears as one of the choices in the print dialog box (Figure I-4). To send a document as a fax right from a computer, the user selects "Print," which opens the "Print" dialog box. Instead of selecting a printer, however, the user selects "Send with eFax Messenger Plus." Instead of printing the document, eFax Messenger Plus converts it to a proprietary .efx format that will be sent to the recipient as an e-mail attachment. A fax can be sent from the print menu of any Windows application.

After being guided through the steps with the eFax Messenger Plus wizard, the user clicks "Finish" and next sees the "Document Delivery" window with the "Fax" tab selected (Figure I-5). From the fax number field, the user enters the fax

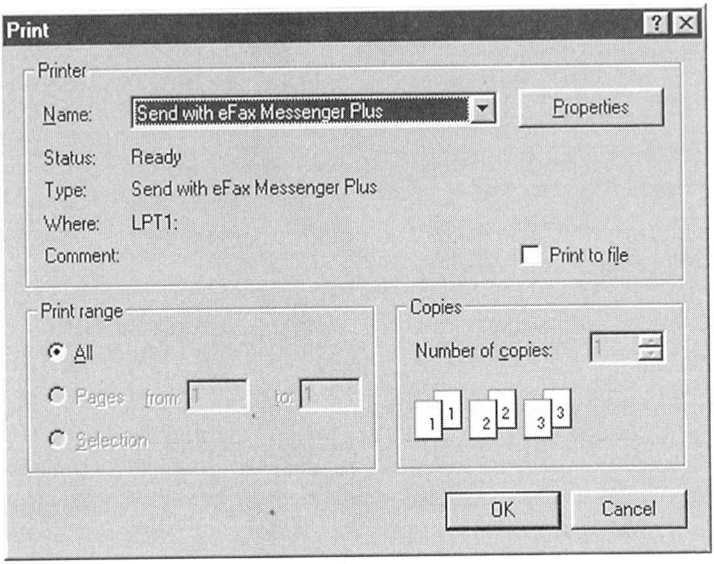

Figure I-4 To send a document as a fax from the computer, the user selects "Print," which opens the "Print" dialog box. Instead of selecting a printer, however, the user selects "Send with eFax Messenger Plus."

number to which the document should be sent. Clicking the "Send Fax" button launches the user's e-mail software with the .efx file attached to it. Clicking on "Send" actually mails the attachment. The document travels over the Internet to the eFax service center. There it is converted to a traditional fax and sent over phone lines to the receiving fax machine.

The person receiving the fax is notified via an e-mail message. Clicking on the eFax icon opens the eFax viewer, allowing the person to see the imaged document (Figure I-6).

The company offers a fax number and free usage of that number to individuals so that they can receive private faxes at their own e-mail account. Users go to www.efax.com and set up their personal fax account with minimal registration information. They get an eFax.com number instantly. This number is like any other fax number with an area code and

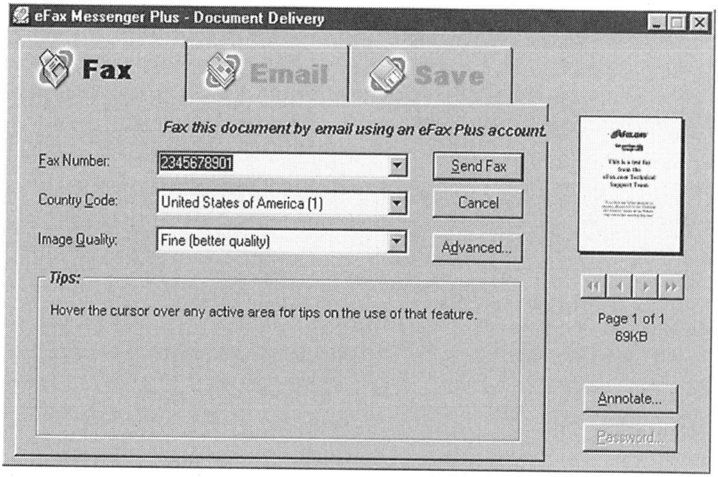

Figure I-5 At the "eFax Messenger Plus" dialog box, the user enters the phone number of the target fax machine and the country code (if any) and sets the image quality for the document.

seven digits. Once users give this number to associates, family, or friends, they can receive documents sent from any standard fax machine as e-mail attachments.

eFaxes look just like regular e-mails with attachments. Attachments are opened with an eFax viewer so that users can read them on screen, print them, or forward them to other e-mail addresses. The small proprietary viewer is included with the first fax users receive for installation on their computers. Most faxes sent using eFax.com are compressed two to three times more than standard digital documents, making the download process extremely fast. Like regular faxes, eFaxes maintain all original formatting, including text, graphics, and handwritten notes or signatures.

Windows Fax Software

There are some inexpensive stand-alone alternatives to faxing over the Internet. Typically, these are installed as print

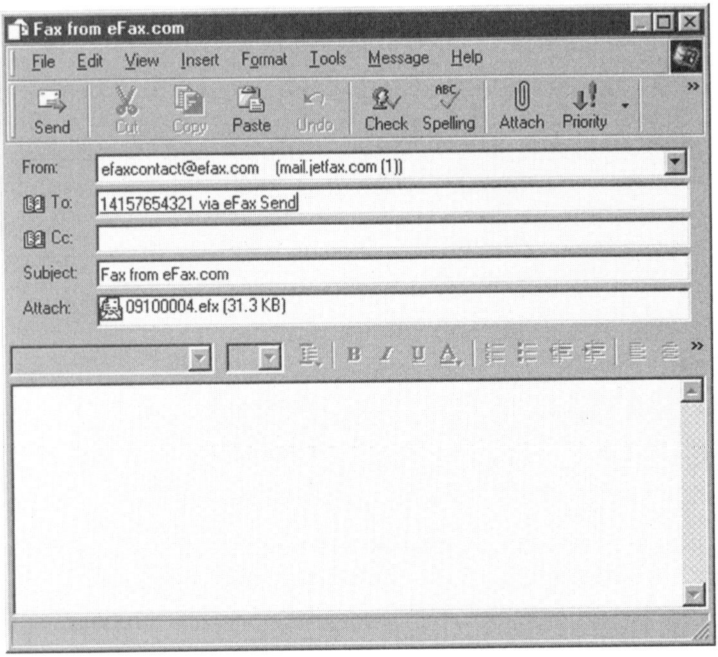

Figure I-6 On opening the eFax viewer, the recipient will be able to see the imaged document as if it arrived on a conventional fax machine. With eFax Messenger Plus, however, the sender can add audio clips and various stamps to the document.

drivers on desktop computers, allowing documents to be faxed from any Windows application. The fax software itself usually supports directories, offline queuing, status messages, and the creation of cover sheets. Some allow received faxes to be redirected to another location. At the destination end, the fax is received via e-mail as an attachment.

Attachment support is usually limited with stand-alone products, and some of these products are not capable of reaching conventional fax machines unless the document goes through a mail-to-fax gateway service. The recipient opens his or her e-mail as usual and uses an appropriate image viewer to read or print out the attachment.

Mainstream fax software, such as Symantec Corp.'s WinFax PRO (starting with version 7.5), supports fax transmissions over the Internet. After preparing the document for faxing, the user is given a choice of delivery methods—through a normal phone line or through the Internet. Once the user has chosen to send a fax over the Internet, WinFax PRO compresses and encrypts the fax and sends it through an Internet fax service provider. During transmission, WinFax PRO provides real-time status to the user.

Do-It-Yourself

Anyone with an Internet connection, e-mail software, and a scanner can send faxes over the Internet without the aid of extracost services or off-the-shelf fax software. Documents are simply scanned, saved into a graphic format that can be opened by the recipient, and sent as e-mail attachments. This homegrown approach does not provide the bells and whistles of professional services and products, but the quality of the received faxes is the same. A possible hindrance to this method is that each page must be individually placed on a scanner to create an image of the document. This can be time-consuming for sending large volumes of faxes daily or when documents have a large number of pages. This problem could be overcome if the user has a scanner with an automatic sheet feeder.

Summary

Cost savings is the principal reason for turning to Internet-based fax solutions. Various methods are available to accomplish this. The choice will depend on several factors, including geographic reach, feature requirements, the number of faxes sent per month, cost per fax (if any), and ease of use.

See also

Electronic Mail

Internet
Unified Messaging

INTERNET SERVICE PROVIDERS

Internet service providers (ISPs) provide consumers and businesses with access to the Internet. For a monthly fee, subscribers are given a software package, user name and password, and access phone number. Equipped with a modem or router, users can then log on to the Internet and access all the services it supports over a dial-up or dedicated connection to the ISP. The ISPs themselves are connected to one another through network access points (NAPs). There are almost 10,000 ISPs in the United States, many of them very small, serving subscribers in their local communities. Large ISPs, however, are national in scope and typically pursue revenue in at least four key business areas:

- Narrowband access, which involves monthly fees charged to customers for dial-up Internet access and one-time setup fees
- Web hosting, which consists of providing services to companies and individuals wishing to have a Web or e-commerce presence
- Broadband access, which consists of high-speed, high-capacity access services including DSL, cable, fixed wireless, and dedicated circuits
- Content, commerce, and advertising revenues, which come from sales of banner and other online ads, fees generated through revenue-sharing arrangements with online retailers, and the sale of advertising and content space on the ISP's various online properties

Services

ISPs offer Internet access software that incorporates a telephone dialer and e-mail program with several third-party

Internet access tools, including Web browsers. The software provides a functional, easy-to-use Internet access solution for Windows and Macintosh platforms. The software automatically installs these and other software applications on customer computers. The simple point-and-click functionality of the software, combined with its easy-to-use multimedia registration and installation system, permits online credit card registration, allowing both novice and experienced customers to quickly set up access to the Internet.

The ISP typically supplies local phone numbers that users can dial to access the service. Large ISPs like Earthlink provide customers with the means to access the Internet from any location through both wired and wireless non-PC devices and appliances. ISPs also provide or support broadband connections via DSL and cable. Large ISPs offer businesses dedicated Internet access connections over T-carrier lines at speeds of up to 45 Mbps and optical fiber at speeds of 2.5 Gbps and beyond. Some ISPs offer "burstable" dedicated Internet access, which gives business customers more bandwidth granularity at a more affordable price but allows them to automatically burst to a higher speed when applications require greater bandwidth. The customer is billed for the higher bandwidth only when it is actually used. Other Internet access methods include ISDN PRI, frame relay, ATM, and metropolitan area Ethernet service.

The large ISPs route customer traffic over physically diverse fiber backbones with congestion management and automatic rerouting capabilities. The connections are proactively monitored on a 24×7 basis from a network operations center. Their status as a tier 1 ISP enables them to expedite the handling of business traffic through major peering points to ensure minimum latency. Complementary services may be offered by the ISP, such as applications hosting, help desk services, and collocation space. Often ISPs will pursue partners that can provide these and other services.

In conjunction with their Internet services, some ISPs offer virtual private networks (VPNs) for businesses looking for secure, reliable, and affordable enterprise-wide IP networking and remote access solutions. A VPN provides point-to-point connectivity through the public Internet via "tunnels" that are set up between the routers at each location. Access to the VPN can be controlled with various authentication techniques and the traffic encrypted so that it cannot be intercepted on the way to its destination. The ISP can support VPNs with dial-up connections as well as dedicated connections.

The value proposition for business customers includes a choice of dedicated Internet access services in a variety of speeds to suit their specific application requirements, network monitoring to the customer premises, and Web-based bandwidth usage reports. For companies that want to eliminate the hassles and overhead of setting up and managing a complicated hosting infrastructure, some ISPs offer a choice of dedicated Web hosting packages, differing by platform and capacity. The server can be used as a Web server, FTP server, e-mail server, or a combination of all three. The customer can remotely administer the server via a Web browser and Secure Sockets Layer (SSL) connection.

Customer Service and Technical Support

ISPs provide customer service and technical support as a means to retain existing customers and attract new ones. At a minimum, ISPs provide

- Toll-free account setup and technical and billing assistance
- E-mail–based assistance
- Help sites and Internet guide files on the ISP's Web site
- Printed or CD reference material

The ISP may provide premium-level support for businesses with dedicated access connections and Web sites. In addition, the ISP may maintain newsgroups on the Internet where subscribers can post requests for help, and other subscribers, as well as its own support personnel, can respond. The ISP also may contract with call center service vendors whose agents are trained to provide additional technical support assistance.

Competition

ISPs operate in a very competitive market that comprises the following categories of companies:

- Established online services, such as America Online, Earthlink, Microsoft Network, and Prodigy
- Thousands of local, regional, and national ISPs
- National telecommunications companies, such as AT&T and Sprint
- Regional Bell operating companies, such as BellSouth and SBC Communications
- Online cable services, such as AT&T Broadband and Cox
- Free and low-cost ISPs, such as NetZero and Juno
- Municipal utility and cable companies

Competition is likely to increase as large, diversified telecommunications and media companies acquire ISPs and otherwise provide ISP services and as ISPs consolidate into larger, more competitive companies. Diversified competitors may continue to bundle other content, services, and products with Internet connectivity services, potentially increasing competition even more.

In a slow-growth economy, the online industry faces significant challenges. Free ISPs have not sustained business and cannot afford to keep even active customers onboard while advertising dries up. The DSL market, which had been so strong initially, is struggling as independent companies go out of business. And for the first time ever, PC sales in 2001 declined on a year-over-year basis in the United States.

Regulation

The FCC does not regulate the Internet or ISPs. Under FCC rules, ISPs are considered enhanced service providers (ESPs). The FCC does not regulate the rates that enhanced service providers charge to their subscribers. Although ISPs purchase local phone lines so that their customers can call them for access to the Internet, the FCC considers ISPs as end users when they purchase services from local telephone companies. Thus ISPs pay the same rates as any other business customer, and these rates are set separately in each state. By contrast, long-distance companies are considered "carriers," and they pay interstate access charges regulated by the FCC.

The FCC's responsibilities include protecting consumers against telephone fraud and the bad business practices of telephone companies. Since the FCC does not regulate the Internet or ISPs, however, consumers must contact their state consumer protection office or, if there is possible fraud involved, the Federal Trade Commission or the DoJ's Internet Fraud Complaint Center.

Summary

ISPs provide consumers and businesses with access to the Internet. A number of ISPs provide a wide range of online services for personal and business use, such as weather, entertainment news, movie listings and reviews, sports coverage, stock quotes, financial services, parental control or screening capabilities, and more. ISPs compete on the basis of price and value. Consumers generally gravitate to ISPs that offer the lowest price, while businesses tend to choose an ISP on the basis of value—the ability of the ISP to provide a range of services companies need, custom solutions, technical support, and service level guarantees.

See also

Application Service Providers

Electronic Commerce
Internet
Virtual Private Networks
World Wide Web

INTERNET TELEPHONY

The ability to place telephone calls over the Internet was considered a hobby for Internet enthusiasts only a few years ago. Voice quality was diminished by variable delay, clipped speech that resulted from dropped packets, and confusion about whose turn it was to talk at any given time. Continued advancements in digital signal processing (DSP) technology, the emergence of standards for interoperability, the development of scalable IP switches, and the availability of IP/PSTN gateways that enable ordinary phones to be used for Internet calls have changed all this. Now, traditional carriers, cable companies, national ISPs, and their competitors offer commercial voice over IP (VoIP) service, while businesses of all types and sizes have started to view their IP infrastructures as a means to save on long-distance call charges and implement multimedia applications.

Humble Beginnings

Voice over IP (VoIP) was first demonstrated in the early 1980s when Bolt, Beranek and Newman (BBN) in Cambridge, Massachusetts, set up the "voice funnel" to communicate with team members on the West Coast as part of its work with the Advanced Research Projects Agency (ARPA). The voice funnel digitized voice, arranged the resulting bits into packets, and sent them through the Internet. Further development of the technology had to wait until the 1990s, when improvements in microprocessors, digital signal processing (DSP), codec technology, and routing protocols all came together to make feasible easy-to-use products for consumer and business use.

Since 1995, IP telephony has developed rapidly to become a successful commercial service offered even by AT&T, Sprint, and WorldCom, as well as cable operators and numerous smaller companies like Net2Phone that specialize in IP telephone service. Carrier-class gateway platforms that connect to legacy systems via T1/E1 or analog interfaces have matured to reliably handle authentication, call management, and billing. VoIP service providers and customers alike can now access call detail, billing history, and account information over a secure Web site in real time.

Now that VoIP is a proved technology and has many benefits to commend it, users will be looking for more advanced capabilities, such as IP centrex, and more complex features, such as unified voice, e-mail, and fax messaging. These and other innovations can be implemented quickly and economically through a flexible "soft switch" architecture that uses servers to control data calls on IP networks much like circuit switches do for voice calls on the PSTN. In addition, soft switches provide a full range of IP-based communications services that are virtually indistinguishable in quality and ease of use from services on traditional circuit voice networks.

Today's IP switches closely match the capabilities of legacy Class 4 and Class 5 telephony switches. They are designed to meet the rigorous requirements of public network service providers, including complete redundancy of all system elements, toll-quality voice, interoperability with the SS7 network, and scalability to hundreds of thousands of calls.

Latency

There is a significant difference in the quality of calls placed over the Internet versus those carried over the PSTN. The key difference is latency and jitter (variable delay). Whereas telephone conversations via satellite have 0.5 to 1.0 second of delay, it is common to have up to 5.0 seconds or more of delay on the Internet, depending on the distance of the call and the amount of traffic traversing the net at any given time. The Internet, however, is not expected to handle commercial IP telephony services. Instead, managed high-capacity IP backbones handle the long-haul portion of the call, greatly reducing delay.

A number of technologies are employed to maintain consistent call quality over IP networks, including the use of native IP switches, ATM on high-capacity fiber backbones, and routing protocols that give preference to real-time traffic. Although delay will not be eliminated entirely, neither will it remain an ongoing problem. As managed IP backbones branch out to reach major metropolitan areas, delay will become less of an issue to the point where consumers will not even know their calls are being carried over a packet network.

Voice Quality

Although the poor voice quality offered by first-generation Internet telephony products condemned them to hobby status, voice quality over IP has continued to improve. The use of server-based gateways rather than sound cards in users' computers provides the processing power needed to minimize compression/decompression time, while improvements in DSPs provide high-quality sound. The mean opinion score used to rate the quality of speech codecs gives toll-quality voice over the PSTN a top score of 4.4.

The voice compression algorithms used in IP telephony applications—known as G.723.1—bring the bandwidth requirement down to 6.3 Kbps or 5.3 Kbps, depending on the encoding scheme used. The mean opinion score for these levels of compression are 3.9 and 3.5, respectively, which provide voice quality comparable to that of cell phones. Now that the industry has seemingly coalesced around the international

G.723.1 standard for VoIP networks, there is the added benefit of interoperability among the products of different vendors.

Increasingly, PBX vendors are recognizing the importance of providing customers with easier and more economical ways of supporting voice over IP and, in the process, helping them transition to voice and data convergence. Vendors provide solutions with the flexibility to support voice, video, and data traffic over the Internet, intranets, extranets, public switched networks, and ATM. When used as a gateway, such systems convert voice traffic to packets for reliable transmission over IP networks. The quality of service for each call is monitored, so that if the IP network's performance is not acceptable for voice or fax calls, the switch will reroute the call over an alternative network, if available.

Remote User Support

Enterprises with a large number of telecommuters and mobile professionals can opt for a VoIP solution that supports remote log-in so these employees can have the same capabilities as their desktop telephone sets, including hold, call forward, transfer, speed dial, and conference, as well as multiple call appearances and call displays on their laptop computers while working remotely. Such systems offer the means for remote users to log into the corporate switch so they can take advantage of all these features and work virtually from any location. Remote users can also have access to voice mail.

This solution can even be applied to call center operations, helping companies take full advantage of their data networks by delivering call signaling and phone features to a call center agent's PC through an IP connection. Among other things, this allows agents working at home to provide the same high level of customer care as agents working in a traditional call center environment. For the company, the solution lowers the cost of support operations and helps them attract qualified staff by offering them the means to work at home.

Network Management

Network management vendors are offering tools that help IT administrators and network managers monitor voice and data traffic performance on IP nets. These tools provide call monitoring, adaptive voice and data traffic prioritization, QoS, bandwidth management, and accounting and billing metrics for voice-related traffic over IP networks.

Network managers can even monitor information such as callers and destinations, call duration, time of day distribution, and associated costs. Voice and data traffic are dynamically prioritized in real time, on the basis of actual bandwidth availability at the circuit level. Voice sessions can be accounted and billed for according to factors such as priority, user type, and time of day. Corporations can use these detailed accounting features to bill-back departments, while service providers may use them to bill customers on a per-use and/or priority basis.

Summary

VoIP technology is ready to deploy today, but works best on managed IP-based networks where performance can be closely monitored and fine tuned by an enterprise or a service provider. According to various industry estimates, 50 percent of all companies that have a private intranet are already running or experimenting with integrated voice-data applications and IP telephony. The technology has progressed to the point that there is little or no reason for companies to delay the phased implementation of VoIP solutions, especially for intracompany communications.

See Also

Cable Telephony

LAN Telephony

Transmission Control Protocol/Internet Protocol

Voice Compression
Voice-Data Convergence
Voice over IP

INTRANETS

An intranet is a private Transmission Control Protocol/Internet Protocol (TCP/IP) network that usually supports the same protocols and services as the public Internet, including e-mail, news, chat rooms, and Web pages. Businesses build intranets to improve internal communication, facilitate information distribution, broaden access to corporate resources, enable group scheduling, and provide a browser front end to various corporate databases and services.

Reasons for Intranets

There are a number of practical reasons for setting up a corporate intranet. The biggest reason is to improve internal communications and facilitate decision making. For example, an intranet makes it possible for employees to access information without documents having to be printed and distributed in paper form. Posting the documents on a department Web site provides direct access to the information from any location so that employees can get what they need when they need it without involving anyone else. This empowers employees to make decisions on their own, without causing them to experience information overload.

Another reason to have an intranet is that it reduces the cost of internal operations. With employees able to access information directly, schedule conferences and collaborate with each other using automated tools, submit timesheets directly to accounting, and communicate across departmental boundaries with e-mail, chat, IP fax, and IP telephony, there is no need for a middle management layer in the organization to act as the facilitating agent. The result is across-the-board improvements in productivity, as well as cost savings from streamlined business operations. The applications to do all this are very inexpensive, and the browsers are free. The protocols run over the existing corporate LANs and WANs, eliminating the need to invest in a separate network. Rollout of the intranet can be gradual, modular, and minimally disruptive.

The cross-platform nature of TCP/IP provides another reason to establish an intranet. Most organizations are heterogeneous on the client side, having a mix of Macintosh computers, UNIX workstations, Windows PCs, and even some OS/2 machines. Intranets are the easiest way to get these devices talking. Since all the operating systems have TCP/IP stacks already built into them, the clients are "intranet ready," requiring no extra costs to network them together. There may be additional costs associated with servers and routers, but these are often incremental expenses because these systems are typically in place to support other applications on the LAN and WAN. It is just a matter of taking advantage of the TCP/IP stacks already embedded in the operating systems of these devices to support the intranet.

Another facet of intranets is that they are fast—much faster than the public Internet. The reason is that a company is in sole control of such critical elements as bandwidth, the technologies and protocols, and the applications and devices on the intranet. All these elements impact performance. When under control of a single company, steps can be taken to optimize the performance and safeguard the integrity of the intranet end to end. For example, the company can implement quality of service (QoS) mechanisms, traffic prioritization schemes, and network caching and add bandwidth wherever it is needed and even partition it among the applications. This is not possible when relying on the public Internet, which has no central management authority to make these decisions and see that they are carried out. With an intranet, a company can push the envelope in terms of applications and make adjustments to ensure peak performance.

Finally, setting up an intranet is a risk-free proposition. The underlying technologies and protocols that are used to implement corporate intranets have been in use on the public Internet since the beginning and have proved to be reliable and robust. Even when new capabilities are added, such as IP telephony and streaming video, which the original Internet was never intended to support, the protocols necessary for implementing the new capabilities are designed to work within the TCP/IP framework.

Infrastructure Availability

The decision to implement an intranet is relatively easy for large companies because they typically have the necessary components already in place. For example, they have LANs and use TCP/IP on the WAN in support of e-mail, file transfers, remote database access, and other routine communications needs. They usually have the technical expertise to install and configure the necessary components—including a heterogeneous client base, plus servers, routers, switches, and gateways—and manage these and other network elements via an enterprise-level management system that also supports SNMP. They also may have people who implement and maintain client-server technology over LANs that also provide connectivity to legacy host systems. For these companies, it is relatively simple to add a graphical front end to this environment in the form of browsers and offer extra functionality, such as a structured query language (SQL) query capability, from Web servers distributed on the TCP/IP network.

Even for companies that do not already have an existing TCP/IP-based infrastructure, it does not take much to learn how to take advantage of Internet technology and adapt it for internal use. These companies, as well as very small companies that lack any kind of technical expertise, can avail themselves of numerous vendors and service providers who are eager to educate potential customers on the benefits of corporate intranets and offer their own ideas concerning intranet implementation.

If a company does not want to build and run its own intranet, there are service providers that handle this as well. In fact, every aspect of designing, provisioning, and managing a corporate intranet can be outsourced to a carrier or third-party firm—including creating the intranet Web page, selecting the equipment and software, hosting one or more intranet Web sites, and procuring and managing the access and transport facilities of the network.

Address Management

Critical to keeping an intranet running smoothly is IP address administration, which can become unwieldy as intranets lead to a proliferation of devices requiring IP addresses. Intranet-driven IP administration can be facilitated by Dynamic Host Configuration Protocol (DHCP) software. For managers of large IP networks, DHCP reduces the work necessary to administer a large number of IP addresses. It does this by automatically assigning IP addresses to clients as they log onto the TCP/IP network.

Running on a server, the DHCP software also reclaims unused IP addresses and maintains a pool of reusable addresses. These features greatly simplify the workload of network managers, who would otherwise have to issue static IP addresses to every device on the network and manually assign an address to any device that is changed, moved, or added on the network. DHCP is also good for the organization because there is less chance of running out of IP addresses and having to justify the request for additional addresses.

Security

Perhaps the most serious issue related to intranet implementation is security. Increasing the number of people who have access to important data or systems can make a company's IT infrastructure vulnerable to attack if the right precautions are not taken. A comprehensive security solution addresses internal as well as external threats and should include policies and procedures and the ability to monitor and enforce them, as well as robust security tools that work well together and do not leave any gaps in protection. The following basic functions are necessary for broad security coverage:

- Access-control software allows varying degrees of access to applications and data.
- Secure transmission mechanisms such as encryption prevent outside parties from intercepting, eavesdropping, or changing data sent over the network.
- Authentication software validates that the information that appears to have been originated and sent by a particular individual actually was sent by that person.
- Disaster-recovery software and procedures assist in recovering data from a server that experiences a major fault.
- Antivirus software detects and removes viruses before they cause problems.
- Packet filtering controls what information can pass between internal subnets and between the intranet and Internet on the basis of such criteria as source and destination addresses, specific applications, users or groups of users, and even time of day.
- Intrusion detection identifies hacking attempts before they progress far enough to do any damage.

Costs and ROI

The cost of developing a corporate intranet varies considerably on a case-by-case basis. Large companies often can build sophisticated intranets using existing TCP/IP networks, equipment, and management tools. In such cases, the intranet is treated as just another set of applications that is added to meet business needs. For such companies, the startup cost for intranet development can be incremental.

The startup cost for an intranet that supports 400 to 500 people can be as low as $25,000. This includes browsers for the client, a Web server, content development tools, and the communications hardware and software. The recurring cost of facilities and services can be obtained from the various carriers and compiled into an annual figure. Companies that do not have in-house technical expertise also should plan to spend 10 percent of the total startup cost of equipment and software on integration services.

Fortune 100 companies with worldwide locations that must be tied into the intranet can expect to pay quite a bit more, especially if they intend to offer a high level of interactivity, engage in electronic commerce, and Web-enable various business processes. Here, security is extremely important and constitutes a significant cost to factor into the budget. Such companies should plan to spend at least $10 million.

As companies put together budgets for intranet development and management, eventually they will have to address the issue of return on investment (ROI), as they typically do for any other major capital expenditure. The extent to which this can be done with any degree of accuracy often depends on how the proposed intranet will be used.

For example, if the intranet will be used to publish staff handbooks, telephone directories, forms, office notices, and other administrative documentation, the annual cost of printing, distributing, updating, and storing these materials contributes to the ROI of the intranet. Although harder to quantify, there is also the significant cost of staff time for filing, updating, and referring to paper-based material that also would be eliminated. A publishing application can garner an annual ROI of as much as 30 percent. Other applications, such as database access and inventory management, may yield 70 and 50 percent annual returns, respectively.

Reliance on electronic publishing would improve overall productivity, which is a "soft dollar" benefit that can be used to cost-justify the intranet, especially when the intranet includes a search engine or SQL query capability that allows users to key in on desired information quickly.

If the company plans to use the intranet for transaction processing, ROI can be fairly easy to calculate. For example, the company can post all its business forms on the intranet, including various health insurance forms, travel authorization and expense reimbursement forms, vacation schedule forms, Worker's Compensation forms, 401(k) plan forms, and purchase order forms—just to name a few. These and other forms can be called up on the intranet with a browser, filled in by the employee, and sent to the appropriate department via e-mail. Employees need not waste time tracking down the paper forms they need, and since the employee-supplied information is submitted in electronic form, departments can process it faster and readily integrate it into various databases. The savings in time and improved form processing constitutes another element that can be factored into the intranet's return on investment.

Summary

Corporate intranets are changing businesses in a profound way. They empower employees by providing them with a high degree of autonomy, encouraging creativity, enhancing decision making, and improving productivity. The result is that the company improves customer service and responds more effectively to changing market conditions.

See also

Extranets

Internet

Transmission Control Protocol/Internet Protocol

INVERSE MULTIPLEXERS

Inverse multiplexers allow users to put together increments of bandwidth and use it as a high-speed channel to support a given application. Originally, inverse multiplexing specifically addressed the bandwidth needs of videoconferencing, but the concept now applies to other applications and to providing scalable bandwidth for Internet access. Inverse multiplexers may be used to combine bandwidth on multiple dial-up or dedicated connections. The connections may be in the local loop or on the network (i.e., interoffice) side.

In the dial-up scenario, inverse multiplexing might come into play when the user wants to access the Internet at speeds greater than 56 kbps but does not have broadband services such as DSL or cable available. With the right software and a multiport modem, up to three dial-up connections can be established to the ISP to achieve a data rate of 168 kbps. In the case of dedicated connections, up to eight T1 lines can be bonded together by an inverse multiplexer to achieve up to 12 Mbps of access bandwidth to the Internet or to a carrier service such as ATM, which would be more economical that having to step up to a T3 line.

On the network side, inverse multiplexing could be used to dial up long-distance bandwidth within the public switched telephone network (PSTN) in support of videoconferencing or document collaboration among distributed corporate locations. In this case, multiple 56/64-kbps channels would be dialed by the inverse multiplexer within the carrier's network. The company pays for the number of interexchange channels only when they are set up to handle the conference call. On completion of the call or as different locations drop off the conference, the channels are taken down and carrier billing stops. This method of access obviates the need for overprovisioning the corporate network to support infrequently used applications.

Implementation

Inverse multiplexing can be implemented in customer premises equipment (CPE) or as a carrier-provided service. Either way, the advantages of inverse multiplexing include the immediate availability of extra bandwidth when needed, which eliminates the need for standby leased lines that are billed to the user whether fully used or not. This adds up to significant cost savings for the organization. Inverse multiplexers can support virtually any type of traffic: voice, video, data, and IP for scalable bandwidth to the Internet.

Some inverse multiplexers can be configured to support multiple applications simultaneously. For example, an inverse multiplexer that can be used to link (1) multiple applications at a single site to the public network via a T1 or ISDN access facility, (2) a PBX to a VPN, (3) a router to a fractional T1 network, or (4) a video codec to a switched digital service. This capability appeals to users who want to spread the cost of a T1 access line across multiple applications. Some products allow users to switch multiple applications on a call-by-call basis over different carriers' services simultaneously. While some inverse multiplexers interface only to switched services, others can access both switched and dedicated communications facilities.

Another capability of some inverse multiplexers is the transport of bandwidth-intensive data across multiple T1 circuits to achieve a fractional T3 circuit. T3-level inverse multiplexers are intended for applications that require transport between the T1 and T3 rates of 1.544 and 44.736 Mbps. As many as eight T1 circuits can be aggregated to achieve the desired increment of bandwidth without the organization being forced to lease a more expensive T3 facility, much of which would go unused. If the organization needs more bandwidth than eight T1 circuits would provide, then it is more economical to lease a T3 line.

System Management

The system management interface usually consists of a microcomputer equipped with software that allows the network manager to define and monitor traffic flow, bandwidth requirements, access line quality, and various configuration parameters.

Through this interface, administrative functions are also performed, such as the creation of call profiles. A "call profile" is a file that contains the parameters of a particular data call so that a similar call can be quickly reestablished at another time simply by loading the call profile. Usually the call profile function includes a factory-loaded profile that acts as the template for creating and storing user-defined call profiles. Because each data call may involve as many as 25 separately configurable parameters, the use of call profiles can save a lot of time. Users typically load or edit a call profile using keyboard commands to the management software on the microcomputer.

Inverse multiplexer management interfaces often support remote devices. This capability allows a network administrator at a central location to configure, test, and otherwise manage other inverse multiplexers at remote locations in much the same way as is currently offered by the in-band management systems of some T1 multiplexers. This is accomplished by the management interface reserving a certain amount of the network bandwidth, usually not more than 2 percent, as a subchannel to implement remote management.

Most inverse multiplexers can be remotely monitored and controlled via SNMP. This is usually accomplished with SNMP agent software included with the product. The agent collects detailed error statistics, utilization ratios, and performance histories that can be retrieved for analysis.

Standards

The Bandwidth on Demand Interoperability Group (BONDING), formed in late 1991, defined interoperability standards for inverse multiplexers. The BONDING specification is often used for non-Internet applications, such as videoconferencing. There is also a set of international standards for bandwidth-on-demand services called Global Bandwidth on Demand (GloBanD). The BONDING specification describes four modes of inverse multiplexer operation:

- *Mode 0* Enables inverse multiplexers to receive two 56-kbps calls from a video codec and initiate dual 56-kbps calls to support a videoconference.
- *Mode 1* Enables inverse multiplexers to spread a high-speed data stream over multiple switched 56/64-kbps circuits. Because this mode does not provide error checking, the inverse multiplexers operating in this mode have no way of knowing if one of the circuits in a multicircuit call has failed. In this case, it is up to the receiving node to detect that it has not received the full amount of data and request more bandwidth.
- *Mode 2* Adds error checking to each 56/64-kbps circuit by stealing 1.6 percent of the bandwidth from each circuit for the passage of information that detects circuit failures and reestablishes links.
- *Mode 3* Uses out-of-band signaling for error checking, which may be derived from a separate dial-up circuit or the unused bandwidth of an existing circuit.

When establishing calls, inverse multiplexers at both ends first determine whether they can interoperate using the vendor's proprietary protocol. If not, this means that the inverse multiplexers of different vendors are being used and that they should use the BONDING protocol to support the transmission.

Summary

The inverse multiplexer allows network managers to match bandwidth to the application. These devices (or a carrier-provided service) provide a degree of configuration flexibility that cannot be matched in efficiency or economy by any other technology. With inverse multiplexers, organizations no longer have to overprovision their networks to handle peak traffic or run occasional high-bandwidth applications. Instead, they can order bandwidth only when it is needed and, in the process, save on line costs. Another type of inverse multiplexer is the integrated access device (IAD), which supports multiple protocols, bandwidth contention among different applications, and QoS features to ensure optimal performance for all applications.

See also

Integrated Access Devices

Multiplexers

J

JAVA

Java is a network programming language that was developed at Sun Microsystems in 1991 by James Gosling (Figure J-1). Java was designed with a principal objective of eliminating a problem that software developers confront when working in traditional software environments: the need to create and distribute different versions of their programs for different operating system and hardware platforms. To accomplish this goal, the Java technology is specifically designed to give developers the ability to build software applications that can run on multiple platforms.

The cross-platform nature of Java enabled it to become one of the fastest-growing programming languages, particularly for the thin client model of computing. Java is actually a scaled-down version of the C++ programming language that omits many seldom-used features while adding an object orientation. Java provides a cleaner, simpler language that can be processed faster and more efficiently than C or C++ on nearly any microprocessor.

Whereas C or C++ source code is optimized for a particular model of processor, Java source code is compiled into a universal format. It writes for a virtual machine in the form

Figure J-1 James Gosling, developer of the Java programming language, is currently vice president and fellow at Sun Microsystems. As chief scientist of the Java Software Division at Sun, Gosling's main responsibilities include reviewing and guiding the ongoing development of the Java programming language and the Java run-time environment, which includes the Java virtual machine and the Java class libraries.

of simple binary instructions. Compiled byte code is executed by a Java run-time interpreter, performing all the usual activities of a real processor, but within a safe, virtual environment instead of a particular computer platform. This allows the same Java applications to run on all platforms and networks, eliminating the need to "port" an application to different client platforms. In fact, Java applications can run anywhere the virtual machine software is installed, including any Java-enabled browser, such as Microsoft's Internet Explorer.

The use of Java enables remote users, mobile professionals, and network managers to access corporate networks, systems, and legacy data through Java applets that are downloaded from the server to the remote computer only when needed. An applet is a piece of a larger application that resides on the server. The function of the applet is to extend the capabilities of the larger application to the remote user. This is the fundamental principle of "network computing."

In most cases, the applets are stored in cache on a hard disk at the client location or in cache memory. Either way, the applet does not take up permanent residence on the client machine. Since applets are delivered to the client only as needed and all software maintenance tasks are performed at the server, users are assured of access to the latest application release level. This not only saves on the cost of software, it also permits companies to get away with cheaper computers, since every computer need not be equipped with the resources necessary to handle every conceivable application. At the same time, there is no sacrifice in the capabilities of users to do their work while away from the office.

Rapid Applications Development

The acceptance of Java has spawned a steady stream of visual development tools that aid in rapid application development (RAD). Among the second-generation RAD tools is Borland International's JBuilder. When opened, the tool displays the main window and AppBrowser from which the user can access all the usual development functions through three major panes: Navigation pane, Content pane, and Structure pane (Figure J-2).

The Navigation pane shows a list of projects with associated files, which may include Java, HyperText Markup Language (HTML), text, or image files. The Content and Structure panes display information about the selected file. For example, if a Java file is selected, the Structure pane shows such information as imported packages, the classes

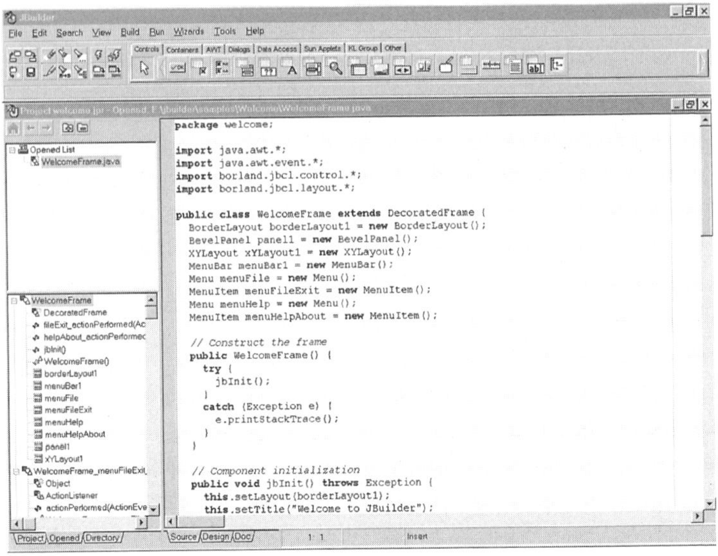

Figure J-2 JBuilder's multipaned AppBrowser, showing the Navigation pane (*above, left*), the Structure pane (*below, left*), and the Content pane (*right*).

and/or interfaces in the file, any ancestor classes and/or interfaces, and variables and methods. With the AppBrowser in Project Browser mode, the user can manipulate the files in a project.

JBuilder includes an Object Gallery that contains shortcuts that create skeletal instances of many objects, letting the user quickly manufacture such things as applets, applications, frames, dialogs, panels, data modules, classes, and HTML files (Figure J-3).

To guide the user through the major tasks necessary to create Java programs, JBuilder also comes equipped with wizards to create new projects and files and modify existing ones. Among the wizards is an application wizard that creates a new Java application shell containing a frame and an applet wizard that creates a minimal applet and HTML file containing the applet. There is a wizard that wraps an

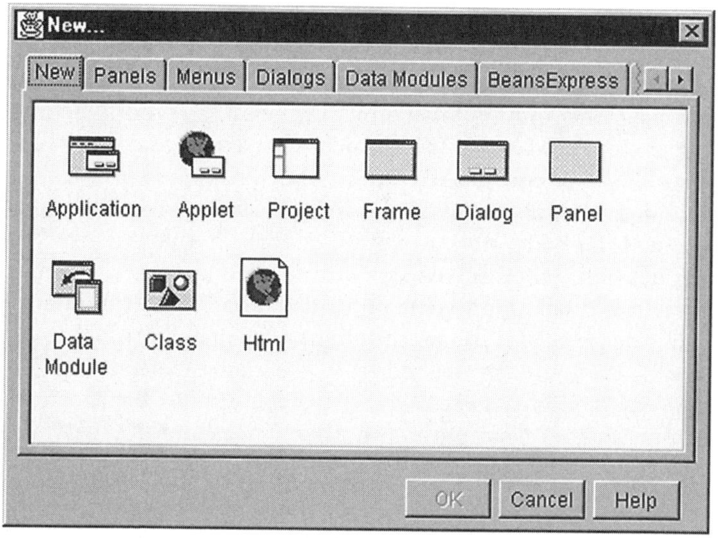

Figure J-3 JBuilder's Object

existing applet in a JavaBean class, making it look like a JavaBean to other applets and applications. The wizards can even work with existing projects to convert an application to an applet or vice versa. Other wizards can be added to JBuilder as they become available.

JBuilder was designed with business application development needs in mind. In addition to JavaBeans Express for easy JavaBean creation and deployment, it includes drag-and-drop database components and tools, complete Java data base connectivity (JDBC) connectivity, more than 100 JavaBeans—including grid with source code, charting, numerous wizards, and command-line tools—and a Local InterBase for offline structured query language (SQL) database development.

Summary

The early success of the C++ programming language owes a great deal to its ability to access legacy code written in C.

Similarly, Java preserves much of C++ and offers a number of compelling benefits: It is object-oriented, portable, and relatively easy to master and maintain. Once written, Java applications and applets can run unchanged on any operating system that has a Java interpreter. These and other benefits of Java can greatly speed the development cycle for Web-based applications, including those for integrated computer telephony and remote network management. The applications themselves are accessed only when needed, with the most updated version downloaded to the client's cache as an applet. When the client disconnects from the network, the applet is flushed from the cache, conserving limited system resources. This is the basis for network computing, a new paradigm that, in essence, treats the Internet as the computer.

See also

World Wide Web

L

LAN TELEPHONY

Local area network (LAN) telephony integrates voice and data over the same medium, enabling automated call distribution, voice mail, and interactive voice response, as well as voice calls and teleconferencing, between workstations on a LAN. The benefit of LAN-based telephony is that it can eliminate the costly, proprietary Private Branch Exchange (PBX) and replace it with a standards-based Ethernet/Internet Protocol (IP) solution. By carrying voice conversations in the form of IP packets, local calls can traverse the Ethernet LAN, while long-distance calls can go out to the wide area IP-based intranet or even the public Internet. Through the use of IP/PSTN gateways, calls also can reach conventional telephones off the IP network.

With LAN telephony, users working away from their offices—at home or in a hotel—can use a single phone line to carry both data and voice traffic. The user dials up to access the corporate intranet, which would be equipped and engineered to carry real-time voice traffic. Such a system provides an integrated directory view, enabling remote users to locate individuals within the corporation for voice- or e-mail connection in a unified way. Likewise, phone callers (internal or external to the corporation) can locate the mobile workers connected to any part of the intranet. Thus LAN telephony allows users to work seamlessly from any location.

By using the LAN-based conferencing standards, transparent connectivity of different terminal equipment can be achieved; the media used by any conference participant would be limited only by what is supported by his or her terminal equipment. Connectivity to room-based conference systems or analog telephones can be achieved by means of gateways, which would perform the required protocol and media translations.

IP PBX

The traditional PBX is a circuit switch that provides organizations with access to communications services and call handling features. It sets up a communication path between the calling and called party, supervises the circuit for various events (e.g., answer, busy, and disconnect), and tears down the path when it is no longer required. In many ways, the PBX mimics a telephone company's central office switch, except that it is smaller in scale and is privately owned or leased.

PBXs based on IP can transport intraoffice voice over an Ethernet LAN, a managed IP network, and, via a gateway, the public switched telephone network (PSTN) to reach off-net locations. Full-featured digital phone sets link directly to the Ethernet LAN via a 10BaseT interface without requiring connection to a desktop computer. Phone features can be configured using a Web browser. Existing analog devices, such as phones and fax machines, can be linked to the LAN via a gateway. In addition to IP nets, calls can be placed or received using T1, ISDN PRI, or traditional analog telephone lines.

All the desktop devices have access to the calling features offered through the IP PBX management software running on a LAN server. The call management software allows client devices on the network, such as phones and computers, to perform functions such as call hold, call transfer, call forward, call park, and calling party ID. In addition, advanced PBX functions, such as multiple lines per phone or multiple phones per line, can be offered through the management software. The software also offers directory services. Unified messaging capabilities allow voice-mail messages to be sent to an existing electronic mailbox, along with e-mail and faxes.

Standards

The building block of LAN telephony is the international H.323 Standard, which specifies the visual telephone system and equipment for packet-switched networks. H.323 is an umbrella standard that covers a number of audio and video encoding standards. Among these standards is H.225 for formatting voice into packets. H.225 is based on the Internet Engineering Task Force's (IETF) Real Time Protocol (RTP) specification and the H.245 protocol for capability exchange between workstations.

On the sending side, uncompressed audio/video information is passed to the encoders by the drivers and then given to the audio/video application program. For transmission, the information is passed to the terminal management application, which may be the same as the audio/video application; the media streams are carried over RTP/UDP, and call control is performed using H.225-H.245/Transmission Control Protocol (TCP).

Gateways provide the interoperability between H.323 and the PSTN, as well as networks running other teleconferencing standards such as H.320 for the Integrated Services Digital Network (ISDN), H.324 for voice, and H.310/H.321 for Asynchronous Transfer Mode (ATM). An example H.323 deployment scenario involves H.323 terminals interconnected in the same local area by a switched LAN. Gateways, routers, or integrated gateway/router devices provide access to remote sites. The gateways provide communication with H.320 and H.324 terminals remotely connected to the ISDN and PSTN, respectively. H.323-to-H.323 communication between two remote sites can be achieved using routers that directly carry IP traffic over the Point-to-Point Protocol (PPP) running on ISDN. For better channel efficiency, gateways can translate H.323 streams into H.320 streams to be carried over ISDN lines, and vice versa.

In addition to H.323, there is the Session Initiation Protocol (SIP) defined by the IETF. SIP offers mechanisms for call routing, call signaling, capabilities exchange, media control, and supplementary services. It is a newer protocol that offers scalability, flexibility, and ease of implementation in building complex systems, while H.323 is an older protocol valued for its manageability, reliability, and interoperability with the PSTN. Standards bodies are working on procedures to allow seamless internetworking between the two protocols.

Summary

A LAN-based PBX eliminates the need for IP telephony software to be loaded on each client PC. It also allows organizations to avoid having to set up and manage separate LAN and PBX infrastructures. A unified backbone to the desktop allows common delivery of voice and data for reduced wiring and maintenance costs. Using a switched 100-Mbps Ethernet, network engineers can design telephone networks with essentially unlimited capacity. When the need arises for more workstations (i.e., extensions), another Ethernet switch is added. Administering these systems is done locally through a Windows graphical user interface or remotely through a Web browser.

See also

Call Centers

Ethernet
Internet Telephony
Transmission Control Protocol/Internet Protocol (TCP/IP)
Voice over IP

LATENCY

Latency is the amount of delay that affects all types of communications links. Delay on telecommunications networks is usually measured in milliseconds (ms), or thousandths of a second.

A rule of thumb used by the telephone industry is that the round-trip delay for a telephone call should be less than 100 ms. If delay exceeds 100 ms, participants perceive a slight pause in the conversation and use it as the opportunity to begin speaking. But by the time their words arrive at the other end, the other speaker has already begun the next sentence and feels that he or she is being interrupted. When telephone calls go over satellite links, the round-trip delay is typically about 250 ms, and conversations become full of awkward pauses and accidental interruptions.

Latency affects the performance of applications on data networks as well. On the Internet, for example, excessive delay can cause packets to arrive at their destination out of order, especially during busy hours. The reason packets may arrive out of sequence is that they can take different routes on the network. The packets are held in a buffer at the receiving device until all packets arrive and are put in the right order. While this does not affect e-mail and file transfers, which are not real-time applications, excess latency does affect the performance of multimedia and real-time applications.

If the packets containing voice or video do not arrive within a reasonable time, they are dropped. When packets containing voice are dropped, a condition known as "clipping" occurs, which is the cutting off of the first or final syllables in a conversation. Dropped packets of video cause the image to be jerky. Excessive latency also causes the voice and video components in a videoconference to arrive out of synchronization with each other, causing the video component to run slower than the voice component. For example, a person's lips will not match what he or she is saying.

The effects of latency can be overcome by assigning an appropriate quality of service (QoS) to each application and prioritizing the traffic offered to the network. QoS parameters can be programmed into the operating systems of routers, switches, and integrated access devices (IADs). When traffic is set to go onto the network, prioritization ensures that mission-critical applications obtain the bandwidth before routine applications.

Summary

The availability of policy-based network management tools from a variety of vendors has made it easier for large enterprises to implement QoS policies and traffic prioritization schemes with enough granularity to ensure that all applications are served in an appropriate manner without the company having to constantly shell out for more bandwidth and associated resources in a futile effort to stay ahead of the performance curve.

See also

Ping

LOCAL EXCHANGE CARRIERS

Local exchange carriers (LECs) provide residential, business, and interexchange access services. In addition to centrex, many of the larger LECs are developing and/or offering value-added services such as voice and data messaging via cellular and Personal Communications Services (PCS) networks. The LECs, which are commonly called "telephone companies" (or "telcos"), include the 22 former Bell Operating Companies (BOCs) divested from AT&T in 1984, as well as Cincinnati Bell, Southern New England Telephone (SNET), and the telephone companies of GTE and Sprint. These companies are now referred to as ILECs—incumbent local exchange carriers—to distinguish them from competitors in the local market and the hundreds of smaller telephone companies serving largely rural areas.

In addition to providing local phone service and allowing interexchange carriers (IXCs) to access the local loop, the LECs provide billing services. Phone bills that come from a LEC actually can represent charges from a number of services and providers. A telephone bill can comprise many basic elements, including charges for local message units (MSUs), special service offerings, directory service, 911 emergency service, cellular calls, and Internet access. These and other charges are identified in a consolidated monthly invoice. In addition, charges for long-distance calls carried by IXCs and cellular service providers may appear on the monthly invoice as well.

Another type of LEC is the competitive local exchange carrier (CLEC). This type of service provider offers business and residential users lines and services on a resale basis or from its own facilities-based network, enabling customers to save money on their communications bill. Regional teleports, metropolitan fiber carriers, and CATV operators are among the types of companies that are now involved in providing competitive local exchange services. Typically, these alternative access carriers offer service in major markets, where traffic volumes are greatest and, consequently, users are hardest hit with high local service charges.

There are also data local exchange carriers (DLECs), which specialize in providing data services such as Digital Subscriber Line (DSL) for high-speed Internet access. The

DLEC usually provisions its service over the same line that provides telephone service to the subscriber. Since the voice and data use different frequencies, the subscriber can talk on the phone and surf the Web at the same time. A DSL access multiplexer in the central office combines all the user traffic onto a fiber link that is ultimately connected to the Internet backbone via a network access point (NAP). DSL services are offered by ILECs and CLECs as well, sometimes in partnership with the DLECs.

Building local exchange carriers (BLECs) specialize in setting up integrated voice and data services in office buildings. Typically, the BLEC targets buildings with 10 or more tenants and over 100,000 square feet of office space. It decides on what buildings to approach based on tenant profile and anticipated demand, the economic opportunity in the building, and access to broadband circuits. The BLEC selects the buildings and portfolio owners in the target markets in which it desires to secure rights and negotiates an arrangement with the real estate owner that will benefit both parties. The real estate owners may be paid either a fixed rental fee per month or 5 percent of the revenue generated in the building. The typical lease or license agreement with an owner is for a term of 10 or more years. A typical network costs between $175,000 and $200,000 per building, with the BLEC picking up the entire amount.

Summary

With passage of the Telecommunications Act of 1996, new entities are allowed into the market for local telephone service, including cable operators, electric utilities, Internet service providers (ISPs), and entertainment companies. Until late 1999, the ILECs were restricted to providing local service within their assigned serving areas, called LATAs—local access and transport areas. The first ILEC to obtain Federal Communications Commission (FCC) permission to offer long-distance service in its own territory was Bell

Atlantic, which changed its name to Verizon. Ultimately, all the ILECs will receive permission to offer long-distance service in their respective territories. The objective is to offer customers bundled services consisting of local and long-distance service and Internet access at very attractive prices and, in the process, limit competition from carriers that are not in a position to offer such bundles.

See also

Building Local Exchange Carriers
Competitive Local Exchange Carriers
Incumbent Local Exchange Carriers
Interexchange Carriers

LOCAL LOOP

The local loop is an unbundled network element (UNE) that the FCC defines as a transmission facility between a distribution frame, or its equivalent, in a central office and the network interface device at the customer premises. This definition includes, for example, two- and four-wire analog voice-grade loops and two- and four-wire loops that are conditioned to transmit the digital signals needed to provide services such as ISDN, DSL and DS1 signals.[1]

ILECs are required to provide access to these transmission facilities only to the extent technically feasible. If it is not technically feasible to condition a loop facility to support a particular service, the ILEC need not provide unbundled access to that loop. For example, a local loop that exceeds the maximum length allowable for the provision of high-bit-rate

[1]Carriers traditionally have defined local loops in more detailed terms than discussed here. Likewise, the definition of the term differs among state public utility commissions (PUCs). The FCC has taken a general approach in an effort to minimize complex and resource-intensive disputes between ILECs and competitors over whether a particular function qualifies as a "loop."

Digital Subscriber Line (HDSL) service could not feasibly be conditioned for such service.

The FCC's definition of loops in some instances requires the ILEC to take affirmative steps to condition existing loop facilities to enable requesting carriers to provide services not currently provided over such facilities. For example, if a competitor seeks to provide a digital service such as DSL, and the loop is not currently conditioned to carry digital signals, but it is technically feasible to do so, the ILEC must condition the loop to permit the transmission of digital signals. Thus the FCC rejects the arguments of some ILECs that competitors must take the networks as they find them with respect to unbundled network elements. The requesting carrier must, however, bear the cost of compensating the ILEC for such conditioning.

The FCC further requires that ILECs provide competitors with access to unbundled loops regardless of whether the incumbent LEC uses integrated digital loop carrier (IDLC) technology or similar remote concentration devices for the particular loop sought by the competitor. IDLC technology allows a carrier to aggregate and multiplex loop traffic at a remote concentration point and deliver that multiplexed traffic directly into the switch without first demultiplexing the individual loops. If the FCC did not require ILECs to unbundle IDLC-delivered loops, end users served by such technologies would not have the same choice of competing providers as end users served by other loop types. Further, such an exception would encourage ILECs to "hide" loops from competitors through the use of IDLC technology.

In most cases it is technically feasible to unbundle IDLC-delivered loops. One way to unbundle an individual loop from an IDLC is to use a demultiplexer to separate the unbundled loop(s) prior to connecting the remaining loops to the switch. There are other ways to separate out individual loops from IDLC facilities, including methods that do not require demultiplexing. For example, IDLC loops can be moved onto other loop carrier links or, alternatively, can be removed from the multiplexed signal through a grooming process. Again, the costs associated with these mechanisms must be borne by requesting carriers.

The FCC also requires ILECs to offer unbundled access to the network interface device (NID) at the customer premises. When a competitor deploys its own loops, the competitor must be able to connect its loops to customers' inside wiring in order to provide competing service, especially in multitenant buildings. In many cases, inside wiring is connected to the ILEC's loop plant at the NID. In order to provide service, a competitor must have access to this facility. Therefore, a requesting carrier is entitled to connect its loops, via its own NID, to the ILEC's NID. The new entrant bears the cost of connecting its NID to the incumbent LEC's NID.

Summary

The purpose of requiring incumbent LECs to make available unbundled local loops is to facilitate market entry and improve consumer welfare. Without access to unbundled local loops, new entrants would need to invest immediately in duplicative facilities in order to compete for customers. Such investment and building likely would delay market entry and postpone the benefits of local telephone competition for consumers. Moreover, without access to unbundled loops, new entrants would be required to make a large initial capital investment in loop facilities before they had a customer base large enough to justify such expenditures. This would increase the risk of entry and raise the new entrant's cost of capital. By contrast, the ability of a new entrant to purchase unbundled loops from an ILEC allows the new entrant to build facilities gradually and to deploy loops for its customers where it is efficient to do so.

See also

Digital Subscriber Line Technologies

Integrated Services Digital Network

LOCAL MULTIPOINT DISTRIBUTION SERVICE

Local Multipoint Distribution Service (LMDS) is a two-way millimeter microwave technology that operates in the 27- to 31-GHz range. This broadband service allows communications providers to offer a variety of high-bandwidth services to homes and businesses, including broadband Internet access. LMDS offers greater bandwidth capabilities than a predecessor technology called Multichannel Multipoint Distribution Service (MMDS) but has a maximum range of only 7.5 miles from the carrier's hub to the customer premises. This range can be extended, however, through the use of optical fiber links.

Applications

LMDS provides enormous bandwidth—enough to support 16,000 voice conversations, plus 200 channels of television programming. Figure L-1 contrasts LMDS with the bandwidth available over other wireless services.

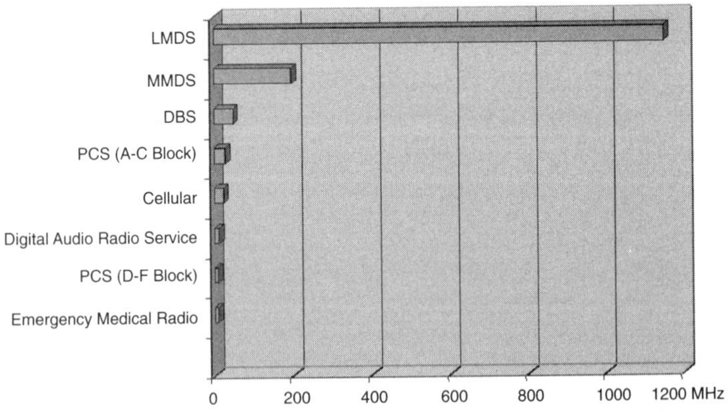

Figure L-1 Local Multipoint Distribution Service (LMDS) operates in the 27- to 31-GHz range and offers 1150 MHz of bandwidth capacity, which is over 2 times more than all other auctioned spectrum combined.

CLECs can deploy LMDS to completely bypass the local loops of the ILECs, eliminating access charges and avoiding service-provisioning delays. Since the service entails setting up equipment between the provider's hub location and customer buildings for the microwave link, LMDS costs far less to deploy than installing new fiber. This allows CLECs to very economically bring customer traffic onto their existing metropolitan fiber networks and, from there, to a national backbone network.

The strategy among many CLECs is to offer LMDS to owners of multitenant office buildings and then install cable to each tenant who subscribes to the service. The cabling goes to an on-premises switch, which is run to the antenna on the building's roof. That antenna is aimed at the service provider's antenna at its hub location. The line-of-sight wireless link between the two antennas offers a broadband "pipe" for multiple voice, data, and video applications. Subscribers can use LMDS for a variety of high-bandwidth applications, including television broadcast, videoconferencing, LAN interconnection, broadband Internet access, and telemedicine.

Operation

LMDS operation requires a clear line of sight between the carrier's hub station antenna and the antenna at each customer location. The maximum range between the two is 7.5 miles. However, LMDS is also capable of operating without having a direct line of sight with the receiver. This feature, highly desirable in built-up urban areas, may be achieved by bouncing signals off buildings so that they get around obstructions. At the receiving location, the data packets arriving at different times are held in queue for resequencing before they are passed to the application. This scheme does not work well for voice, however, because the delay resulting from queuing and resequencing disrupts two-way conversation.

At the carrier's hub location there is a roof-mounted multisectored antenna. Each sector of the antenna receives/

transmits signals between itself and a specific customer location. This antenna is very small, some measuring only 12 inches in diameter. The hub antenna brings the multiplexed traffic down to an indoor switch, which processes the data into 53-byte ATM "cells" for transmission over the carrier's fiber network. These individually addressed cells are converted back to their native format before going off the carrier's network to their proper destinations—the Internet, PSTN, or the customer's remote location.

At each customer's location, there is a rooftop antenna that sends/receives multiplexed traffic. This traffic passes through an indoor NIU that provides the gateway between the radiofrequency (RF) components and the in-building equipment, such as a LAN hub, PBX, or videoconferencing system. The NIU includes an up/down converter that changes the frequency of the microwave signals to a lower intermediate frequency (IF) that the electronics in the office equipment can manipulate more easily (and inexpensively).

Spectrum Auctions

In May 1999, the FCC held the last auction for LMDS spectrum. Over 100 companies qualified for the auctions, bidding against each other for licenses in select basic trading areas (BTAs).[2] The FCC auctioned two types of licenses in each market: An A-block license permits the holder to provision 1150 MHz of spectrum for distribution among its customers, while a B-block license permits the holder to provision 150 MHz. Most of the A-block licenses in the largest BTAs were won by major CLECs, while the B-block licenses were taken by smaller companies, ISPs, universities, and government

[2] Basic trading area (BTA) is a term used in the geographic definition of economic activity based on data compiled by Rand McNally. Most large cities are metropolitan trading areas (MTAs), and most of the larger U.S. towns are classed as BTAs. These are not the same as local access and transport areas (LATAs), which have defined the local service boundaries of the former BOCs since their divestiture from AT&T in 1984.

agencies.[3] The licenses are granted for a 10-year period, after which the FCC can take them back if the holder does not have service up and running.

Development History

Bernard Bossard is generally recognized as the inventor of LMDS. Bossard, who had worked with microwaves for the military, believed he could make point-to-multipoint video work in the 28-GHz band. Not interested in sending high-powered, low-frequency signals over long distances, Bossard focused instead on sending low-powered, high-frequency signals over a short distance. The result was LMDS. In 1986 he received funding and formed CellularVision with his financial backers. CellularVision then spun off the technical rights to the technology into a separate subsidiary, CT&T, which licenses it to other companies.

CellularVision was awarded a pioneer's preference license by the FCC for its role in developing LMDS. CellularVision began operating a commercial LMDS in metropolitan New York, providing video programming to subscribers in the Brighton Beach area. In 1998, CellularVision changed its name to SPEEDUS.COM. The company has a network operations center and recently has been expanding the number of operating cells in the New York area and now claims more than 12,000 residential and business subscribers.

SPEED is delivered via 14 fully functional Internet broadcast stations in operation under SPEEDUS.COM's FCC license covering metropolitan New York. SPEED subscribers are able to browse the Web using the company's SPEED modem, which is capable of downstream speeds of up to 48 Mbps, about 30 times faster than a full T1 line.

[3] The ILECs, such as the BOCs, were forbidden to enter the LMDS market for 3 years. In 2002, they were expected to use LMDS, among other technologies, to bypass each other's local loops to extend services to target markets. By mid-2002, however, the technical limitations of LMDS (line-of-sight requirement and weather-related impairments), plus return-on-investment concerns, prevented it from being rolled out in many markets.

In the SPEED.COM system, cable programming is downlinked from satellites to the company's head-end facility, where local broadcast transmissions are also received. At the company's master control room, the programming signals are then amplified, sequenced, scrambled, and up-converted to 28 GHz. The SPEED.COM transmitters and repeaters then broadcast a polarized FM signal in the 28-GHz band over a radius of up to 3 miles to subscribers and to adjacent cells for transmission. A 6-inch-square, highly directional flat-plate, window, roof, or wall-mounted antenna receives the scrambled signal and delivers it to the addressable set-top converter, which decodes the signals. The subscriber receives 49 channels of high-quality video and audio programming, including pay-per-view and premium channels.

Potential Problems

A potential problem for LMDS users is that the signals can be disrupted by heavy rainfall and dense fog—even foliage can block a signal. In metropolitan areas where new construction is a fact of life, a line-of-sight transmission path can disappear virtually overnight. For these reasons, many IT executives are leery of trusting mission-critical applications to this wireless technology. Service providers downplay this situation by claiming that LMDS is just one local access option and that fiber links are the way to go for mission-critical applications. In fact, some LMDS providers offer fiber as a backup in case the microwave links experience interference.

There is controversy in the industry about the economics of the point-to-multipoint architecture of LMDS, with some experts claiming that the business model of going after low-usage customers is fundamentally flawed and will never justify the service provider's cost of equipment, installation, and provisioning. With an overabundance of fiber in the ground and metropolitan area Gigabit Ethernet services coming online at a competitive price, the time for LMDS may have come and gone. In addition, newer wireless technologies such as free-air laser hold a significant speed advantage over LMDS, as does submillimeter transmission in the 60- and 95-GHz bands.

Another problem that has beset LMDS is that the major license holders have gotten caught up in financial problems, some declaring Chapter 11 bankruptcy. These carriers built their networks quickly, incurring massive debt, without lining up customers fast enough. This strategy worked well as long as the capital markets were willing to continue funding these companies. But once the capital markets dried up in 2000, so did the wireless providers' coffers and their immediate prospects. The uncertain future of these financially strapped carriers has discouraged many companies from even trying LMDS.

Summary

Fiberoptics is the primary transmission medium for broadband connectivity today. However, of the estimated 4.6 million commercial buildings in the United States, 99 percent are not served by fiber. Businesses are at a competitive disadvantage in today's information-intensive world unless they have access to broadband access services, including high-speed Internet access. These businesses, including many data-intensive high-technology companies, can be served adequately with LMDS. Despite the financial problems of LMDS providers, the technology has the potential to become a significant portion of the global access market, which will include a mix of many technologies, including DSL, cable modems, broadband satellite, and fiberoptic systems.

See also

Asynchronous Transfer Mode
Cable Telephony
Digital Subscriber Line Technologies
Multichannel Multipoint Distribution Service

M

MODEMS

A modem, or modulator-demodulator, converts the digital signals generated by a computer into analog signals suitable for transmission over dial-up telephone lines or voice-grade leased lines. Another modem, located at the receiving end of the transmission, converts the analog signals back into digital form for manipulation by the data recipient. Although long-distance lines are digital, most local lines are not, which explains why modems are often required to access the Internet, transfer files, access bulletin boards, send electronic mail, and connect to host computers from remote locations.

Modems are packaged as cards for PCs, stand-alone devices, or rack-mounted models for use by carriers and Internet service providers (ISPs). Card modems insert into the vacant slot of a desktop computer or notebook. There are also modem cards for communications servers, which can be shared by multiple users. External desktop modems connect to a computer's RS-232 serial port, while rack-mount modems are modules that are housed in an equipment frame.

Modem manufacturers are continually redesigning their products to incorporate the latest standards, enhance existing features, and add new ones. The advancement of modulation techniques, error correction, data compression, and diagnostics is one of the continuing activities of modem manufacturers.

Modulation Techniques

Modems use modulation techniques to encode the serial digital data generated by a computer onto the analog carrier signal. The simplest modulation techniques employ two signal manipulations to transmit information: frequency shift keying and phase shift keying. Frequency shift keying (FSK) is similar to the frequency modulation technique used to broadcast FM radio signals. By forcing the signal to shift back and forth between two frequencies, the modem is able to encode one frequency as a 1 and the other as a 0. This modulation technique was used widely in the early 300-bps modems. At higher rates, FSK is too vulnerable to line noise to be effective.

Phase shift keying (PSK), another early modulation technique, makes use of shifts in a signal's phase to indicate 1s and 0s. The problem with this method is that "phase" refers to the position of a waveform in time; therefore, the data terminal clocks at both ends of the transmission must be synchronized precisely.

Another method, known as differential phase shift keying (DPSK), uses the phase transition to indicate the logic level. With this scheme, it is not necessary to assign a specific binary state to each phase; it only matters that some phase shift has taken place. The telephone bandwidth is limited, however, so it is only possible to have 600 phase transitions per second on each channel, thus limiting transmission speeds to about 600 bps. To increase speed, it is possible to expand DPSK from a two- to a four-state pattern represented by four 2-bit symbols (known as "dibits") as follows:

- Maintain the same state (0, 0).
- Shift counterclockwise (0, 1).

- Shift clockwise (1, 0).
- Shift to the opposite state (1, 1).

Other modulation techniques, such as trellis encoding, are much more sophisticated and capable of moving data at much higher speeds. Trellis encoding entails the use of a 32-bit constellation with "quintbits" to pack more information into the carrier signal and offer more immunity from noise. The use of quintbits offers 16 extra possible state symbols. These extra transition states allow dial-line modems to use transitions between points, rather than specific points, to represent state symbols. The receiving modem uses probability rules to eliminate illegal transitions and obtain the correct symbol. This gives the transmission greater immunity to line impairments. Additionally, the fifth bit can be used as a redundant bit, or checksum, to increase throughput by reducing the probability of errors.

By increasing the number of points in the signal constellation, it is possible to encode greater amounts of information to increase the modem's throughput. This is because ever-slighter variations in the phase-modulated signal may be used to represent coded information, which translates into higher throughput. Some modem manufacturers use constellations consisting of 256 or more points.

Transmission Techniques

Modems support two types of transmission techniques: asynchronous and synchronous. The user's operating environment determines whether an asynchronous or synchronous modem is required. During asynchronous transmission, start and stop bits frame each segment of data during transfer to distinguish each bit from the one preceding it. Synchronous transmission transfers data in one continuous stream; therefore, the transmitting and receiving devices must be synchronized precisely in order to distinguish each character in the data stream. Some types of modems support both types of transmission.

Modem Speed

For many users, the most important modem characteristic is data rate. The quality of the connection has a lot to do with the actual speed of the modem. If the connection is noisy, for example, the modem may have to step down to a lower speed to continue transmitting data. Some modems are able to sense improvements in line quality and can automatically step up to higher data rates as line quality improves.

In 1997, a class of modems became available that offered data transmission rates of up to 56 kbps. The modems are based on technology that exploits the fact that for most of its length, an analog modem connection is really a digital connection. When an analog signal leaves the user's modem, it is carried to a phone company central office, where it is digitized. If it is destined for a remote analog line, the signal is converted back to analog at the central office nearest the receiving user. The conversion is made at only one place—where the analog line meets the central office. During the conversion, noise is introduced that cuts throughput. But the noise is less in the other direction, from digital to analog, allowing the greater downstream throughput (Figure M-1).

It is also possible to bond two or three 56-kbps dial-up lines to achieve a combined data rate of up to 168 kbps[1] (Figure M-2). When two or three modems in a modem pool device are used for Internet access (for example, they call the Internet service provider simultaneously and share the downloading of Web pages, resulting in the greater throughput rate), download time can be cut by as much as two-thirds. Although the Internet service provider (ISP) does not have to do anything different as far as hardware is concerned—except have enough 56 kbps modems—it must permit users to establish multiple sessions with a single user ID and password.

[1]Since the quality of each line may differ at any given time, the aggregate speed of three dial-up lines actually will be much less than 168 kbps.

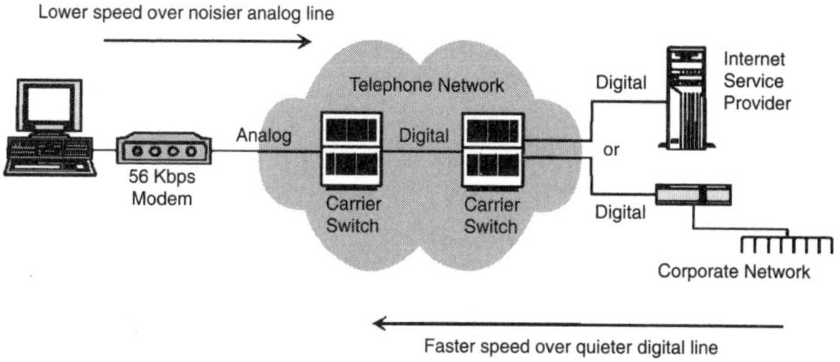

Figure M-1 Today's 56-kbps modems can send data at top speed, but only from a digital source. Since it is already in digital form, the traffic is free of impairments from noise introduced when an analog modem signal is converted to digital within the carrier network. From an analog source, the top speed is quite a bit lower than 56 kbps, since the traffic is subject to impairment from noise.

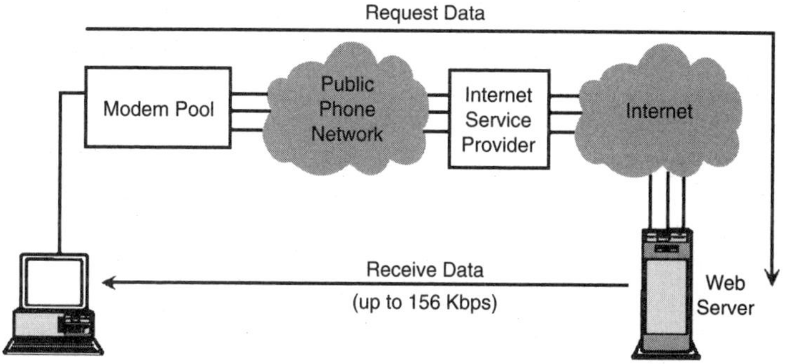

Figure M-2 The modem pool device initiates two or three simultaneous calls, divvying up the separate TCP/IP sessions and making Web page downloads go much faster. The Web server sends the requested page to the modem pool device, and the page is rendered on the user's screen.

Modem Features

Most modems come equipped with the same basic features, including error correction and data compression. In addition, they have features associated with the network interface, such as flow control and diagnostics. There are also various security features that are implemented by modems.

Error Correction Networks often contain disturbances that modems must deal with or, in some cases, overcome. These disturbances include attenuation distortion, envelope delay distortion, phase jitter, impulse noise, background noise, and harmonic distortion—all of which negatively affect data transmission. To alleviate the disturbances encountered when transferring data over leased lines (without line conditioning) and dial-up lines, most products include an error-correction technique in which a processor puts a bit stream through a series of complex algorithms prior to data transmission.

The most prominent error-correction technique has been the Microcom Networking Protocol (MNP),[2] which uses the cyclic redundancy check (CRC) method for detecting packet errors and requests retransmissions when necessary.

Link Access Procedure B (LAP-B), a similar technique, is a member of the High Level Data Link Control (HDLC) protocol family, the error-correcting protocol in X.25 for packet-switched networks. LAP-M is an extension to that standard for modem use and is the core of the International Telecommunication Union (ITU) error-correcting standard, V.42. This standard also supports MNP stages 1 through 4. Full conformance with the V.42 standard requires that both LAP-M and MNP stages 1 through 4 be supported by the modem. Virtually all modems currently made by major manufacturers conform to the V.42 standard.

The MNP is divided into nine classes. Only the first four deal with error recovery, which is why only those four are referenced in V.42. The other five classes deal with data compression. The MNP error recovery classes are as follows:

- MNP classes 1 to 3 packetize data and ensure data integrity.

[2]Microcom no longer exists as an independent company; Compaq Computer acquired the company in 1997.

- MNP class 4 achieves up to 120 percent link throughput efficiency via adaptive packet assembly and data phase optimization, which automatically adjusts packet size relative to line conditions and reduces protocol overhead.

Data Compression With the adoption of the V.42bis recommendation by the ITU in 1988, the data-compression standard known as Lempel-Ziv was adopted. This algorithm compresses most data types, including executable programs, graphics, ASCII text, or binary data streams. Compression ratios of 4 to 1 can be achieved, although actual throughput gains from data compression depend on the types of data being compressed. Text files are the most likely to yield performance gains, followed by spreadsheet and database files. Executable files are most resistant to compression algorithms because of the random nature of the data.

Diagnostics and Other Features Most modems perform a series of diagnostic tests to identify internal and transmission-line problems. Most modems also offer standard loopback tests, such as local analog, local digital, and remote digital loop back. Once a modem is set in test mode, characters entered on the keyboard are looped back to the screen for verification.

Most modems also include call-handling features such as automatic dial, answer, redial, fallback, and call progress monitoring. Calling features simplify the chore of establishing and maintaining a communications connection by automating the dialing process. Telephone numbers can be stored in nonvolatile memory.

Other standard modem features commonly offered include fallback capability and remote operation. Fallback allows a modem to automatically drop, or fall back, to a lower speed in the event of line noise and then revert to the original transmission speed after line conditions improve. Remote operation, as the name implies, allows users to activate and configure a modem from a remote terminal.

Security

Modems that offer security features typically provide two levels of protection: password and dial-back. The former requires the user to enter a password, which is verified against an internal security table. The dial-back feature offers an even higher level of protection. Incoming calls are prompted for a password, and the modem either calls back the originating modem using a number stored in the security table or prompts the user for a telephone number and then calls back.

Security procedures can be implemented before the modem handshaking sequence rather than after it. This effectively eliminates the access opportunity for potential intruders. In addition to saving connection establishment time, this method uses a precision high-speed analog security sequence that is not detectable even by advanced line-monitoring equipment.

Transmission Facilities

Modems support two types of lines: leased or dial-up. The primary difference between the two is the procedure for establishing a connection as opposed to the line itself.

Dial-up Lines Dial-up lines are used for typical telephone service. These lines usually connect to a small modular wall jack called an RJ-11 and a companion plug, which is inserted into the jack to establish a connection to the telephone or modem. When a voice-grade analog line is used with a modem, a short cord with an RJ-11 connector on both ends is inserted into the jack and into the modem.

Although the RJ-11 modular jack connection is common for low-speed modems, problems may arise with regard to the consistency of the signals transmitted over the line. To ensure a consistently high signal level, a special data-line jack, such as an RJ-41 or an RJ-45, can be installed. These data-line jacks are designed specifically to operate with modem circuitry and are often used in leased-line environments to help maintain the quality of the transmitted signal.

Leased Lines Leased lines are available in two- and four-wire versions. Four-wire leased lines differ from their two-wire counterparts in terms of cost (four-wire lines are more expensive) and in the mode of modem operation supported. Not all modems can support leased lines. An effective way of determining whether a modem can support leased-line connections is to examine the way the telephone line is connected. Modems designed for two- and four-wire leased-line operation have two sets of terminal screws with which to attach the two pairs of lines.

To sustain the optimal performance of leased-line modems, the lines may be specifically selected for their desirable characteristics. This is an extra-cost service called "line conditioning" that is provided by the carriers on a best-effort basis. AT&T's D6 line conditioning, for example, addresses phase jitter, attenuation distortion, and envelope delay distortion—all of which can impair transmission at data rates approaching 19.2 kbps.

The monthly charges and installation cost of D6 conditioning are higher than for other levels of conditioning if only because there are fewer wire pairs available that exhibit the higher immunity from generic noise and nonlinear distortion in high-density locations. Immunity from noise and nonlinear distortion occur on copper pairs more by chance than by design because they may be caused externally and be beyond the control of the carrier. This means that numerous wire pairs must be tested before those having the desired characteristics can be identified and put into service as "conditioned" lines.

Wireless Links

Wireless modems are required to transfer data over public wireless services and private wireless networks. These modems come in a variety of hardware configurations:

stand-alone, built-in, and removable PC card. The newer modems are programmable and therefore capable of being used with a variety of wireless services using different frequencies and protocols. There are even modems that mimic wireline protocols, allowing existing applications to be run over the wireless network without modification.

Single-Frequency Modems Private wireless networks operate in a range of unique frequency bands to ensure privacy. Using radio modems operating over dedicated frequencies within these frequency bands also permits the transmission of business-critical information without interference problems. Furthermore, the strategic deployment of radio modems can provide metropolitan area coverage without the use of expensive antenna arrays.

Such modems are designed to provide a wireless, protocol-independent interface between host computers and remote terminals located as far away as 30 miles. Most provide a transmission rate of at least 19.2 kbps point to point in either half- or full-duplex mode. Some radio modems even support point-to-multipoint radio network configurations, serving as a virtual multidrop radio link that replaces the need for expensive, dedicated lines (Figure M-3). In this configuration, one modem is designated as the master, passing

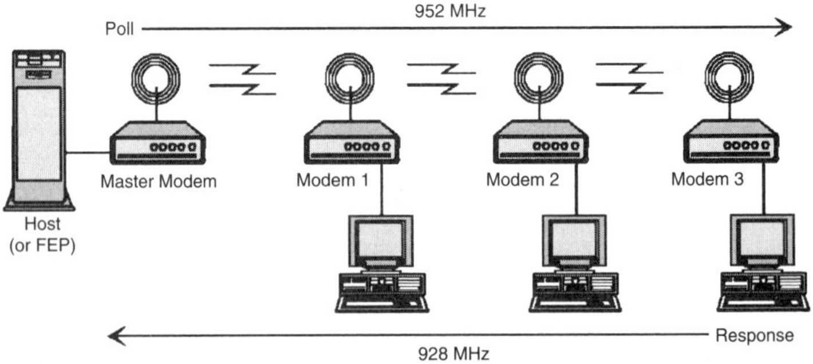

Figure M-3 A typical multidrop radio modem configuration.

polling information and responses between the host and terminals over two different frequencies.

In multidrop configurations, a given radio network is capable of supporting one type of asynchronous or synchronous polling protocol. Since such modems perform no processing or interpreting of the protocol, the host (or front-end processor) must generate all required protocol framing, line discipline, node addressing, and data encapsulation. And depending on vendor, these modems may be optionally equipped with an integral repeater to maintain signal integrity over longer distances.

Multifrequency Modems Regardless of the transmission technology or the hardware configuration used, the modem must be tuned to the frequency of the service provider's wireless network to operate properly. Until recently, modems were offered in different versions, according to the wireless network to which the modem would connect. This delayed product development and inflated the cost of manufacturing, which was passed onto users in the form of higher prices for equipment. To overcome these problems, chip manufacturers have developed programmable chipsets that are not limited to a specific network's radiofrequency. Newer wireless modems are computer-configurable—within specified frequency ranges, the transmitting and receiving frequencies are independently selectable via software.

Multimedia Modems Not only can modems be programmed for multifrequency use, they also can provide seamless integration of multiple media—wireline and wireless—through a common programmable interface. This is accomplished with a chipset that supports both wireline and wireless communications. Special software used with the chipset provides a method for connecting cellular phones to modems, which is important because cellular phones lack dial tones and other features used by modems on the wireline phone network. The software makes it appear that those features exist.

Multifunction Modems

Multifunction modems use programmable digital signal processing (DSP) technology to turn a computer into a complete desktop message center, allowing the user to control telephone, voice (recording and playback), fax, data transfers, and e-mail. Typical features include multiple mailboxes for voice mail, caller ID support, call forwarding, remote message retrieval, phone directory, and contact database.

In some cases, the modem is actually on a full-duplex sound card. By plugging in speakers and a subwoofer, the user can even enjoy a stereo sound speakerphone. A separate connection to a CD-ROM player allows the user to work at the computer while listening to music. With DSP, the modem can be upgraded easily to the latest communications standards, and new capabilities can be added simply by loading additional software. For example, a 33.6-kbps modem can be upgraded to 56 kbps by installing new software instead of having to buy new hardware, often at no extra charge from the vendor.

Multimedia modems use digital simultaneous voice and data (DSVD), which enables the user to send voice and data at the same time over a single telephone line. The biggest advantage of DSVD is that users no longer need to interrupt telephone conversations or install a separate line to transmit data or receive faxes. Multimedia modems typically include full-duplex speakerphone, fax, modem, and 16-bit stereo audio capabilities.

Another new way vendors are packaging modems is by integrating them with Integrated Services Digital Network (ISDN) terminal adapters. This allows users to communicate with conventional dial-up services and also take advantage of ISDN when possible—all without cluttering the desktop or having to use up scarce slots in the PC.

Soft Modems

PC analog telephony modems are available today in three technology implementations: controller-based, host-controlled, and soft modems. Controller-based modems use DSP chips and microprocessor controllers; host-controlled modems (also called DSP-accelerated), use DSPs for modulation but rely on the host computer to provide the controller functions; and soft modems rely on the host computer to perform both modulation (data pump) and control functions.

Controller-based modems contain all the chips necessary to carry out their functions with minimal reliance on the host computer, while host-controller modem designs take advantage of low-cost dedicated silicon for computationally intensive data-pump functions and use the host PC microprocessor for system-control functions. Host controllers are optimal for internal add-in card modems and motherboard modems.

Soft modem design puts the highest load on the host processor, typically consuming 50 to 60 MHz on a conventional microprocessor, thereby reducing resources available to other programs. Soft modems are a reasonable choice for those who use analog phone connections infrequently or who tend not to run multiple applications while connected online.

Summary

Although today's modems max out at 56 kbps, they can theoretically go faster with continued innovations in signal processing. Each telephone line channel contains 4 kHz of analog bandwidth, which is equivalent to 64 kbps of digital bandwidth. Under perfect conditions, a telephone line could support 64 kbps if signal-processing rates could be bumped up a bit. It is powerful signal processing in the latest V.90 and V.92 modems that currently allows for 56 kbps over this channel. Further improvements to increase throughput will depend not only on progress in signal-processing technologies but also on the priority given to their development by the industry. It seems that the industry is focused on other technologies such as Digital Subscriber Line (DSL), which provides multimegabit-per-second speeds over ordinary telephone lines.

See also

Cable Telephony

Digital Subscriber Line Technologies

Integrated Services Digital Network

MULTICHANNEL MULTIPOINT DISTRIBUTION SERVICE

Multichannel Multipoint Distribution Service (MMDS) is a microwave technology that traces its origins to 1972 when it was introduced to provide an analog service called Multipoint Distribution Service (MDS). For many years, MMDS was used for one-way broadcast of television programming, but in early 1999 the Federal Communications Commission (FCC) opened up this spectrum to allow for two-way transmissions, making it useful for delivering telecommunication services, including high-speed Internet access to homes and businesses.

This technology, which has now been updated to digital, operates in the 2- to 3-GHz range, enabling large amounts of data to be carried over the air from the operator's antenna towers to small receiving dishes installed at each customer location. The useful signal range of MMDS is about 30 miles, which beats Local Multipoint Distribution Service (LMDS) at 7.5 miles and DSL at 18,000 feet. Furthermore, MMDS is easier and less costly to install than cable service.

Operation

With MMDS, a complete package of TV programs can be transmitted to homes and businesses. Since MMDS operates within the frequency range of 2 to 3 GHz, which is much lower than LMDS at 28 to 31 GHz, it can support only up to 24 stations. However, operating at a lower frequency range means that the signals are not as susceptible to interference as those using LMDS technology.

Most of the time the operator receives TV programming via a satellite downlink. Large satellite antennas installed at the head end collect these signals and feed them into encoders that compress and encrypt the programming. The encoded video and audio signals are modulated, via amplitude modulation (AM) and frequency modulation (FM), respectively, to an intermediate frequency (IF) signal. These IF signals are up-converted to MMDS frequencies and then amplified and combined for delivery to a coax cable, which is connected to the transmitting antenna. The antenna can have an omnidirectional or sectional pattern.

The small antennas at each subscriber location receive the signals and pass them via a cable to a set-top box connected to the television. If the service also supports high-speed Internet access, a cable also goes to a special modem connected to the subscriber's PC. MMDS sends data as fast as 10 Mbps downstream (toward the computer). Typically, service providers offer downstream rates of 512 kbps to 2.0 Mbps, with burst rates up to 5 Mbps whenever spare bandwidth becomes available.

Originally, there was a line-of-sight limitation with MMDS technology. But this has been overcome with a complementary technology called Vector Orthogonal Frequency Division Multiplexing (VOFDM). Because MMDS does not require an unobstructed line of sight between antennas, signals bouncing off objects en route to their destination require a mechanism for being reassembled in their proper order at the receiving site. VOFDM handles this function by leveraging multipath signals, which normally degrade transmissions. It does this by combining multiple signals at the receiving end to enhance or recreate the transmitted signals. This increases the overall wireless system performance, link quality, and availability. It also increases service providers' market coverage through non-line-of-sight transmission.

Channel Derivation

MMDS equipment can be categorized into two types based on the duplexing technology used: Frequency Division Duplexing (FDD) or Time Division Duplexing (TDD). Systems based on FDD are a good solution for voice and bidirectional data because forward and reverse use separate and equally large frequency bands. However, the fixed nature of this scheme limits overall efficiency when used for Internet access. This is so because Internet traffic tends to be "bursty" and asymmetric. Instead of preassigning bandwidth with FDD, Internet traffic is best supported by a more flexible bandwidth-allocation scheme.

This is where TDD comes in; it is more efficient because each radio channel is divided into multiple time slots through Time Division Multiple Access (TDMA) technology, which enables multiple channels to be supported. Because TDD has flexible timeslot allocations, it is better suited for data delivery—specifically, Internet traffic. TDD enables service providers to vary uplink and downlink ratios as they add customers and services. Many more users can be supported by the allocation of bandwidth on a nonpredefined basis.

Summary

MMDS is being used to fill the gaps in market segments where cable modems and DSL cannot be deployed because of distance limitations and cost concerns. Like these technologies, MMDS provides data services and enhanced video services such as video on demand, as well as Internet access. MMDS will be another access method to complement a carrier's existing cable and DSL infrastructure, or it can be used alone for direct competition. With VOFDM technology, MMDS is becoming a workable option that can be deployed cost-effectively to reach urban businesses that do have line-of-sight access and in suburban and rural markets for small businesses and telecommuters.

See also

Digital Subscriber Line Technologies
Local Multipoint Distribution Service

MULTIPROTOCOL LABEL SWITCHING

With the explosive growth of the Internet in recent years, there is growing dissatisfaction with its performance. New techniques are available to improve performance, such as Multiprotocol Label Switching (MPLS), which delivers quality of service (QoS) and security capabilities over Internet Protocol (IP) networks, including virtual private networks (VPNs).

MPLS attaches tags, or labels, to IP packets as they leave the edge router and enter the MPLS-based network. The labels eliminate the need for intermediate router nodes to look deeply into each packet's IP header to make forwarding and class-of-service handling decisions. The result is that packet streams can pass through an MPLS-based wide area network (WAN) infrastructure very quickly, and time-sensitive traffic can get the priority treatment it requires.

The same labels that distinguish IP packet streams for appropriate class-of-service handling also provide secure isolation of these packets from other traffic over the same physical links. Since MPLS labeling hides the real IP address and other aspects of the packet stream, it provides data protection at least as secure as other layer 2 technologies, including frame relay and Asynchronous Transfer Mode (ATM).

Operation

To enhance the performance of IP networks, the various routes are assigned labels. Each node maintains a table of label-to-route bindings. At the node, a label switch router

(LSR) tracks incoming and outgoing labels for all routes it can reach, and it swaps an incoming label with an outgoing label as it forwards packet information (Figure M-4). Since MPLS

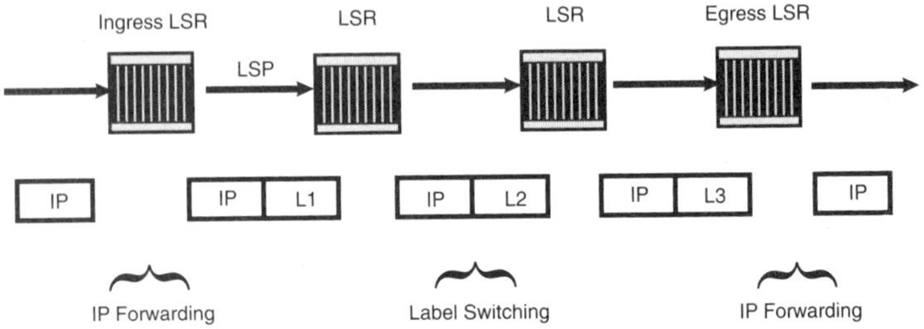

Figure M-4 A label-switched route is defined by fixed-length tags appended to the data packets. At each hop, the LSR strips off the existing label and applies a new label, which tells the next hop how to forward the packet. These labels enable the data packets to be forwarded through the network without the intermediate routers having to perform a complex route lookup based on destination IP address.

routers do not need to read as far into a packet as a traditional router does and perform a complex route lookup based on destination IP address, packets are forwarded much faster, which improves the performance of the entire IP network.

Although MPLS routers forward packets on a hop-by-hop basis, just like traditional routers, they operate more efficiently. As a packet arrives on an MPLS node, its label is compared to the label information base (LIB), which contains a table that is used to add a label to a packet, while determining the outgoing interface to which the data will be sent. After consulting the LIB, the MPLS node forwards the packet toward its destination over a label-switched path (LSP). The LIB can simplify forwarding and increase scalability by tying many incoming labels to the same outgoing label, achieving even greater levels of efficiency in routing. The LSPs can be used to provide QoS guarantees, define and enforce service-level agreements, and establish private user groups for VPNs.

MPLS provides a flexible scheme in that the labels could be used to manually define routes for load sharing or to establish a secure path. A multilevel system of labels can be used to indicate route information within a routing domain (interior routing) and across domains (exterior routing). This decoupling of interior and exterior routing means MPLS routers in the middle of a routing domain would need to track less routing information. That, in turn, helps the technology scale to handle large IP networks.

MPLS could provide a similar benefit to corporations that have large ATM-based backbones with routers as edge devices. Normally, as such networks grow and more routers are added, each router may need additional memory to keep up with the increasing size of the routing tables. MPLS alleviates this problem by having the ATM switches use the same routing protocols as routers. In this way, the routers on the edge of the backbone and the ATM-based label switches in the core would maintain summarized routing information and only need to know how to get to their nearest neighbor—not to all peers on the network.

MPLS also offers benefits to ISPs and carriers. It allows layer 2 switches to participate in layer 3 routing. This increases network scalability because it reduces the number of routing peers that each edge router must deal with. It also enables new traffic tuning mechanisms in router-based networks by integrating virtual circuit capabilities available previously only in layer 2 fabrics. With label switching, packet flows can be directed across the router network along predetermined paths, similar to virtual circuits, rather than along the hop-by-hop routes of normal routed networks. This enables routers to perform advanced traffic management tasks, such as load balancing, in the same manner as ATM or frame relay switches.

Finally, MPLS can be applied not only to the IP networks but also to any other network-layer protocol. This is so because tag switching is independent of the routing protocols employed. While the Internet runs on IP, a lot of campus backbone traffic is transported on protocols such as internetwork packet exchange (IPX), making a pure IP solution inadequate for many organizations.

Summary

MPLS came about as a result of Cisco's tag-switching concept, which was given over to the Internet Engineering Task Force (IETF) for further development and standardization. In 1996, the framework document published by the IETF presented MPLS as a label-switching architecture suitable for any protocol. The label process takes place without referencing the content of the data packet, eliminating the need for protocol-specific handling. By having the data-handling layer of MPLS separate from the control layer, multiple control layers—one for each protocol—could be supported. The IETF, however, has focused on MPLS as a means of improving IP networking, where the commercial opportunity is greatest. MPLS many encourage more service providers to migrate core infrastructures from ATM to IP. Now that MPLS provides IP with high speed, QoS, and security, there may be less reason for service providers to build an ATM infrastructure, which provides these advantages, but at a much high cost than IP.

See also

Asynchronous Transfer Mode

Quality of Service

Routers

Transmission Control Protocol/Internet Protocol (TCP/IP)

Virtual Private Networks

MULTISERVICE NETWORKING

Under the concept of multiservice networking, voice-enabled routers, local area network (LAN) and WAN switches, systems network architecture- (SNA-) to-LAN integration solutions, dial and other access products, Web site management tools, Internet appliances, and network management software are brought together within the same network infrastructure in an effort to better implement and manage current and emerging business applications. By definition, multiservice networks inherently support any type of traffic and, therefore, any type of applications networking requirement. The multiservice network typically consists of many components that all work together to ensure consistent QoS for the various applications, each of which may have its own performance requirements in terms of bandwidth, response time, and network availability.

Voice over IP

A key capability of the multiservice network is voice over IP (VoIP), which can be implemented in a variety of ways. A complete solution might consist of IP telephones, call manager software installed on a communications server, and a WAN gateway—all of which are attached to an existing LAN/WAN infrastructure. Locally, the infrastructure is usually a shared or switched Ethernet that provides the bandwidth and connectivity between the attached devices—IP phones or multimedia computers. IP address assignment is provided through the Dynamic Host Control Protocol (DHCP) installed on a communication server.

The call-management software provides the intelligence necessary to implement public branch exchange (PBX)–like features. This application is usually installed, along with DHCP, on a Windows NT/2000 server and provides basic call processing, signaling and connection services to IP phones and "soft" phones, voice-over-IP gateways, and other local and remote devices. This includes the management and control of various signaling protocols such as Q.931 for ISDN WAN control and H.225/H.245 for IP packet control.

The call-management software also implements supplementary and enhanced services such as hold, transfer, forward, conference, multiple-line appearances, automatic route selection, speed dial, last-number redial, and other features, which are extended to IP phones and gateways via parameters stored in a configuration database. Microsoft's Internet Information Server (IIS), for example, could be installed at the communications server to provide a browser interface to the configuration database. With administrator privileges, users can configure their own phones through a Web interface.

IP/PSTN Gateways

Gateways convert voice from the packet domain to the circuit-switched domain. Specifically, this type of device converts the voice packets that have been placed into an Ethernet frame into the format that can be accepted by the public switched telephone network (PSTN). Gateways—digital or analog—also pass signaling information, including dial tone and network signaling, as well as caller ID.

A digital gateway supports G.711 audio encoding at 64 kbps, offers Ethernet access, supports ISDN PRI, and provides integrated digital signal processor (DSP) functions. Each T1 interface card supports 24 channels with line echo cancellation and packet-to-circuit conversion for voice or fax calls. In addition to H.323/SIP compliance,[3] which allows interoperability with other H.323/SIP client applications and gateways, the gateway also supports supplementary services such as call forward, transfer, and hold. The gateway is configured using a Web interface.

---

[3]Both H.323 and Session Initiation Protocol (SIP) define mechanisms for call routing, call signaling, capabilities exchange, media control, and supplementary services. The H.323 from the ITU is valued for its manageability, reliability, and interoperability with the PSTN. SIP is a newer protocol from the IETF that offers scalability, flexibility, and ease of implementation when building complex systems. Standards bodies are working on procedures to allow seamless internetworking between the two protocols.

An analog gateway supports G.711 (64 kbps) or G.723 (dual-rate 5.3 or 6.3 kbps) audio compression and comes with integrated DSPs and modular analog circuit-switched interfaces. The analog system not only connects to local analog telephone company lines but also provides connectivity to devices such as fax machines, voice mail, and analog phone systems.

Multiservice Access Routers

With companies spending billions of dollars each year on internal phone calls and faxes between their own offices, there is ample incentive to reduce these costs by integrating voice, fax, and data onto a single multiservice network infrastructure. With voice and fax over IP, companies can deploy integrated, scalable networks without sacrificing voice and fax quality. In addition, the deployment of these multiservice capabilities can be done without changing the way phone calls are made or the way faxes are sent.

There are routers that enable traditional telephony traffic such as voice and fax to be integrated with traditional data traffic such as IP, IPX, and SNA. This integration is achieved with traditional telephony interfaces so that PBXs, key systems, traditional phones, IP phones, fax machines, and even the PSTN can physically connect to the router. Once these connections are established, the voice and fax traffic is processed by the DSPs and placed into IP packets or frame relay cells for transfer to other network locations. In keeping as much voice and fax traffic as possible on the data network, toll charges can be eliminated or greatly reduced. These routers also let users take advantage of advanced applications such as secure Internet access, managed network services, virtual private networks (VPNs), and electronic commerce.

Workstations on the remote LAN can be assigned IP addresses dynamically by either a communications server at the corporate location or the access routers using DHCP. The access routers also support network address translation

(NAT), which effectively creates a "private network" composed of IP addresses that are invisible to the outside world. NAT enables network administrators to assign IP addresses normally reserved for the Internet for private use over a remote LAN. For businesses that want to allow select access to the network, NAT can be configured to allow only certain types of data requests, such as Web browsing, e-mail, or file transfers.

Security features implemented through the router's operating system (OS) protect the privacy of company communications and commerce transactions over the Internet. The OS also provides the means to build custom security solutions, including standard and extended access-control lists (ACLs), dynamic ACLs, router and route authentication, and generic routing encapsulation (GRE) for tunneling. Perimeter security features control traffic entry and exit between private networks, intranets, or the Internet. To protect the corporate LAN from unauthorized access, the routers also may support token cards, the Password Authentication Protocol (PAP), the Challenge Handshake Authentication Protocol (CHAP), and other security features available through an optional firewall. Access routers can be managed with a software application that provides configuration and security management, as well as performance and fault monitoring. Centralized administration and management can be applied via the Simple Network Management Protocol (SNMP), Telnet, or local management through the router's console port.

Multiservice Concentrator

A multiservice concentrator is basically a wire-speed T1 router and serial data device that has voice, video, and ATM capabilities. It includes Ethernet LAN and data capabilities as well as IP and SNA suites. This type of router can be deployed over private or public networks to reduce equipment and connection costs, simplify network management, and improve application performance. Through the concentrator's operating system, these systems perform multiprotocol routing and bridging. They also can be tightly integrated with smaller access routers at branch office locations.

In addition to providing 24 channels of voice through the T1 port, the multiservice concentrator provides echo cancellation for all voice channels and achieves further cost savings through the use of voice activity detection (VAD). This feature halts voice traffic during the silent periods of a conversation, allowing the idle bandwidth to be used for data. Further bandwidth efficiencies can be achieved with voice compression at 8 kbps (G.729, G.729a) or 32 kbps (G.711, ADPCM).

The multiservice concentrator connects to any standard PBX switch, key system, or telephone. At smaller corporate sites, the concentrator can be used as a local voice system, which possibly can obviate the need for centrex, key system, or PBX. It offers an ISDN BRI voice interface and supports an array of call-handling capabilities for voice connections. It can be used in tie-line and ring-down modes. It also can support dual-tone multifrequency (DTMF) digit-based per-call switching using dialed digits to select destination sites and network calls.

The multiservice concentrator supports transparent common channel signaling (CCS) and the Q.SIG voice signaling protocol. Q.SIG is a form of CCS that is based on ISDN Q.931, the signaling method used by the D-channel for call setup and teardown. Q.SIG provides transparent support for supplementary PBX services so that proprietary PBX features are not lost when connecting PBXs to networks composed of multiservice concentrators.

In addition to voice calls, the multiservice concentrator also supports both circuit- and packet-mode video. Circuit video is transported bit by bit through circuit emulation over a constant-bit-rate (CBR) ATM connection. Packet video can be supported over a variable-bit-rate (VBR) ATM connection or over the LAN, through the router engine, and over an unspecified-bit-rate (UBR) connection.

The multiservice concentrator is also compatible with the drop-and-insert capability of digital cross-connect systems (DCS) used on the PSTN. Drop and insert refers to the software-defined capability of the DCS to exchange channels from one digital facility to another, either to implement appropriate routing of the traffic, to reroute traffic around failed facilities, or to increase the efficiency of all the available digital facilities. Accordingly, the multiservice concentrator allows some time slots of a T1 facility to be used for on-net traffic and services, while the rest can be dropped/inserted off-net for transport over the PSTN when necessary.

Multiservice LAN Switches

A multiservice LAN switch implements telephony in the LAN and provides seamless integration with campus and WAN systems. In addition to a range of connectivity options and network services, this type of switch also provides an IP telephone and call manager to enable organizations to build corporate intranets for multicast, mission-critical, and voice applications. These switches also offer redundancy and topology resiliency for high availability and Gigabit Ethernet and ATM interfacing for high performance. In addition, they reduce complexity with application awareness and policy classification, which eliminates the need for network managers and administrators to engage in detailed configuration of QoS parameters.

Such switches provide layer 3 routing capabilities over Fast and Gigabit Ethernets, T1 to OC-48 ATM, and packet over SONET (PoS) uplinks. The ATM switching capabilities provide campus backbones with the means to integrate data, voice, and video traffic via such methods as ATM LAN emulation (LANE), multiprotocol over ATM (MPOA), or multiprotocol labeling-switching (MPLS)/tag-switching networks.

LANE is a layer 2 bridging protocol that makes a connection-oriented ATM network appear to higher-layer protocols and applications as just another connectionless LAN segment. Implemented at a central connection point, such as a communications server or LAN switch, LANE provides a means to migrate today's legacy networks toward ATM networks without requiring that existing protocols and applications on the LAN be modified to make them "ATM aware." The scheme supports backbone implementations, directly attached ATM servers and hosts, and high-performance, scalable computing workgroups. By defining multiple emulated LANs across an ATM network, switched virtual LANs can be created using MAC addresses for improved security and greater configuration flexibility. Additional benefits include minimal latency for real-time applications and QoS for emulated LANs.

MPOA preserves the benefits of LAN emulation while allowing intersubnet, internetwork communication over ATM virtual circuits without requiring routers in the data path. This framework synthesizes bridging and routing with ATM in an environment of diverse protocols, network technologies, and virtual LANs. MPOA is capable of using both routing and bridging information to locate the optimal exit from the ATM cloud. It allows the physical separation of internetwork layer route calculation and forwarding, a technique known as "virtual routing." This separation allows efficient intersubnet communication. It enhances manageability by decreasing the number of devices that must be configured to perform internetwork layer route calculation. It also reduces the number of devices participating in internetwork layer route calculation and eliminates the need for edge devices to perform internetwork layer route calculation.

MPLS, based on Cisco Systems' tag switching, is an IETF standard for IP service delivery. MPLS labels or "tags" provide the ability to differentiate service classes for individual data flows. The tags work like address labels on packages in an express delivery system—they expedite packet delivery on large corporate enterprise networks, allowing for the creation of faster, lower-latency intranets that can effectively support data, voice, and video on a common network infrastructure.

Configuring and deploying QoS policies are achieved through a policy manager. The policy manager's graphical user interface enables network administrators to define traffic classification and QoS enforcement policies. This system includes a rules-based policy builder, integrated policy validation and error reporting, and a policy distribution manager that downloads policies to the various network devices. This enables network administrators to quickly apply a mix of QoS policy objectives that protect business-critical application performance. In the process, organizations can more easily make the transition from unconstrained bandwidth utilization toward more consistent application performance over currently available bandwidth. This is not only increasingly necessary to eliminate unpredictable performance for business-critical applications running in corporate networks, but it is required to integrate data, voice, and video over a common network infrastructure.

Multiservice Routers

Multiservice routers provide organizations with the flexibility to meet the constantly changing requirements at the core and distribution points of their networks. Typically, these routers deliver up to 1 Gbps of throughput over a midplane that provides the ability to switch DS0 time slots between multichannel T1 interfaces, much like a TDM multiplexer. At the same time, such routers provide digital voice connectivity via an ATM circuit emulation service module. Together these capabilities allow the router to be connected to an ATM network on one side and to the TDM network on the other side.

Integral networking software provides routing and bridging functions for a wide variety of protocols and network media, including any combination of Ethernet, Fast Ethernet, Token Ring, fiber distributed data interface (FDDI), ATM, serial, ISDN, and high-speed serial interface (HSSI). Port and service adapters are connected to the router's peripheral component interconnect (PCI) buses, enabling connection to external networks.

Multiservice routers generally are equipped with the most advanced reliability features. For example, software-defined configuration changes take effect without rebooting or interrupting network applications and services. Port adapters and service adapters can be inserted and removed while the system is online. An automatic reconfiguration capability enables seamless upgrades to higher density and new port adapters without the need for rebooting, taking the system offline, or manual intervention. Dual hot-swappable, load-sharing power supplies provide system power redundancy; if one power supply or power source fails or is taken off line, the other power supply maintains system power without interruption. Alerts are issued when potentially problematic system fluctuations occur before they become critical, thereby enabling resolution while the system remains online.

Multiservice Edge Switches

Multiservice edge switches provide data, voice, and video integration over the wide area. They extend the WAN backbone to branch offices, providing high service levels regardless of location. They internetwork between existing routers and LAN switches to provide seamless traffic flow between LAN/campus and WAN domains. By combining ATM's dynamic bandwidth management with queuing techniques, the edge switches minimize recurring WAN bandwidth costs while ensuring fairness and high QoS for individual applications. The switches support legacy applications, large-scale packet voice, frame relay, and ATM and provide integrated LAN interfaces. Internetworking and management functions between the edge switch and LAN switching, routing, and branch-office devices improve end-to-end network performance and QoS.

Port interfaces ranging from 1.2 kbps to 155 Mbps are typically supported, as are network or trunk interfaces ranging from T1 to OC-3. Advanced traffic-management features and multiplexing techniques deliver high levels of bandwidth utilization and efficiency. The voice compression, silence suppression, and fax relay capabilities deliver additional bandwidth savings.

In being able to consolidate multiple WAN infrastructures, organizations have the flexibility to deploy voice over IP, voice over ATM, or voice over frame relay. Full control over network resources is exercised in a variety of ways—with per virtual circuit (VC) queuing, per VC rate scheduling, and multiple classes of service (CoS)—guaranteeing QoS levels for the individual applications. This enables all applications to be supported according to their specific requirements using advanced traffic-management and CoS features.

Access devices can be connected to the backbone through a leased line or though public frame relay or ATM WANs. Depending on the modules used to connect to the access device, different levels of internetworking are possible. Data applications can be based on Ethernet or Token Ring LANs, as well as frame relay, ATM, and legacy protocols. Voice applications can use different transport technologies. In addition to VoIP, the edge switch supports voice over frame relay (VoFR) and voice transport over ATM (VToA). Dial-up frame relay, SNA networking, and frame forwarding are also supported. For locations that need ATM broadband speeds, the systems offer frame relay to ATM service internetworking.

Frame-to-ATM service internetworking entails segmenting and mapping variable-length frame relay frames into fixed-length ATM cells. This enables the switch to provide transparent connectivity between large ATM and small frame relay locations. Small sites that do not have enough traffic to justify a dedicated access connection can connect to a frame relay network on a demand basis using switched (dial-up) lines.

Through frame forwarding, the edge switch can transport frame-based protocols, such as SDLC, X.25, or any other HDLC-based protocol, at speeds ranging from 9.6 kbps to 16 Mbps. With frame forwarding, all valid HDLC frames are forwarded or tunneled from one frame relay port to another port without frame relay header processing or local management interface (LMI) control. This method of transport results in efficient bandwidth usage and low latency on the corporate network.

Circuit data services are provided by the edge switch for transport of asynchronous or synchronous circuit data or video, which is transparently carried through a fixed-delay, fixed-throughput, zero-discard, point-to-point data connection across the WAN. These capabilities enable the transport of both legacy and TDM traffic and facilitate the migration to future ATM networks. The reliability of multiservice edge switches is enhanced by common equipment that can be configured for redundancy. New software releases can be remotely downloaded onto the redundant processor for background installation while traffic continues to run. Advanced distributed intelligence algorithms enable the network to automatically route new connections and, if necessary, reroute traffic around failures in network facilities.

Summary

Multiservice networking has emerged as a strategically important issue for both companies and carriers. It entails the convergence of all types of communications—data, voice, and video—over a single packet- or cell-based infrastructure. The benefits of multiservice networking include reduced operational costs; higher performance; greater flexibility, integration, and control, as well as faster deployment of new applications and services than can otherwise be achieved over traditional voice-oriented PSTNs or even TDM-based leased-line private networks. Organizations are interested in data, voice, and video integration for short-term cost savings. But this integration also meets medium-term requirements for the support of emerging applications and leads to the long-term objectives of reducing complexity and network redundancy through technology convergence.

See also

Internet Telephony
Inverse Multiplexers

N

NETWORK AGENTS

Network agents are special programs that accomplish specified tasks by executing commands remotely. Network managers can create and use intelligent agents to execute critical processes, including performance monitoring, fault detection and correction, and asset management. The agent-manager concept is not new. The manager-agent relationship is intrinsic to most standard network management protocols, including the Simple Network Management Protocol (SNMP) used to manage Transmision Control Protocol/Internet Protocol (TCP/IP) networks. In fact, SNMP agents are widely available for all kinds of network devices, including bridges, routers, hubs, multiplexers, and switches.

In the SNMP world, agents respond to polls from a management station on the operational status of the various devices on the network. Depending on the information returned, agents can then be directed by the management station to get more data, set performance variables, or generate traps when specified events occur. However, to retrieve the collected data, the agents must be polled by central management software, a process that increases network traffic. On wide area networks (WANs), which are being increasingly burdened with multimedia and other delay-sensitive applications, traffic from continuous polling and the resulting data transfers can degrade network performance. So-called intelligent agents address this problem.

What makes these agents so smart is the addition of programming code containing rules that tell them exactly what to do, how to do it, and when to do it. In essence, the intelligent agent plays the dual role of manager and agent. Under this rules-based scheme, polling is localized, events and alarms are collected and correlated, various tasks are automated, and only the most relevant information is forwarded to the central management station (Figure N-1) for analysis.

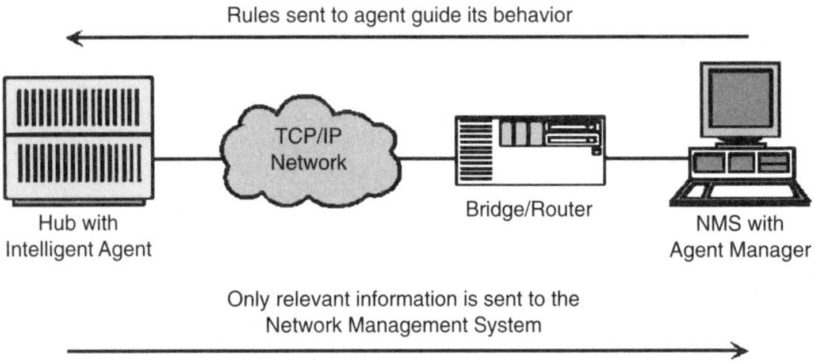

Figure N-1 Rules sent to the agent tell it what to do and when to do it so that only relevant information is sent to the network management system.

In the process, network traffic is greatly reduced, and problems are resolved faster.

Network Agent Applications

Agent technology has been available for several years and still represents one of the fastest growing areas in network management. In a global economy that encourages the expansion of networks to reach new markets and discourages the addition of personnel to minimize operating costs, it makes sense to automate as many management tasks as possible through the use of intelligent agents. In recognition of these new business realities, the list of tasks that are being handled by network agents is continually growing.

Performance Management Network performance monitoring can help determine network service-level objectives by providing measurements to help managers understand typical network behavior in normal periods. The following capabilities of intelligent agents are particularly useful for building a network performance profile:

- *Baselining and network trending* Identifies the true parameters of the network by defining typical and normal behavior. Baselining also provides long-term measurements to check service-level objectives and show out-of-norm conditions that, if left unchecked, can diminish the productivity of network users.
- *Application usage and analysis* Identifies the overall load of network traffic, what times of the day certain applications load the network, which applications are running between critical servers and clients, and what their load is throughout the day, week, and month.
- *Client-server performance analysis* Identifies which servers may be overutilized, which clients are hogging server resources, and what applications or protocols they are running.
- *Internetwork performance* Identifies traffic rates between subnets so that the network manager can determine which nodes are using WAN links to communicate. This information can be used to define typical throughput rates between interconnected devices.
- *Data correlation* Allows peak network usage intervals to be selected throughout the day to determine which nodes were contributing most to the network load. Traffic source and associated destinations can be determined with seven-layer protocol identification.

Fault Management When faults occur on the network, problems must be resolved quickly to decrease the negative impact on user productivity. The following capabilities of intelligent agents can be used to gather and sort the data needed to quickly identify the cause of faults on the network:

- *Packet interrogation* Isolates the actual session that is causing the network problem, allowing the network manager to assess the nature of the problem quickly.
- *Data correlation* Since managers cannot always be on constant watch for network faults, it is important to have historical data available that provides views of key network metrics at the time of the fault. Such metrics can be used to answer questions like: What is the overall error/packet rate and the types of errors that occurred? What applications were running at the time of the fault? Which servers were most active? Which clients were accessing these active servers, and which applications were they running?
- *Identification of top error generators* Identifies the network nodes that are generating the faults and contributing to problems such as bottlenecks caused by errors and network downtime.
- *Immediate fault notification* With immediate notification of network faults, managers can instantly learn when a problem has occurred before users do. Proactive alarms help detect and solve the problem as it is happening.
- *Automated resolution procedures* The intelligent agents can be configured to automatically fix a problem when it occurs. The agent can even be programmed to automatically e-mail or notify help desk personnel with on-screen instructions on how to solve the problem.

Capacity Planning and Reporting Capacity planning and reporting allow for the collection and evaluation of information to make informed decisions about how to respond to network growth. For this purpose, the following capabilities of intelligent agents are useful:

- *Baselining* This capability allows the network manager to determine the true operating parameters of the network against which future performance can be measured.
- *Load balancing* Load-balancing capabilities allow the network manager to compare internetwork service objectives from multiple sites at once to determine which subnets are over- or underutilized. It also helps the network manager discover which subnets can sustain increased growth and which require immediate attention for possible upgrade.
- *Protocol/application distribution* Protocol and application distribution capabilities can help the network manager understand which applications have outgrown which domains or subnets. For example, these capabilities can find out if certain applications are continuously taking up more precious bandwidth and resources throughout the enterprise. With this kind of information, the network manager can better plan for the future.
- *Host load balancing* Allows the network manager to obtain a list of the top network-wide servers and clients using mission-critical applications. For example, the information collected from intelligent agents might reveal if specific servers always dominate precious local area network (LAN) or WAN bandwidth or spot when a CPU is becoming overloaded. In either case, an agent on the LAN segment, WAN device, or host can initiate load balancing automatically when predefined performance thresholds are met.
- *Traffic profile optimization* To ensure adequate service-level performance, the ability of network managers to compare actual network configurations against proposed configurations is very valuable. From the information gathered and reported by intelligent agents, traffic profiles can be developed that allow "what if" scenarios to be put together and tested before incurring the cost of physically redesigning the network.

A growing number of capacity planning and reporting tools have become available in recent years. One is Nortel Networks' Optivity Enterprise, a family of network management products that includes the Optivity Design and Analysis suite of network design and optimization applications for Ethernet and Token Ring environments. Among the tools available in this suite is DesignMan, which performs simulation activities using live traffic information gathered by embedded management agents on the network (Figure N-2).

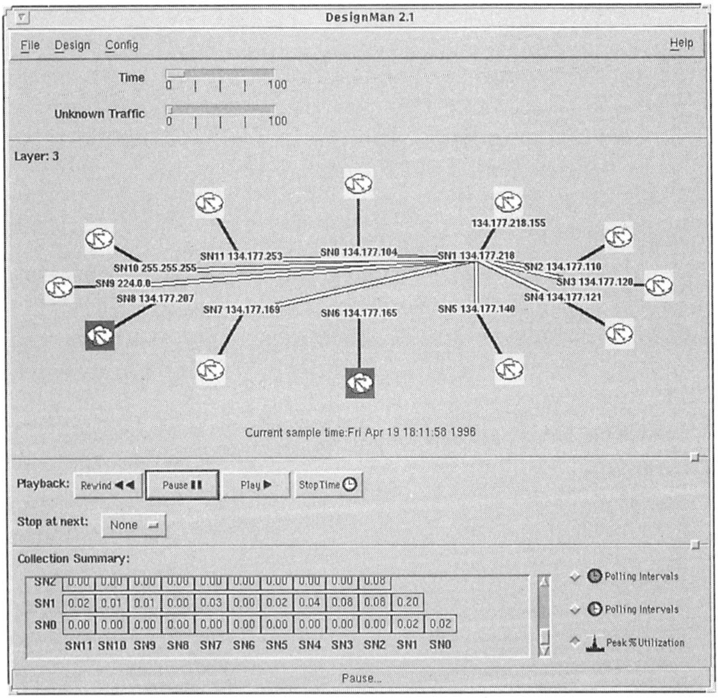

Figure N-2 Operating at layer 3 (network), Nortel Networks' DesignMan application shows the traffic flow between logically connected subnets. The user can apply what-if scenarios to the traffic data collected by intelligent agents to see the effect of moving a server, for example, from one subnet to another. The application uses a VCR metaphor, allowing the user play, pause, stop, and rewind the scenario to view its impact on the entire network.

Security Management A properly functioning and secure corporate network plays a key role in maintaining an organization's competitive advantage. Setting up security objectives related to network access must be considered before mission-critical applications are run over "untrusted" networks, particularly the Internet. The following capabilities of network agents can help discover holes in security by continuously monitoring access attempts:

- *Monitor effects of firewall configurations* By monitoring firewall traffic, the network manager can determine if the firewall is functioning properly. For example, if the firewall was just programmed to disallow access to a corporate host via Telnet but the program's syntax is wrong, the intelligent agent will report this fact immediately.
- *Show access to and from secure subnets* By monitoring access from internal and external sites to secure data centers or subnets, the network manager can set up security service-level objectives and firewall configurations based on the findings. For example, the information reported by the intelligent agent can be used to determine what external sites should have access to the company's database servers or legacy hosts.
- *Trigger packet capture of network security signatures* Intelligent agents can be set up to issue alarms and automatically capture packets on the occurrence of external intrusions or unauthorized application access. This information can be used to track down the source of security breaches. Some network agents even have the ability to initiate a trace procedure to discover a breach's point of origination.
- *Show access to secure servers and nodes with data correlation* This capability reveals which external or internal nodes are accessing potentially secure servers or nodes and identifies which applications they are running.
- *Show applications running on secure nets with application monitoring* This capability evaluates applications

and protocol use on secure networks or traffic components to and from secure nodes.

- *Watch protocol and application use throughout the enterprise* This capability allows the network manager to select applications or protocols for monitoring by the intelligent agent so that the flow of information throughout the enterprise can be viewed. This information can identify who is browsing the Web, accessing database client-server applications, or running Notes, for example.

Intelligent agents can be used for a variety of other tasks, including Internet-related tasks. They can monitor information logged by servers on the Web, for example. When the log entries exceed a designated threshold, it may indicate a high demand for applications and impending congestion if the logging rate continues. An intelligent agent can act on this information to redirect traffic to another server to balance the load across the available Web servers.

Applications Management There are now client-side agents that continuously monitor the performance and availability of applications from the end user's perspective. A "just in time" application's performance management capability captures detailed diagnostic information at the precise moment that problem or performance degradations occur, pinpointing the source of the problem so that it can be resolved immediately. Such agents are installed on clients as well as on application servers. They monitor every transaction that crosses the user desktop, traversing networks, application servers, and database servers. They monitor all distributed applications and environmental conditions in real time, comparing actual availability and performance with service-level thresholds.

Via a management console, a window is provided into application availability and performance throughout the enterprise. Via the console, IT personnel can identify which users are experiencing problems and then drill down to view successive layers of problem and diagnostic detail. Through the console, the IT administrator also can define service-level performance thresholds, specify automated corrective action plans, and fine-tune data collection and reporting. A repository stores all exception and historical end-user application usage data in a standard stuctured query language (SQL) database.

Remote User Support With the increasing number of mobile professionals and telecommuters—most having no permanent connections to a LAN or WAN—IT administrators are faced with the challenge of managing this growing base of unattached computers. They are tasked with ensuring that each system is working properly, configured to corporate standards, running the right versions of the right software, and functioning reliably. However, typical management solutions are not proactive and often involve long-distance calls to the help desk, user downtime, and shipping costs for sending a system in for upgrades or repairs.

To deal with the problems of providing remote users with support, there is agent software that can be installed on the client that gives administrators a presence on each machine, regardless of its location. With this capability, an administrator can define profiles for these out-of-reach PCs and how they should be configured, what software must be on the hard drive, and how often they should run programs like diagnostics and virus checks.

The server component communicates with each and every PC, comparing the correct configuration profile with the computer's current profile. It notes exactly which files and scripts a particular client system needs, requests only those files, and downloads and installs them. This exchange between the client and server continues until the target system exactly fits the defined configuration. The agent software even checks periodically to make sure the configuration stays the same, alerting the administrator of any unauthorized changes or system problems.

Communication between the server component and the remote clients takes place through e-mail. Since remote users typically check e-mail several times a day, the agent can report on the status of the client several times a day as well. The agent can even be instructed to log onto a File Transfer Protocol (FTP) server and download files. If the transmission is interrupted, the agent can pick up the installation right where it left off during the next logon.

The agent does not hog limited bandwidth to perform its tasks. It breaks the data packets into smaller pieces and sends them encrypted, one by one, to optimize bandwidth. The IT administrator can determine the size of the packets before they are compressed and transmitted sequentially. The agent gathers all the packets at the client side, whether they arrive over one e-mail session or multiple sessions. It decrypts and reassembles them into standard-sized packets before implementation. Thus a large file, such as a word processing program, can be sent over a period of days before being installed.

This method of providing remote support is nonintrusive—the user is not interrupted and may never even know what is going on in the background. The agent takes remote users out of the loop so that IT administrators will not have to deal with resistance or noncompliance from busy users.

WAN Service-Level Management Despite the migration from private WANs to public WANs in recent years, network managers are still accountable for overall network performance. Consequently, they must ensure that their carrier is equally concerned about service quality. An effective approach for ensuring service quality is to implement the relatively new concept of WAN service-level management, a collaborative effort between subscriber and service provider to manage the service quality of public network services. In this arrangement, both the carrier and the customer work together to plan, monitor, and troubleshoot WAN service quality.

WAN service-level management offers a number of benefits. Subscribers can increase network availability and performance, reduce the need for recurring support, and ensure that business needs are met at the lowest possible cost. Service providers can reduce operational support costs, prioritize response to alarms, issue trouble tickets, set expectations for service quality, and help justify recommendations to upgrade bandwidth.

To ensure the success of WAN service-level management requires historical data and the collection of WAN service quality information to arrive at baseline performance metrics. Normally, these are difficult tasks, but today's network agents automate the collection, interpretation, and presentation of WAN service-level information, making it easier to monitor and verify the performance of carrier-provided services.

Summary

Over the years, intelligent agents have proved to be indispensable tools for providing network management assistance. Problems can be identified and resolved locally by network agents rather than by harried operators at a central management console or by sending technicians to remote locations—both of which are expensive and time-consuming. In many cases, intelligent agents can implement restorations automatically in response to specified events. These actions can be as simple as resetting a device by turning it off and then back on. Other times the restoration might consist of balancing the load across multiple lines or servers to avoid impending congestion. Agents will become even more indispensable as networks continue to expand to international locations. In today's global economy, having the capability to effectively monitor remote systems and networks becomes even more important, especially when organizations are under pressure to minimize staff to reduce operating costs.

See also

Simple Network Management Protocol

NETWORK DIRECTORY SERVICES

Network directory services provide an easy way for users to access resources and find other people on the network. Applications such as e-mail, facsimile, personal information managers (PIMs), personnel management systems, messaging products, and numerous others all come with directories to facilitate user communication. When an individual or a company or needs a new application, it probably will need a directory to manage users, user groups, routing, security, and other information. With each type of product having its own directory structure, the only way to keep the content consistent is to manually enter the same information into each one. This wastes valuable time, increases the chance of error, and interferes with productivity.

Maintaining directory information for multiple applications is a costly and burdensome chore for most administrators and organizations. The Lightweight Directory Access Protocol (LDAP) is intended to provide a common method of accessing server directories, and it enables directories to be extended across intranets and the Internet, allowing them to be accessed by e-mail applications and Web browsers.

LDAP is based on the standards contained within the international X.500 Standard but is significantly simpler. And unlike X.500, LDAP supports TCP/IP, which is necessary for any type of Internet access. Because it is a simpler version of X.500, LDAP is sometimes referred to as "X.500-lite."

To enable the LDAP to run directly over the TCP/IP stack, it had to shed many of X.500's overhead functions. However, LDAP makes up for this loss of power in the following ways:

- Whereas X.500 requires special network access software, LDAP was designed to run over TCP, making it ideal for Internet and intranet applications.
- LDAP has simpler functions, making it easier and less expensive for vendors to implement.

- LDAP encodes its protocol elements in a less complex way than X.500, thereby streamlining coding/decoding of requests.
- LDAP servers return only results or errors, which lightens their processing burden.
- LDAP servers take responsibility for "referrals" by handing off the request to the appropriate network resource. X.500 returns this information to the client, which must then issue a new search request.

Although LDAP enjoys widespread industry support, there is incompatibility among LDAP-compliant applications because the standard does not specify a consistent naming scheme for accessing directories by such fields as name, address, phone number, and e-mail address. Thus vendors have been using different ways for storing and maintaining this information. This problem has been addressed by the Lightweight Internet Person Schema (LIPS).

Lightweight Internet Person Schema

LIPS is designed to ensure easier implementation of LDAP through the definition of common terms for attribute names and content. For example, a messaging client may want to browse an LDAP directory to retrieve a name and phone number. Without LIPS, one server could define "phone number" as a field called PHONE with a length of 10 characters, and another vendor could define the field as BUS_PHNE with a length of 20 characters to accommodate international numbers. LIPS solves this problem by defining the field name, size, and acceptable characters (syntax) for 37 common attributes.

This is not intended to be an exhaustive list of attributes; in fact, most directories have far more than 37 fields. LIPS presents a baseline schema containing only the minimum number of common fields that loosely define an individual.

By adhering to these standardized attributes, client software vendors can build server-independent products using the LDAP standard. To be fully compliant, a vendor must expose all the LIPS attributes with the given field names and minimum sizes (larger values are allowed). However, there is no requirement that the attributes contain any data.

LIPS is not designed to be a server-to-server synchronization solution; it only defines how the data are presented to a client. There is no facility for initiating a server-to-server connection and replicating information, nor is that planned in the future. However, products that employ LDAP and LIPS can be used to perform server-to-server directory synchronization.

XML for Directory Access

An eXtensible Markup Language (XML)–based standard for directory access is available that defines how applications running on the Web or mobile devices can access a directory without needing a special client, as required with LDAP. This means that a cell phone or personal digital assistant (PDA) can use XML to access a directory instead of requiring it to have a bulkier LDAP implementation on that client.

Directory Services Markup Language (DSML) provides a standard way for a client application to read, query, update, and search a directory. DSML also simplifies application creation because developers can write exclusively in the XML without having to know LDAP.

DSML eliminates the special-purpose client code. Although it standardizes basic directory functions such as query and update, it does not address user identification and authorization or chaining, the act of stringing directories together. The specification supports referrals, however, which lets one directory refer queries to another directory. The advantage of wrapping XML around LDAP is that vendors and companies do not have to reinvent their current LDAP products. Every directory vendor supports LDAP today, and DSML merely adds a more efficient way to deliver queries to their directories. This results in a broader reach of directory services to a new level of client applications.

Summary

LDAP offers a method of accessing directories, making it possible for almost any application running on virtually any computer platform to obtain directory information, such as e-mail addresses and public keys. Because LDAP is an open protocol, applications do not have to be tailored to the specific type of server hosting the directory. DSML improves directory services in that it does not require the client device, which may be memory constrained, to run an LDAP client. For these devices, DSML provides a more efficient way to deliver queries to directories on the network.

See also

Electronic Mail

NETWORK INTERFACE CARDS

A network interface card (NIC) is an adapter that plugs into a computer, enabling it to connect to a LAN for the purpose of communicating with other computers and devices. The NICs are network-specific—there are adapters for Ethernet, Token Ring, fiber distributed data interface (FDDI), Asynchronous Transfer Mode (ATM), and other types of networks. NICs are also media-specific—there are adapters for shielded and unshielded twisted-pair wiring, thick and thin coaxial cabling, and single-mode and multimode optical fiber. NICs also are bus-specific—there are adapters for the Industry Standard Architecture (ISA), Extension to Industry Standard Architecture (EISA), Micro-Channel Architecture (MCA), and Peripheral Component Interconnect (PCI/PCI-X) architectures. NICs are also available in the PC card form factor for connecting mobile notebook users to the LAN. Major vendors offer software that adds management capabilities of their NICs.

MAC Addresses

Devices on conventional LANs like Ethernet and Token Ring use Media Access Control (MAC) addresses. These are the 6-byte hardware-level addresses of the NICs that provide workstations and other devices with the means to interconnect with each other through a hub or switch. An example of a MAC address is

00 00 0C 00 00 01

The first 3 bytes contain a manufacturer code (the one above is for Cisco Systems); the last 3 bytes contain a unique station ID that are burned into the NIC's firmware. Manufacturer IDs are assigned by the Institute of Electrical and Electronics Engineers (IEEE).

These addresses provide the means to implement virtual LANs (VLANs). This technology lets the same server link—via one NIC or a team of NICs—carry traffic for up to 64 logical subgroups created via software. This capability provides additional bandwidth-management and security features and helps reduce the administrative overhead required to manage workstation moves and changes.

Client NICs

Client NICs provide the means to connect desktop computers, printers, and other devices to the LAN. Today's Ethernet and Token Ring NICs have an autosensing capability that allows the NIC, when connected to a switch or hub port, to automatically sense and connect at the highest network speed. NICs are available at different speeds. With Ethernet, for example, there are 10BaseT, 100BaseT, and 1000BaseT cards. Some cards support 10/100BaseT or combine all three Ethernet speeds onto the same card. By simultaneously performing multiple processing tasks, some NICs provide the fastest data transfer speeds available for the peripheral component interconnect (PCI) bus.

NICs that feature 32-bit multimaster concurrency technology permit the card to communicate directly with the computer's CPU, bypassing sluggish interrupts and I/O channels. NICs that feature an onboard boot ROM socket allow for remote workstation boot-up from the server. Light-emitting diodes (LEDs) on the card report link status, packet activity, transmission speed, and transmission mode (half or full duplex).

Many NICs are optimized to work in specific operating environments. For example, NICs for Windows environments—specifically PCs running Windows 95/98/XP or NT/2000—and are plug-and-play compliant. The installation software allows connection of the PC to Novell NetWare networks as well. Windows-based diagnostics and configuration utilities facilitate installation and troubleshooting. Different NICs are available for the Macintosh and some UNIX environments.

Server NICs

Server NICs include the functions of client NICs but have additional functionality and provide higher bandwidth. For example, NICs may be configured in a way to increase the fault tolerance of the server's LAN link. If one NIC fails, the fail-over software deactivates the faulty NIC and switches LAN traffic to an alternate card. The rerouting takes place almost instantaneously, without human intervention. The software also gives Simple Network Management Protocol (SNMP) alerts on the failed NICs. When the failed NIC begins working again, the software brings it back into the array automatically and starts balancing traffic across it again.

Many server NICs provide asymmetric port aggregation—also referred to as "asymmetric load balancing" or "asymmetric trunking." This technology distributes outbound server traffic between two or more cards, providing a wider data pipe. The NICs operate together and appear as a single device with one network address. Asymmetric port aggregation is especially useful for Web servers, e-mail servers, and other applications where most of the traffic flows in one direction from the server to the client PCs. This method also provides fault tolerance in that if one of the NICs fails, the others take its load.

A companion technology is symmetric port aggregation (or symmetric load balancing or symmetric trunking). This method combines two or more connections into a wider pipe that can transmit data in both directions. Since combining several 100-Mbps connections does not require replacing hubs or switches, a user can put off having to invest in upgrades to gigabit LAN technologies. A dual homing capability allows the server NIC to connect to different switches for additional redundancy, ensuring that the server remains available even if one of the attached switches fails.

There are also server NICs that allow Token Ring traffic to run over a 100-Mbps Fast Ethernet backbone connection, eliminating congestion across the Token Ring network backbone without the costs associated with ATM and FDDI. The NIC tunnels Token Ring traffic in Fast Ethernet frames, delivering high-speed performance to Token Ring clients. When installed in a Fast Ethernet server, the NIC allows Token Ring clients to communicate with the server at Fast Ethernet speeds via a special module installed in a Token Ring switch.

Summary

NICs are used to connect computers and other devices to the LAN. Although many users can get by with inexpensive "dumb" NICs costing under $20 (U.S.), enterprise networks require "intelligent" NICs that ensure high availability to support mission-critical applications and management software to facilitate monitoring and control from a central location.

See also

Transceivers

NETWORK SECURITY

Protecting vital information has always been a high-priority concern among companies. While access to distributed data networks improves productivity by making applications, processing power, and mass storage readily available to a large and growing user population, it also makes those resources more vulnerable to abuse and misuse. Among the risks are unauthorized access to mission-critical data, information theft, and malicious file tampering that can result in immediate financial loss and, in the long term, loss of customer confidence and damage to competitive position. However, various protective measures can be taken to safeguard information in transit as well as information stored at various points on the network, including servers and desktop computers.

Physical Security

Protecting data in distributed environments starts with securing the premises. Such precautions as locking office doors and wiring closets, restricting access to the data center, and having employees register when they enter sensitive areas can greatly reduce risk. Issuing badges to visitors, installing electronic locks on doors, providing visitor escorts, and having a security guard station in the lobby can reduce risk even further.

Other measures such as keyboard and disk-drive locks are also effective in deterring unauthorized access to unattended workstations. These are important security features, especially since some workstations may provide management access to wiring hubs, LAN servers, bridge-routers, and other network access points. In addition, locking down workstations to desks can help protect against equipment theft. Some companies even collect unattended laptops in the office after business hours when building maintenance crews start their shift.

Access Controls

Access controls can prevent unauthorized local access to the network and control remote access through dial-up ports. Network administrators can assign multiple levels of access to different users based on need: public, private, and shared access. Public access allows all users to have read-only access to file information. Private access gives specific users read-and-write file access, while shared access allows all users to read and write to files.

When a company offers network access to one or more databases, it should restrict and control all user query operations. Each database should have a protective "key," or series of steps, known only to those individuals entitled to access the data. To ensure that intruders cannot duplicate the data from the system files, users should first have to sign on with passwords and then prove that they are entitled to access the data by responding to a challenge with a predefined response known only to that person. This is the basis of a security procedure known as "personal authentication."

Logon Security

Network operating systems or add-on software can offer effective logon security, which requires that the user enter a logon ID and password to access local or remote systems. Passwords not only can identify the user, but they also can associate the user with a specific workstation, as well as a designated shift, workgroup, or department. The effectiveness of this measure hinges on users' ability to maintain password confidentiality.

A user ID should be suspended after a certain number of passwords have been entered to thwart trial-and-error attempts at access. Changing passwords frequently—especially when key personnel leave the company—and using a multilevel password-protection scheme can enhance security. With multilevel passwords, users can gain access to a designated security level, as well as all lower levels. With specific passwords, on the other hand, users can access only the intended level and not the others above or below. Finally, users should not be allowed to make up their own passwords; they should be assigned using a random password generator or compared to a dictionary to weed out guessable passwords. Although such schemes entail an increased administrative burden, the effort is usually worthwhile.

The effectiveness of passwords can be enhanced by using them in combination with other control measures, such as keyboard lock or card reader. Biometric devices also may be used, which identify an authorized user based on such characteristics as a handprint, voice pattern, or the layout of capillary blood vessels in the retina of the eye. Of course, the choice of control measure will depend on the level of security desired and budgetary considerations.

Data Encryption

To protect data (and voice) as they traverse the network requires that they be scrambled with an encryption algorithm. One of the most effective encryption algorithms is that offered by Pretty Good Privacy (PGP), a method that uses a public key to protect computer and e-mail data. The program generates two keys that belong uniquely to the user. One PGP key is secret and stays in the user's computer. The other key is public and is given out to people the user wants to communicate with. The public key can be distributed as part of the message.

PGP does more than encrypt. It has the ability to produce digital signatures, allowing the user to "sign" and authenticate messages. A digital signature is a unique mathematical function derived from the message being sent. A message is signed by applying the secret key to it before it is sent. By checking the digital signature of a message, the recipient can make sure that the message has not been altered during transmission. The digital signature also can prove that a particular person originated the message. The signature is so reliable that not even the originator can deny creating it.

Firewalls

A firewall is a method of protecting one network from another untrusted network. The actual mechanism whereby this is accomplished varies widely, but in principle, the firewall can be thought of as a pair of mechanisms: one that blocks traffic and another that permits traffic. Some firewalls place a greater emphasis on blocking traffic, while others emphasize permitting traffic.

One way firewalls protect networks is through packet filtering, which can be used to restrict access from or to certain machines or sites. It xcan can also be used to limit access based on time of day or day of the week, the number of simultaneous sessions allowed, service host(s), destination host(s), or service type. In addition to dedicated firewall systems, this kind of functionality can be set up on various network routers, communications servers, or front-end processors.

Transparent proxies are also used to provide secure outbound communication to the Internet from the corporation's internal network. The firewall software achieves this by appearing to be the default router that provides access to the internal network. However, when packets hit the firewall, the software does not route the packets but immediately starts a dynamic, transparent proxy. The proxy connects to a special intermediate host that actually connects to the desired service.

Proxies are often used instead of router-based traffic controls to prevent traffic from passing directly between trusted and untrusted networks. Many proxies contain extra logging or support for user authentication. Since proxies must "understand" the application protocol being used, they also can implement protocol-specific security (e.g., an FTP proxy might be configurable to permit incoming FTP and block outgoing FTP).

Remote Access Security

With an increasingly decentralized and mobile workforce, organizations are coming to rely on remote access arrangements that enable telecommuters, traveling executives, salespeople, and branch offices to dial into the corporate network with an 800 number or a set of regional or local access numbers. Appropriate security measures can prevent unauthorized access to corporate resources from the remote access server. One or more of the following security methods can be employed:

- *Authentication* This involves verifying the remote caller by user ID and password, thus controlling access to the server. Security is enhanced if the ID and password are encrypted before going out over the communications link.
- *Access restrictions* This involves assigning each remote user a specific location (i.e., directory or drive) that can be accessed in the server. Access to specific servers also can be controlled.
- *Time restrictions* This involves assigning each remote user a specific amount of connection time, after which the connection is dropped.
- *Connection attempts* This involves limiting the number of consecutive connection attempts and/or the number of times connections can be established on an hourly or daily basis.

Among the most popular remote access security schemes are Remote Access Dial-n User Service (RADIUS) and Terminal Access Controller Access Control System+ (TACACS+). Of the two, RADIUS is the more popular. Users are authenticated through a series of communications between the client and the server. When the client initiates a connection, the communications server puts the name and password into a data packet called the "authentication request," which also includes information identifying the specific server sending the authentication request and the port that is being used for the connection. For added protection, the communications server, acting as a RADIUS client, encrypts the password before passing it on to the authentication server.

When an authentication request is received, the authentication server validates the request and decrypts the data packet to access the user name and password information. If the user name and password are correct, the authentication server sends back an authentication acknowledgment that includes information on the user's network system and service requirements. The acknowledgment can even contain filtering information to limit the user's access to specific network resources.

The older security system is TACACS, which has been updated by Cisco into a version called TACACS+. Although the protocols are different, the proprietary TACACS+ offers many of the same features as RADIUS but is used mainly on networks consisting of Cisco remote access servers and related products. Companies with mixed-vendor environments tend to prefer the more open RADIUS.

Callback Systems

Callback security systems are useful in remote access environments. When a user dials into the corporate network, the answering modem requests the caller's identification, disconnects the call, verifies the caller's identification against a directory, and then calls back the authorized modem at the number matching the caller's identification. This scheme is an effective way to ensure that data communication occurs only between authorized devices, more so when used in combination with data encryption.

Security procedures can even be implemented before the modem handshaking sequence rather than after it, as is usually the case. This effectively eliminates the access opportunity from potential intruders. This method uses a precision high-speed analog security sequence that is not detectable even by advanced line-monitoring equipment.

While these callback techniques work well for branch offices, most callback products are not appropriate for mobile users whose locations vary on a daily basis. Newer products accept roving callback numbers. This feature allows mobile users to call into a remote access server or host computer, type in their user ID and password, and then specify a number where the server or host should call them back. The callback number is then logged and may be used to help track down security breaches.

To safeguard very sensitive information, there are third-party authentication systems that can be added to the server. These systems require a user password and also a special credit card–sized device that generates a new ID every 60 seconds, which must be matched by a similar ID number-generation process on the remote user's computer.

Link Level Security

When peers at each end of a serial link support the Point-to-Point Protocol (PPP) suite, link level security features can be implemented. This is so because PPP can integrally support the Password Authentication Protocol (PAP) and Challenge Handshake Authentication Protocol (CHAP) to enforce link security. PPP is a versatile WAN connection standard that can be used for tying dispersed branch offices to the central backbone via dial-up serial links. It is actually an enhanced version of the older Serial Line Internet Protocol (SLIP). SLIP is limited to the IP-only environment, while PPP is used in multiprotocol environments. Since PPP is protocol-insensitive, it can be used to access AppleTalk, IPX and TCP/IP networks, for example.

PAP uses a two-way handshake for the peer to establish its identity. This handshake occurs only on initial link establishment. An ID-password pair is repeatedly sent by the peer to the authenticator until verification is acknowledged or the connection is terminated. However, passwords are sent over the circuit in text format, which offers no protection from interception and playback by network intruders.

CHAP periodically verifies the identity of the peer using a three-way handshake. This technique is employed throughout the life of the connection. With CHAP, the server sends a random token to the remote workstation. The token is encrypted with the user's password and sent back to the server. Then the server does a lookup to see if it recognizes the password. If the values match, the authentication is acknowledged; otherwise, the connection is terminated. Every time remote users dial in, they are given a different token. This provides protection against playback because the challenge value changes in every token.

Some vendors of remote-node products support both PAP and CHAP, while low-end products tend to support only PAP, which is the less robust of the two authentication protocols.

Policy-Based Security

With today's LAN administration tools, security goes far beyond mere password protection to include implementation of a policy-based approach characteristic of most mainframe systems. Under the policy-based approach to security, files are protected by their description in a relational database. This means that newly created files are automatically protected not at the discretion of each creator but consistent with the defined security needs of the organization.

Some products use a graphical calendar through which various assets can be made available to select users only during specific hours of specific days. For each asset or group of assets, a different permission type may be applied: permit, deny, or log. Permit allows a user or user group to have access to a specified asset. Deny allows an exception to be made to a permit, not allowing writes to certain files, for example. Log allows an asset to be accessed but stipulates that such access will be logged.

Although the LAN administrator usually has access to a full suite of password controls and tracking features, today's advanced administration tools also provide the ability to determine whether or not a single login ID can have multiple terminal sessions on the same system. Through the console, the LAN manager can review real-time and historical violation activity online, along with other system activity.

Summary

To protect valuable information, companies must establish a sound security policy before an intruder has an opportunity to violate the network and do serious damage. This means identifying security risks, implementing effective security measures, and educating users on the importance of following established security procedures. Despite advancements in security hardware and software, there are some threats no system can protect against, such as insider attacks or people taking sensitive information out of the building with floppy disks. According to some industry reports, 80 percent of attacks on corporate networks originate with employees.

See also

Firewalls
Proxy Servers

NETWORK STATISTICS

Network Statistics ("Netstat") is a utility that displays useful performance information about network connections and activity. It can be run from the command line of any machine with a network operating system that supports TCP/IP (Figure N-3). Netstat includes options that allow the user to specify the type of information to have displayed. Selected options are described in Table N-1.

To launch Netstat from the command line, the user simply types in the word "netstat" or "netstat" followed by one or more options as in the following examples:

```
netstat -a or netstat -ps
```

When `netstat -a` is run, for example, the following type of information is returned:

Proto	Local Address	Foreign Address	State
TCP	ws1-williamsg1:135	0.0.0.0:0	Listening
TCP	ws1-williamsg1:1029	0.0.0.0:0	Listening
TCP	ws1-williamsg1:1034	0.0.0.0:0	Listening
TCP	ws1-williamsg1:1035	0.0.0.0:0	Listening
TCP	ws1-williamsg1:3684	PNTEXCH01.hq.abc.net:1396	Established
TCP	ws1-williamsg1:3689	PNTEXCH01.hq.abc.net:1450	Established
TCP	ws1-williamsg1:3700	PNTEXCH01.hq.abc.net:1396	Established

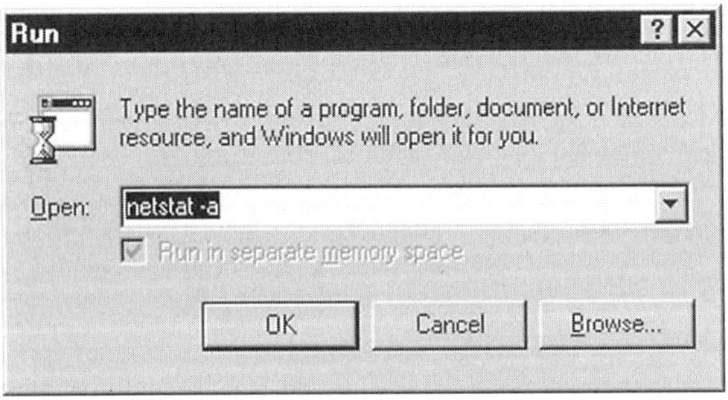

Figure N-3 The Netstat utility with the -a option set, as run from the command line in Windows NT.

Table N-1 Selected Options for the Netstat Utility, Which Can Be Run from the Command Line of Windows NT

Option	Description
-a	Listens for all active ports and displays connection information.
-e	Displays Ethernet statistics. This may be combined with the -s option.
-n	Displays addresses and port numbers in numerical form (rather than attempting name lookups).
-p	Shows connections for the specified protocol specified by Proto; Proto may be TCP or user datagram protocol (UDP). If used with the -s option to display per-protocol statistics, Proto may be TCP, UDP, internet control message protocol (ICMP), or IP.
-r	Displays the contents of the routing table.
-s	Displays per-protocol statistics. By default, statistics are shown for TCP, UDP, ICMP, and IP; the -p option may be used to specify a subset of the default.
Interval	Redisplays selected statistics, pausing interval seconds between each display. Press CTRL+C to stop redisplaying statistics. If this parameter is omitted, Netstat prints the current configuration information once.

The column labeled "Proto" indicates the type of protocol used on the connection. The column labeled "Local Address" refers to the local host and its sockets. The column labeled "Foreign Address" indicates the server connections, if any. The column labeled "State" refers to the status of the port as either listening for a connection or having established a connection.

Summary

Netstat is a useful tool for checking the connections and activity on a network. By setting the appropriate options (more are available than listed above), administrators and technicians can monitor the status of network connections, inspect interface configuration information, examine the routing table, and retrieve operational statistics for various network protocols.

See also

Ping

P

PERFORMANCE BASELINING

Performance baselining is a procedure for understanding the behavior of a properly functioning network so that deviations can help identify the cause of problems that may occur in the future. The only way to know a network's normal behavior is to analyze it while it is operating properly. Later, technicians and network managers can compare data from the properly functioning network with data gathered after conditions have begun to deteriorate. This comparison often points to the right steps that lead to a corrective solution.

Information Requirements

The first step in baselining performance is to gather appropriate information from a properly functioning network. Much of this information may already exist, and it is just a matter of finding it.

Topologic Map For example, many enterprise management systems have the capability of automatically discovering devices on the network and creating a topologic map. This kind of information is necessary for knowing what components exist on the network and how they interact physically and logically. For wide area networks (WANs), this means locations, descriptions, and cable plant maps for equipment such as routers, bridges, and network access devices (LAN to WAN).

In addition, information on transmission media, physical interfaces (T1/E1, V series, etc.), line speeds, encoding methods and framing, and access points to service provider equipment should be assembled. Although it is not always practical to map individual workstations in the local area network (LAN) portions of the WAN or to know exactly what routing occurs in the WAN cloud, knowing the general topology of the WAN can be useful in tracking down problems later.

WAN and LAN Protocols To fully understand how a network behaves, it is necessary to know what protocols are in use. Later, during the troubleshooting process, the presence of unexpected protocols may provide clues as to why network devices appear to be malfunctioning or why data-transfer errors or failures are occurring.

Logs Some network problems begin to occur after new devices or applications are installed. The addition of new devices, for example, can cause network problems that have a ripple effect throughout the network. A new end-user device with a duplicate Internet Protocol (IP) address, for instance, could make it impossible for other network elements to communicate. Or a badly configured router added to the network could produce congestion and connection problems. Other problems occur when new data communications are enabled or existing topologies and configurations are changed. A log of these activities can help pinpoint causes of network difficulty. In addition, previous network trouble—and its resolution—is sometimes recorded, and this too can lead to faster problem identification and resolution.

Statistics Often, previously gathered data can provide a valuable context for newly created baselines. Previously assembled baselines also may contain event and error statistics and examples of decoded traffic based on network location or time of day. These logs may have been gathered over long periods, yielding valuable information about the history of network performance.

Usage Patterns A profile of users and their typical usage patterns also can speed fault isolation. This entails having several types of information, including what kind of LAN traffic is carried over the WAN.

LAN Traffic on the WAN With knowledge of what kind of LAN traffic to expect on the WAN, technicians and network managers will have a better idea of the analysis that might have to be performed later. In addition, knowing how LAN frames might be handled at end stations can help troubleshooters make a distinction between WAN problems and end-station processing problems.

Traffic Content Knowing the WAN traffic type (voice, data, video, etc.) can help troubleshooters estimate when network traffic is most likely to be heavy, what level of transmit errors can be tolerated, and whether it makes sense to even use a protocol analyzer. For example, an analyzer may incorrectly report errored frames and corrupt data when attempting to process voice or video traffic based on data communication protocols.

Peak Usage Knowing when large data transfers will occur—such as scheduled file backups between LANs connected across the WAN—can help network managers predict and plan for network slowdowns and failures. It also can help technicians schedule repairs so that WAN performance is minimally impacted. Some of this information can be obtained from interviews with network administrators or key users. Other times it must be gathered with network analysis tools.

Hard Stats and Decodes

After gathering information on topology, devices, protocols, and typical users of the WAN, hard statistics and examples of decoded network traffic should be gathered. Getting comprehensive baseline data may entail gathering them at regular intervals at numerous points throughout the network.

Statistics Logs To understand usage trends and normal error levels over time, a statistics log is created. Many protocol analyzers let technicians specify the period over which these kinds of data are logged, the interval between log entries, and the type of statistics to log. The log file can be exported to a spreadsheet or other application program for offline analysis.

Frame or Packet Data To see details about typical WAN traffic, frame or packet data can be collected and saved to a file for later examination. Data collection can be done at specific periods during the day or week to find differences between peak and off-peak usage. Saved network traffic also provides insight into device configurations for use later during routine upgrades or repairs.

Targeted Statistics Using configurable traffic filters and counters, selected blocks of data or statistics based on specific network events can be captured, which might include error count thresholds, specific frame types, and in-channel alarms. A comprehensive collection of such data provides a benchmark for comparison if the network begins to malfunction. New protocols on the network, unexpected line and channel utilization levels, and increases in normal errors and in-channel alarms can be isolated according to physical link location, helping narrow the search for the problem.

Applying the Baseline

If network performance and reliability problems occur, the information gathered during baselining can be used to help identify the nature and source of the problem through comparison analysis and historical trends.

Comparison Analysis Baseline information is compared with current information to see network changes. For example, to isolate failing devices or connections, the number of errors recorded during baselining is compared with the current number of errors that occur over a similar time interval.

Historical Trends Current network problems can result from subtly changing conditions that are detected only after examining a series of baselines gathered over time. For example, congestion problems may become apparent only as new users are added to a particular part of the network. Examining historical trends can help isolate these situations.

Summary

Performance baselining provides a profile of normal network behavior, making it easier for technicians and network managers to identify deviations so that appropriate corrective action can be taken. This "snapshot" of the current network can also be used as the input data for subsequent performance modeling. For example, network administrators and operations managers can use the baseline data to conduct "what if" scenarios to assess the impact of proposed changes. A wide variety of changes can be evaluated, such as adding routers, increasing WAN bandwidth or application workloads, and relocating user sites. During analysis, performance thresholds can be customized to highlight network conditions of interest. These capabilities enable users to plan and quantify the benefits of feature migrations, such as different routing protocols, and to make more accurate and cost-effective decisions regarding the location and timing of upgrades.

See also

Network Statistics

Ping
Protocol Analyzers

PERSONAL DIGITAL ASSISTANTS

Personal digital assistants (PDAs) are hand-held computers equipped with operating system and applications software. PDAs can be equipped with communications capabilities for short-text messaging, e-mail, news updates, Web surfing, voice mail, and Internet telephony. Today's PDAs also can act as MP3 players, voice recorders, and digital cameras with the addition of multimedia modules. Some PDAs can even accommodate a module that provides location information via the Global Positioning System (GPS). PDAs are intended for mobile users who require instant access to information regardless of their location at any given time (Figure P-1).

The Newton MessagePad, introduced by Apple Computer in 1993, was the first true PDA. Trumpeted as a major milestone of the information age, the MessagePad was soon joined by similar products from such companies as Hewlett-Packard, Motorola, Sharp, and Sony.

These early hand-held devices were hampered by poor performance, excessive weight, and unstable software. Without a wireless communications infrastructure, there was no compelling advantage to owning a PDA. With the performance limitations largely corrected and the emergence of new wireless personal communications services (PCS)—plus continuing advances in operating systems, connectivity options, and battery technology—PDAs are now well on the way to fulfilling their potential.

Applications

Real estate agents, medical professionals, field service technicians, and delivery people are just a few of the people using PDAs. Real estate agents can use PDAs to conveniently

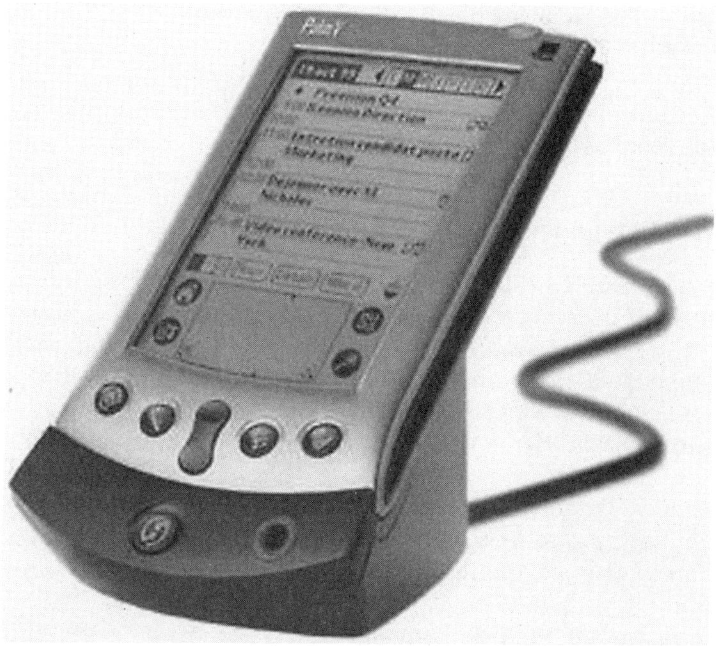

Figure P-1 Palm Computing offers one of the most popular lines of PDAs. This Palm V, shown with cradle charger and HotSync serial cable, weighs in at only 4 ounces.

browse through property listings at client locations. Health care professionals can use PDAs to improve their ability to access, collect, and record patient information at the point of care. Numerous retailers and distributors can collect inventory data on the store and warehouse floor and later export them into a spreadsheet on a PC. Insurance agents, auditors, and inspectors can use PDAs to record data in the field and then instantly transfer that data to PCs and databases at the home office.

For professionals who tote around a laptop computer to give presentations with Microsoft PowerPoint, there is a module for the Springboard Visor that connects the device directly to digital projectors (or other VGA displays) with an interface cable. The user downloads the presentation material from a PC to the Visor and then taps an icon displayed on the Visor screen to start the 1024 × 768 resolution color presentation. The user can even control the presentation from anywhere in the room using the product's infrared remote control.

PDA Components

Aside from the case, PDA components include a screen, keypad, or other type of input device; an operating system; memory; and battery. Many PDAs can be outfitted with fax/modem cards and a docking station to facilitate direct connection to a PC or LAN for data transfers and file synchronization. Some PDAs, such as Palm Computing's Palm VII, have a wireless capability that allows information retrieval from the Internet. Of course, PDAs run numerous applications to help users stay organized and productive. Some PDAs have integral 56-kbps modems and serial ports that allow them to be attached via cable to other devices.

A unique PDA is Handspring's Visor, which uses the Palm OS operating system. What makes the Visor unique is that it is expandable via an external expansion slot, called a Springboard. In addition to backup storage and flash storage modules, the slot lets users add software and hardware modules that completely change the function of the Visor. Springboard modules allow the Visor to become an MP3 player, pager, modem, GPS receiver, e-book, or video game device.

Display The biggest limitation of PDAs is the size of their screens. Visibility is greatly improved through the use of nonglare screens and backlighting, which aid viewing and entering information in any lighting condition. In a dim indoor environment, backlighting is a virtual necessity, but it drains the battery faster. Some PDAs offer user-controllable backlighting, while others let the user set a timer that shuts off the screen automatically after the unit has been idle for a specified period of time. Both features greatly extend battery life. Other PDAs, such as the Visor Prism, feature an active-matrix backlit display capable of displaying over 65,000 colors.

Keyboard Some PDAs have on-screen keyboards, but they are too small to permit touch typing. The use of a stylus speeds up text input and makes task selection easier. Of course, the instrument can be used for handwritten notes. The PDA's handwriting-recognition capability enables the notes to be stored as text for use by various applications, such as date book, address book, and to-do list. As an option, there are foldout full-sized keyboards available that attach to the PDA. They weigh only 8 ounces, making it much easier to respond to e-mail, compose memos, and take notes without having to lug around a laptop.

Operating Systems A PDA's operating system provides the foundation on which applications run. The operating system may offer handwriting recognition, for example, and include solutions for organizing and communicating information via fax or electronic mail, as well as the ability to integrate with Windows and Mac OS-based computers in enterprise environments. The operating system also may include built-in support for a range of modems and third-party paging and cellular communication solutions. Because memory is limited in a PDA, usually between 2 and 64 MB, the operating system and the applications that run on it must be compact.

Some operating systems come with useful utilities. There are utilities that set up direct connections between the PDA and desktop applications to transfer files between them via a cable or infrared connection. A synchronization utility ensures that the user is working from the latest version of a file. Some operating systems offer tools called "intelligent agents" that automate routine tasks. An intelligent agent can be programmed to set up a connection to the Internet, for example, and check for e-mail. To activate this process, the user might only have to touch an icon on the PDA's screen with a pen.

There are two major operating systems in use today—Microsoft's Pocket PC platform and the Palm OS. Pocket PC's predecessor, Windows CE, was too difficult to use and not powerful enough to draw users away from the popular Palm OS. But the Pocket PC's redesigned interface overcomes most of Windows CE's previous problems.

The Pocket PC platform is a version of Windows that preserves the familiarity of the Windows-based desktop and integrates seamlessly with Outlook, Word, and Excel. The platform includes a version of Internet Explorer for browsing the Web over a wireless connection or ordinary phone line and Windows Media Player for listening to digital music and watching digital videos. It also includes Microsoft Reader for reading e-books downloaded from the Internet. Palm OS is a more efficient operating system than Pocket PC. Consequently, it requires less processing power and less memory than equivalent products using the Pocket PC operating system.

Memory Although PDAs come with a base of applications built into ROM—usually a file manager, word processor, and scheduler—users can install other applications as well. New applications and data are stored in RAM. At a minimum, PDAs come with only 2 MB, while others offer up to 64 MB. When equipped with 2 MB of memory, the PDA can store approximately 6000 addresses, 5 years of appointments, 1500 to-do items, and 1500 memos. Some PDAs have a PC card (formerly, PCMCIA) slot that can accommodate storage cards that are purchased separately.

Even though many Pocket PC products come in higher memory configurations, an 8-MB Palm OS product can store as much or even more information than a 16-MB Pocket PC product with little performance degradation. The better performance is due to the efficiencies of the Palm OS, which uses less memory and processing power than equivalent products based on Pocket PC. Since greater memory capacity increases the overall price of the product, vendors such as Handspring believe that 8 MB offers the most utility at the most competitive price.

Power Many PDAs use ordinary AAA alkaline batteries. Manufacturers claim a battery life of 45 hours when users search for data 5 minutes out of every hour the unit is turned on. Of course, using the backlight display will drain the batteries much faster. Using the backlight will reduce battery life by about 22 percent. Other power sources commonly used with PDAs include an ac adapter and rechargeable lithium-ion battery.

The rechargeable battery offers more flexible power management in a smaller space. The components that operate the color screen increase the power draw from the battery. With AAA batteries, the user would have to replace them rather frequently. The rechargeable battery solution enhances the user's experience by providing full power in a pocket-sized package. Lithium-ion rechargeable batteries offer 2 hours of continuous use.

Fax/Modems Some PDAs come with an external fax/modem to support basic messaging needs when hooked up to a telephone line. Others offer a PC card slot (formerly PCMCIA) that can accept not only fax/modems but also storage cards. With fax/modems, PDA users can receive a fax from their office, annotate it, and fax it back with comments written on it in "electronic ink."

There are wireless Ethernet modules available that allow the user to roam about the workplace or campus with secure connections, peer-to-peer links between devices, and high-speed access to the Internet, e-mail, and network resources. Transmissions of up to 11 Mbps are possible, but actual throughput is determined by the speed of the PDA's processor. The modules adhere to the IEEE 802.11b (Wi-Fi) high-rate standard for wireless LANs and support 40- or

128-bit Wired-Equivalent Privacy (WEP) encryption. Transmission ranges of up to 1000 feet (300 meters) in open environments and 300 feet (90 meters) in office environments are supported.

Cradle A cradle allows the PDA to connect to a desktop PC via a standard serial cable or USB cable. The user simply drops the PDA into the cradle and presses a button to automatically synchronize desktop files with those held in the PDA. An alternative to cable is an infrared (IR) connection. With an IR-enabled PDA, users not only can swap and synchronize files with a PC but also can beam business cards, phone lists, memos, and add-on applications to other IR-enabled PDAs. IR-enabled PDAs also can use third-party beaming applications with IR-enabled phones, printers, and other devices.

Summary

Improvements in technology and the availability of wireless communications services, including PCS, overcame many of the limitations of early PDA products, making today's handheld devices very attractive to mobile professionals. In the process, PDAs are finding acceptance beyond vertical markets and finally becoming popular among consumers, particularly those looking for an alternative to notebook computers and younger people who want a versatile device from which they can also play MP3 music files and games as well as read e-books. There are now Bluetooth and Wi-Fi modules available for PDAs, offering users opportunities for personal ad hoc networking.

See also

Bluetooth
Electronic Mail
Unified Messaging

PING

Ping is a simple test function that allows the user to check if a local or remote system on an IP network is currently up and running. Ping can be run on the command line on UNIX machines or within a client application in the Windows or Macintosh environment. The Ping command can be run using plain language domain names or IP addresses. The general command line syntax for implementing Ping is

```
ping abc.com or ping 192.168.100.1
```

This will indicate whether the host at ABC Company is currently online or if Ping is launched from an internal workstation, whether its internal proxy server is in service. The Ping command sends one datagram per second and prints (or displays) one line of output for every echo response returned. No output is produced if there is no response. A count option can be used in the command line syntax to specify the number of requests to be sent. Many implementations of Ping also include an option that measures the roundtrip time of the sent packet in milliseconds (ms) as well as the packet loss between two hosts on the network. When CTRL-C is pressed on the keyboard, Ping provides a brief statistical summary, as in the following example:

```
PING abc.com: 56 data bytes

64 bytes from 132.58.68.1: icmp_seq=0 ttl=251 time=66 ms

64 bytes from 132.58.68.1: icmp_seq=1 ttl=251 time=45 ms

64 bytes from 132.58.68.1: icmp_seq=2 ttl=251 time=46 ms

64 bytes from 132.58.68.1: icmp_seq=3 ttl=251 time=55 ms

64 bytes from 132.58.68.1: icmp_seq=4 ttl=251 time=48 ms


— abc.com ping statistics —
```

5 packets transmitted, 5 packets received, 0% packet loss

round-trip min/avg/max = 45/52/66 ms

Pinging once or twice is generally enough to provide a reliable indication of a remote system's current state. Ping also can be used to continuously monitor the state of the connection. For Windows and Macintosh machines, there are feature-rich graphical utilities offered as shareware that implement Ping (Figure P-2). Among other things, they let the user specify

- The Ping data packet size
- The number of hosts to Ping simultaneously

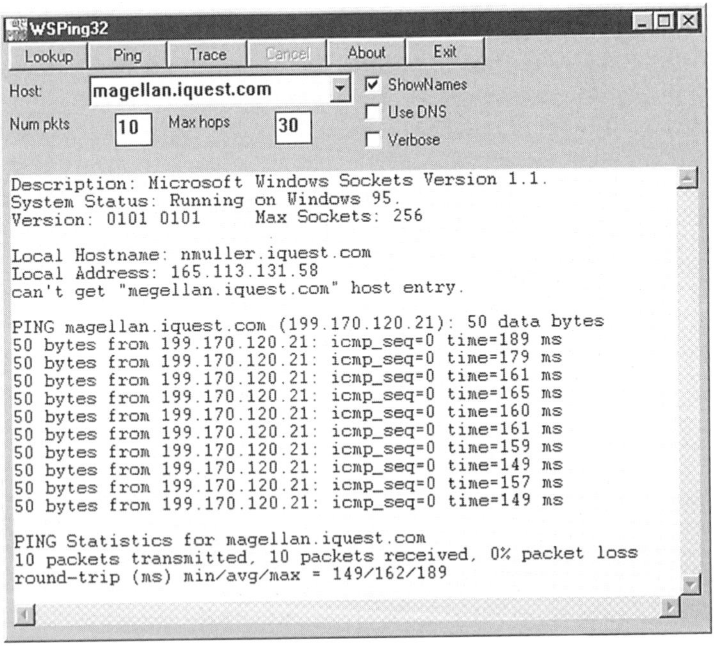

Figure P-2 John A. Junod's ICMP Ping for Windows, a shareware utility.

- The Ping interval
- The amount of milliseconds to wait for echo reply
- Time to wait until next Ping if the last Ping succeeded
- Time to wait until next Ping if the last Ping failed
- The number of failed pings before the utility considers that the host is down

Monitoring network activities is made even easier by setting the Ping utility to take a specific action whenever the remote host changes its state from up to down and vice versa. Among the actions the Ping utility can implement in response to changes in network activity are the playing of audio messages and the running of custom programs.

For example, the user can specify that an audio file be played to indicate when a remote host crashes or recovers from a crash. The Ping utility also can be configured to run a program as soon as the remote host recovers from a crash.

Summary

Ping is a simple but useful troubleshooting tool for tracking down the source of a failure on an internal network or on the Internet. Basically, when Ping is run, packets are issued to one or more designated hosts and echoed back to provide performance information. Various statistics are displayed which tell the user about the state of the local or remote host and the connection. If the target host is out of service, there is no echo. Instead, the Ping function times out, indicating a problem.

See also

Latency
Network Statistics
Quality of Service

POINT-TO-POINT PROTOCOL

The Point-to-Point Protocol (PPP) provides the means to transfer data across any full-duplex (i.e., two-way) circuit, including dial-up links to the Internet via modems, Integrated Services Digital Network (ISDN), and high-speed SONET over fiberoptic lines. PPP is an enhanced version of the older Serial Line Internet Protocol (SLIP). While SLIP is typically used in an IP-only environment, PPP is more versatile in that it can be used in multiprotocol environments. In addition, PPP allows traffic for several protocols to be multiplexed across the link, including IP, IPX, DECnet, ISO, and others. PPP also carries bridged data over complex internets.

Features

PPP supports authentication, link configuration, and link monitoring capabilities via several subprotocols, including

- *Link Control Protocol (LCP)* Negotiates details and desired options for establishing and testing the overall serial link.
- *Authentication protocols* These are the Password Authentication Protocol (PAP) and the Challenge Handshake Authentication Protocol (CHAP). PAP uses a two-way handshake for the peer to establish its identity at the time of link establishment. CHAP periodically verifies the identity of the peer using a three-way handshake, which is employed throughout the life of the connection. Of the two authentication protocols, CHAP is the more robust.
- *Network control protocols* Used to dynamically configure different network layer protocols, such as IP and IPX. For each type of network-layer protocol, there is a network control protocol that is used to initialize, configure, and terminate its use.

- *Link quality monitoring* Provides a standardized way of delivering link quality reports on the quality/accuracy of a serial link.

During Transmission Control Protocol (TCP) sessions, PPP's compression feature can reduce the typical 40-byte TCP header to only 3 to 5 bytes, providing significant savings in transmission time. This is accomplished by sending only the changes in a PPP frame's header values. Since most header information does not change from one frame to the next, the savings can be quite substantial.

Multilink PPP

There is a version of PPP called Multilink PPP (MLPPP), that is used to combine the bandwidth of multiple lines or channels. When used with ISDN, for example, MLPPP is used to combine multiple B-channels of an ISDN link into a single, higher-speed channel. Although B-channels can be added to or subtracted from an established ISDN connection, MLPPP does not offer dynamic control. This is remedied by the Bandwidth Allocation Control Protocol (BACP), which works in conjunction with MLPPP. With BACP, ISDN channels can be added as needed and dropped when no longer required to support the application.

BACP enables bandwidth to change on demand through a standard set of rules while minimizing the need for the end user to be involved in complex connection configuration issues. BACP can even interact with the resource ReSerVation Protocol (RSVP) to provide enhanced functionality. For example, if a bandwidth reservation is queued for lack of bandwidth somewhere on the network, this could trigger the creation of additional channels to support the application. If the application is a videoconference, for example, when the router senses that network load has gone down because participants are dropping out of the session, it starts terminating B-channels. This minimizes the usage charges associated with ISDN.

Summary

While SLIP is typically used to connect computers to an IP network via a dial-up link, it has severe limitations in that it cannot support any other protocol and does not perform error checking. PPP is more versatile in terms of the protocols it can handle and is more functional, particularly with regard to authentication, link configuration, and link monitoring. PPP has displaced SLIP in recent years. Other protocols—such as MLPPP, BACP, and RSVP—are used with PPP to support more sophisticated high-bandwidth applications such as videoconferencing.

See also

Integrated Services Digital Network

PRIVATE BRANCH EXCHANGES

In the simplest terms, a private branch exchange (PBX) is a telephone switch that, through control signaling, performs several basic functions to provide communications services to users within an organization.

In response to a call request, the PBX establishes end-to-end connectivity among its subscribers (on-net) and from its own subscribers to remote subscribers (off-net) through intermediate nodes, which may consist of other PBXs or central office switches on the public switched telephone network (PSTN). The connected path is dedicated to the user for the duration of the call. The PBX also supervises the circuit to detect call request, answer, signaling, busy, and disconnect (hang-up) and "tears down" the path on call termination (disconnect) so that another user can access the resources available over that circuit.

These functions closely parallel those of the central office switch. In fact, the PBX evolved from the operator-controlled switchboards that were used on the public telephone network.

The first of these simple devices was installed in 1878 by the Bell Telephone Company to serve 21 subscribers in New Haven, Connecticut. The operator had full responsibility for answering call requests, setting up the appropriate connections, supervising for answer and disconnect, and tearing down the path on call completion. Interconnectivity among subscribers was accomplished via cable connections at a patch panel.

Basic Features

Many PBX features are under direct user control and can be implemented right from the telephone keypad, including

- *Add-on conference* Allows the user to establish another connection while having a call already in progress.
- *Call forwarding* Allows a station to forward incoming calls to another station. This includes forwarding calls when the station is busy or unattended or as needed.
- *Call hold* Allows the user to put the first party on hold so that an incoming call can be answered.
- *Call waiting* Lets the user know that an incoming call is waiting. While a call is in progress, the user will hear a special tone that indicates that another call has come through.
- *Camp-on* Allows the user to wait for a busy line to become idle, at which time a ring signal notifies both parties that the connection has been made.
- *Last number redial* Allows users to press one or two buttons on the keypad to activate dialing of the previously dialed number.
- *Message waiting* Allows the user to signal an unattended station that a call has been placed. On returning to the station, an indicator tells the person that a message is waiting.
- *Speed dialing* Allows the user to implement calls with an abbreviated number. This feature also allows users to

enter a specified number of speed-dial numbers into the main database. These numbers may be private or shared among all users. Entering and storing additional speed-dial numbers is accomplished via the telephone keypad.

There are also a number of PBX capabilities that operate in the background, transparent to the user. The most common of these system capabilities include

- *Automated attendant* Enables the system to answer incoming calls and prompt the caller to dial an extension or leave a voice message without going through the operator.
- *Automatic call distribution (ACD)* Allows sharing of incoming calls among a number of stations so that the calls are served in order of their arrival. This is usually an optional capability, but it may be integral to the PBX or purchased separately as a stand-alone device.
- *Automatic least-cost routing* Ensures that calls are completed over the most economic route available. This feature may be programmed so that mailroom personnel can access only the cheapest service while executives get to choose whatever service they want.
- *Call detail recording (CDR)* Enables the PBX to record information about selected types of calls for management and cost control.
- *Call pickup* Allows incoming calls made to an unattended station to be picked up by any other station in the same trunk group.
- *Class-of-service restrictions* Controls access to certain services or shared resources. Access to long-distance services, for example, may be restricted by area code or exchange. Access to the modem pool for transmission over analog lines may be similarly controlled.
- *Database redundancy* Enables the instructions stored on one circuit card to be dumped to another card as a protection against loss.

- *Direct inward dialing (DID)* Allows incoming calls to bypass the attendant and ring directly on a specific station.
- *Direct outward dialing (DOD)* Allows outgoing calls to bypass the attendant for completion anywhere over the PSTN.
- *Hunting* A capability that routes calls automatically to an alternate station when the called station is busy.
- *Music on hold* Indicates to callers that the connection is active while the call waits in queue for the next available station operator.
- *Power-fail transfer* Permits the continuance of communication paths to the external network during a power failure. This capability works in conjunction with an uninterruptible power supply (UPS), which kicks in within a few milliseconds after detecting a power outage.
- *System redundancy* Enables sharing of the switching load so that, in the event of failure, another processor can take over all system functions.

IP PBXs

A relatively new development is the IP-based PBX, which transports intraoffice voice calls over an Ethernet LAN and, via an IP-PSTN gateway, over the PSTN. Full-featured digital phone sets link directly to the Ethernet LAN via a 10BaseT interface without requiring direct connection to a desktop computer. Phone features can be configured using a Web browser. Existing analog devices, such as phones and fax machines, can be linked to the LAN via a gateway. In addition to IP nets, calls can be placed or received using T1, PRI ISDN (Primary Rate Interface ISDN), or traditional analog telephone lines.

All the desktop devices have access to the calling features offered through the IP PBX management software running on a LAN server. The call-management software allows hundreds of client devices on the network, such as phones and computers, to perform functions such as call hold, call transfer, call forward, call park, and calling party ID. Other advanced capabilities—such as multiple lines per phone or multiple phones per line—are also implemented in call-management software.

The major vendors, such as Lucent Technologies and Nortel, offer IP interfaces to their conventional PBX systems. Lucent, for example, offers an IP trunk interface for certain models of its Definity product line. The IP interface supports 24 ports and allows businesses to integrate least-cost routing and class-of-service features, giving network managers the ability to add the Internet and corporate intranet as alternative routes for voice and fax services. The ability to add the IP trunk interface directly into the Definity also reduces the cost of obtaining Internet telephony capabilities by eliminating the need to buy a separate IP-PSTN gateway. Nortel also offers a 24-port IP interface for its Meridian communication system, enabling the routing of real-time voice and fax calls over IP data networks rather than the PSTN.

Summary

Despite their humble beginnings over 125 years ago, there is still plenty of room for innovation in PBX systems. Today's PBXs emphasize integration into enterprise networking infrastructures, thereby addressing the applications that will be most in demand for the rest of the 1990s and beyond: LAN interconnectivity, videoconferencing, and multimedia applications. There are PBX add-ons that support in-building mobile communications using wireless technology. PBXs also can be connected to LANs and IP networks, allowing users to send e-mail and run real-time and multimedia applications such as voice calls and videoconferences more economically.

See also

Central Office Switches
Centrex
Communications Services Management

PROTOCOL ANALYZERS

A category of test equipment known as the "protocol analyzer" is used to monitor and diagnose performance problems on LANs and WANs by decoding upper-layer protocols. There are protocol analyzers for all types of communications circuits, including frame relay, X.25, T1 and ISDN, and ATM. There are also protocol analyzers that offer full seven-layer decodes of NetBIOS, SNA, SMB, TCP/IP, DEC LAT, XNS/MS-NET, NetWare, and VINES, as well as the various LAN cabling, signaling, and protocol architectures, including those for AppleTalk, ARCnet, Ethernet, StarLAN, and Token Ring. There are even analyzers for wireless services like Bluetooth, which monitor the frames that are transmitted through the air, as well as capture and analyze Bluetooth serial data as they travel between a host and a host controller.

In the case of a LAN, the protocol analyzer connects directly to the cable as if it were just another node or to the test port of data terminal equipment (DTE) or data communication equipment (DCE) where trouble is suspected (Figure P-3).

Troubleshooting Features

Many protocol analyzers have sophisticated features, such as data capture to RAM or disk, automatic configuration, counters, timers, traps, masks, and statistics. These features can dramatically shorten the time it takes to isolate

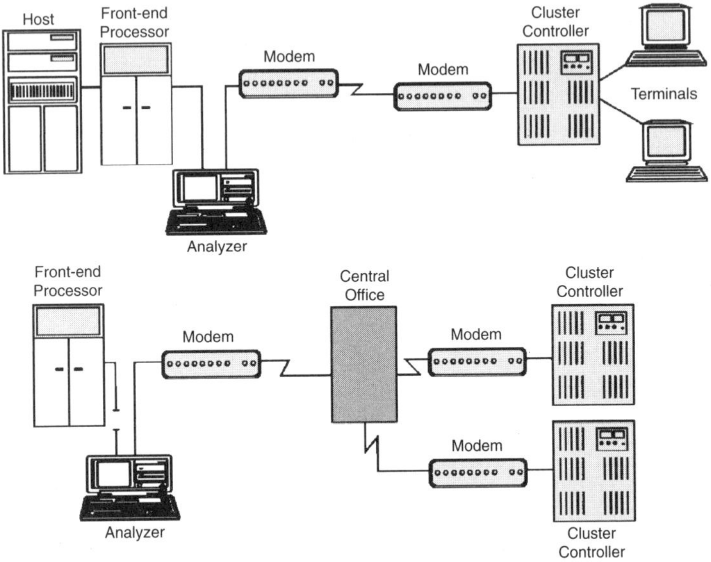

Figure P-3 A protocol analyzer in the monitor mode *(above)* allows the user to check events taking place between front-end processor and a local modem over a synchronous link. A protocol analyzer in the simulation mode *(below)* allows the user to run a program that exhibits proper operation of the suspect front-end processor over a synchronous multidrop link, which includes two cluster controllers.

a problem. Some sets are programmable and offer simulation features.

Monitoring and Simulation Protocol analyzers generally are used in either a passive monitoring application or a simulation application. In the monitoring application, the analyzer sits passively on the network and monitors both the integrity of the cabling and the level of data traffic, logging such things as excessive packet collisions and damaged packets that can tie up an Ethernet LAN, for example. The information on troublesome nodes and cabling is compiled for the network manager. In a monitoring application, the protocol analyzer merely displays the protocol activity and user data (packets) that are passed over the cable, providing a window into the message exchange between network nodes.

In the simulation application, the protocol analyzer is programmed to exhibit the behavior of a network node, such as a gateway, communications controller, or front-end processor (FEP). This makes it possible to replace a suspect device on the network with a simulator that is running a program to simulate proper operation. This also enhances the ability to do fault isolation. For example, a dual-port protocol analyzer can monitor a gateway while running a simulation. More sophisticated protocol analyzers can run simulations designed to stress test individual nodes to verify their conformity to standards. Protocol simulation is used most often to verify the integrity of a new installation.

Trapping The trapping function allows the troubleshooter to command the protocol analyzer to start recording data into its buffer or onto disk when a specific event occurs. For example, the protocol analyzer could be set to trap the first errored frame it receives. This feature permits the capture of only essential information. Some protocol analyzers allow the user to set performance thresholds according to the type of traffic on the network. When these performance thresholds are exceeded, an alarm message is triggered, indicating that there is a problem.

Filtering With the protocol analyzer's filtering capability, the user can exclude certain types of information from capture or analysis. For example, the technician might suspect that errors are being generated at the Data Link Layer, so network-layer packets can be excluded until the problem is located. At the Data Link Layer, the analyzer tracks information such as where the data were generated and whether they contain errors. If no problems are found, the user can set the filter to include only network-layer packets. At this layer, the protocol analyzer tracks information such as where the data are destined and the type of application under which they were generated. If the troubleshooter has no idea where to start looking for problems, then all the packets may be captured and written to disk. Various filters can be applied later for selective viewing.

Bit Error Rate Testing Bit error rate testing is used to determine whether data are being passed reliably over a carrier-provided communications link. This is accomplished by sending and receiving various bit patterns and data characters to compare what is transmitted with what is received. The bit error rate is calculated as a ratio of the total number of bit errors divided by the total number of bits received. Any difference between the two is indicated and displayed as an error. Additional information that may be presented includes sync losses, sync loss seconds, errored seconds, error-free seconds, time unavailable, elapsed time, frame errors, and parity errors.

Packet Generation In being able to generate packets, the protocol analyzer allows the user to test the impact of additional traffic on the network. Using a set of configuration screens, the technician can set the following parameters of the packets:

- The source address the packets will be sent from
- The destination address the packets will be sent to
- The maximum and minimum frame size of the packets
- The spacing between the packets, expressed in microseconds
- The number of packets sent out with each burst

The technician also can customize the contents of the data field section of the packets to simulate real or potential applications. When the packets are generated, the real-time impact of the additional traffic on the network can be observed on the monitor of the protocol analyzer. Packets also may be generated to force a suspected problem to reoccur, thereby expediting identification of the problem.

Load Generation A related capability is load generation, whereby varying traffic rates on the network may be created. By loading the network from 1 to 98 percent, network components such as repeaters, bridges, and transceivers can be stressed for the purpose of identifying any weak links on the network before they become serious problems later.

Mapping Some protocol analyzers have a mapping capability. In automatically documenting the physical location of LAN nodes, many hours of work can be eliminated in rearranging the network map when devices are added, deleted, or moved. The mapping software allows the network manager to name nodes. Appropriate icons for servers and workstations are included. The icon for each station also provides information about the type of adapter used, as well as the node's location along the cable. When problems arise on the network, the network manager can quickly locate the problem by referring to the visual map. Some protocol analyzers can depict network configurations according to the usage of network nodes, arranging them in order of highest to lowest traffic volume.

Programmability The various tasks of a protocol analyzer may be programmed, allowing performance information to be collected automatically. While some analyzers require the use of programming languages, others employ a setup screen, allowing the operator to define a sequence of tests to be performed. Once preset thresholds are met, an appropriate test or sequence of tests is performed automatically. This capability is especially useful for tracking down intermittent problems. An alternative to programming or defining analyzer operation is to use off-the-shelf software that can be plugged into the data analyzer in support of various test scenarios.

Automatic Configuration Some protocol analyzers have an autoconfigure capability that enables the device to automatically configure itself to the protocol characteristics of the line under test. This eliminates the need to go through several manually established screens for setup, which can save a lot of time and frustration.

Decode Capability Some protocol analyzers are unique in their ability to decode packets and display their contents in character notation, in addition to hexadecimal or binary code. Further details about a specific protocol may be revealed through an analyzer's "drill down" capability, which allows the troubleshooter to display each bit field, along with a brief explanation of its status.

Editing Some analyzer software includes a text editor that can be used in conjunction with captured data. This allows the user to delete unimportant data, enter comments, print reports, and even create files in common database formats.

Switched Environments Protocol analyzers were originally designed for shared networks. They pick up and examine all traffic as it is broadcast across a shared wire. With LAN switches growing in popularity, diagnosing problems has become a complex undertaking. Switches break shared networks into segments, and traffic is only broadcast over a particular segment. Although this improves performance by cutting down on contention and devoting more bandwidth to each station, it also makes diagnosing problems more difficult. A protocol analyzer usually can listen to traffic only on the segment to which it is connected. This makes it difficult to obtain an overall picture of what is happening through the switch.

One way to deal with this situation is for the technician to set the switch to operate in promiscuous mode, which sends all Ethernet packets to all ports on the switch, enabling the analyzer to see all traffic. But this results in a measurement that does not reflect real-world switch conditions. Another technique is port mirroring, which copies traffic going through one port to a port where a protocol analyzer is connected. The problem with this approach is that it limits the analyzer's view to one segment at a time.

The functionality of protocol analyzers has been expanded to monitor switches so that network administrators can get traffic statistics across a switch's ports and the switch itself, detect configuration problems in virtual LANs, and track problems between switches and desktop computers. Today's protocol analyzers can discover virtual LAN (VLAN) configurations in a switch, for example, and detect problems in the configuration. Network managers can set thresholds for traffic levels through a switch port. When the threshold is reached, the analyzer takes the traffic going through that port, mirrors it to the port with the analyzer on it, and alerts the help desk.

Another technique for gathering statistics through an entire switch is called "port looping." The analyzer uses port mirroring to look at each port on the switch for only a short time. By sampling traffic through each port, one at a time, the analyzer can build statistics about traffic through the switch.

Summary

Protocol analyzers have long been among the key diagnostic tools of technicians and network managers, helping them quickly isolate trouble spots. However, the use of protocol analyzers was almost always reactive—they captured trace information and network statistics that were later interpreted by network technicians. This required a skilled analyst to interpret the data. Today's protocol analyzers can detect deteriorating conditions that lead to errors and suggest remedial actions to head off problems before they affect performance. There are even tools that track the performance of data offerings covered by service level agreements (SLAs) so that the organization can be assured that it is getting from the carrier a level of performance and availability it is paying for.

See also

Network Statistics

Performance Baselining

Ping

Service Level Agreements

PROXY SERVERS

A proxy server implements a variety of complementary tasks for companies and Internet service providers (ISPs), including caching, filtering Web content, and performing network address translation. The primary function of a proxy server is caching frequently accessed documents to conserve network bandwidth and reduce response times for clients. It also enables network administrators to maintain better control over the use of network resources by blocking access to specific sites by user, document, and other criteria that can be set by a network administrator. The Network Address Translation (NAT) capability allows subnets to be created and protects internal IP addresses from public view on the Internet, conserving IP addresses and enhancing security.

A corporation could deploy a proxy server in a variety of ways. It can deploy a proxy server just behind the firewall to facilitate access to the Internet and reduce response times. It can be used to protect information on the secure Web server behind the firewall and offer load balancing via caching. For companies that have several subnets, a proxy server deployed at each subnet can reduce traffic on the corporate backbone, eliminating the need for more bandwidth. In situations where remote offices are disconnected from the internal network, a proxy server can provide an inexpensive means for quickly replicating content. Outside the United States, where communications bandwidth is typically much more expensive, proxy servers are even more cost-effective for replicating content.

An ISP would deploy one proxy server at each point of presence (POP) and cluster them at the Internet gateway to provide faster, more reliable service and reduce network congestion between the POP and the central Internet gateway. Some ISPs deploy a proxy server only at their gateway to the Internet, which reduces traffic on their link to the Internet but not on their own network from the POP to the Internet gateway.

Caching

A proxy server typically supports HyperText Transfer Protocol (HTTP), File Transfer protocol (FTP), and Gopher for caching. It also may support the Secure Sockets Protocol (SSL) for the transmission of encrypted traffic and SOCKS, which is a generic way of tunneling protocols (such as Telnet) that are not "proxied." A proxy server uses sophisticated statistical analysis to store the documents most likely to be needed.

Among the many features of proxy servers is dynamic caching, which enables an administrator to schedule batch updates to the cache. This includes the ability to preload documents or sites into the cache in anticipation of user demand and the ability to automatically refresh documents that already reside in the cache. Administrators can schedule batch updates to take place at regular intervals and off-peak hours so that network bandwidth is not tied up caching documents during periods of heavy network use. Administrators can check the proxy access logs to determine whether frequently accessed sites are actually desirable for caching.

A proxy server may support the Cache Array Routing Protocol (CARP) and Internet Cache Protocol (ICP), which are proposed standards for distributed caching. CARP provides a mechanism for routing content requests among an array of proxy servers in a deterministic fashion. CARP enables load balancing, fault tolerance, more efficient caching, and easier management for multiple proxy servers. ICP enables a proxy server to send queries to neighbor caches to determine whether they already have a document.

CARP is appropriate for a group of proxy servers that are serving the same audience of downstream clients or proxies and that are all under common administrative control. ICP is appropriate for proxies that are not under common administrative control and that may be serving different clients.

Filtering

Network administrators can grant or limit access to network resources, including specific sites and documents, through the use of user name and password, IP address, host name, or domain name filtering.

A proxy server allows administrators to ban access to particular sites using a list of Uniform Resource Locators (URLs) or wildcard patterns. For example, an administrator could use http://*.playboy.com/* to prevent access to all pages belonging to the Playboy site.

A proxy server also can filter on the basis of content type, such as specific Multipurpose Internet Mail Extensions (MIME) types, and on the basis of content, such as HyperText Markup Language (HTML) tags. In addition, system administrators can implement their own security policies by stopping transmission of Java and JavaScript and ActiveX components. Many proxy servers now include virus-scanning software to prevent damage to client data and applications.

Network Address Translation

A proxy server can enhance firewall security in a variety of ways, including network address translation, which prevents external users on the Internet from being able to view the corporate network's structure and IP addresses. Blocking this information severely limits the chances of attack from hackers via address spoofing.

In Figure P-4, the proxy server gets a packet from station 135.112.56.52 for a destination on the public Internet. The address is rewritten so that it appears to come from 194.70.71.5, and the packet is sent out with this address.

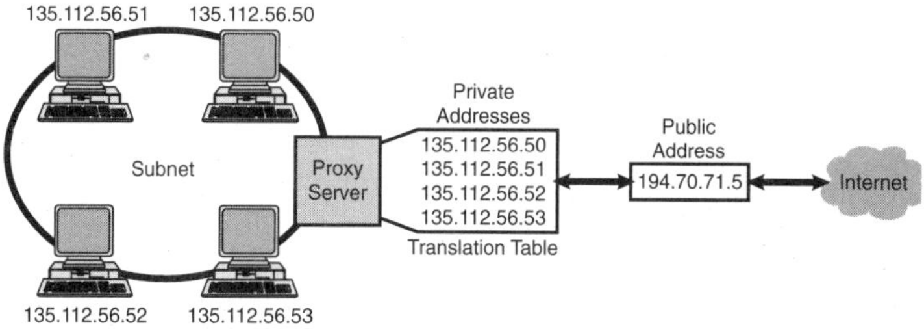

Figure P-4 The network address translation capability of a proxy server allows the creation of subnets with private IP addresses that are administered locally and never exposed to the public Internet. In addition to conserving scarce IP addresses, this capability enhances security by hiding the private IP addresses from public view over the Internet through the use of one or more public IP addresses.

When a reply packet comes back, it will be addressed to the public address 194.70.71.5. The proxy server maintains a database of outstanding requests and will look up the address of the station that made the original request. It then rewrites the address of the return packet to 135.112.56.52.

Both static and dynamic address translations are supported. Static address translations explicitly map an external address to an internal address. For incoming packets that have not been specifically requested, such as e-mail, static mapping is used. With dynamic translations, a pool is allocated, and each new IP address to be translated is dynamically mapped to another IP address from the pool in a round-robin fashion. This real-time assignment of IP addresses is implemented with the Dynamic Host Control Protocol (DHCP).

System Log

A proxy server automatically logs all requests using either the common log-file format or an extended log-file format. The extended log-file format includes the referrer field and user agent. Administrators also can create their own log-file format by selecting which HTTP fields they would like to log. A built-in log analysis program includes reports such as total number of requests, total bytes transferred, most common URLs requested, most common IP addresses making requests, performance during peak periods, cache hit rates, and estimated response time reduction.

Summary

For many companies and ISPs, a proxy server is a key element of their overall Internet gateway strategy because it improves the performance and security of communications across the TCP/IP-based Internet and private intranets while permitting more flexibility in the use of IP addresses. The proxy's disk-based caching feature minimizes use of the external network by eliminating recurrent retrievals of commonly accessed documents. This significantly improves interactive response time for locally attached clients. The resulting performance improvements provide a cost-effective alternative to purchasing additional network bandwidth. And since the cache is disk-based, it can be tuned to provide optimal performance based on network usage patterns.

See also

Firewalls

PUBLIC SWITCHED TELEPHONE NETWORK

Since the invention of the telephone by Alexander Graham Bell in 1876, the public switched telephone network (PSTN) has evolved to become a highly reliable method of communication. For most of this period, spanning 125 years, the PSTN carried voice conversations by telephone. The circuits over which these conversations take place are set up by interconnected switches and remain in place for the duration of the call.

Today, the PSTN increasingly carries data as well, mostly from users with modems who want to access the Internet at speeds of up to 56 kbps. Others may run data over ISDN, whose bearer channels are carried over the PSTN at speeds of 56/64 kbps or 128 kbps. Much of the data destined for the Internet are now diverted to data switches to offload the voice-optimized PSTN of this burden.

The PSTN is so complex that in most network diagrams it is represented merely as a cloud. Inside this cloud are all the discrete components that play a role in setting up and tearing down circuits, implementing services, providing value-added features, and managing the infrastructure.

Customer Premises Equipment (CPE)

CPE includes individual equipment or whole systems that serve a particular customer. The customer can be any subscriber of the telephone service, including an individual, company or government agency. The equipment or system interfaces with the public network.

CPE ranges from simple wiring and telephone sets to advanced PBX and/or Key Telephone Systems (KTS) and all their associated equipment, including automated call distributors (ACDs), voice-mail systems, and fax machines. Modems are also considered CPE. These devices connect to computers and transform data into an acoustical signal, which can be carried over a standard voice-grade line.

Local Loop

The local loop is the connection between the customer premises and the local central office switch. Usually, this connection is a twisted-pair 24 American Wire Gauge (AWG) line that provides the transmission path that allows the CPE to be connected with the major carriers, both local and long distance. Although the local loop is composed mostly of copper wire, new architectures incorporate optical fiber and coaxial cabling into the mix to support broadband data applications. In some cases, the local loop is bypassed through the use of wireless technologies.

Local loop includes the distribution plant that radiates from the central office (CO), consisting mostly of Subscriber Line Carrier (SLC) systems, which use fiberoptic cables to extend the reach of the CO to remote locations. This enables the telephone company to provide service to outlying homes and businesses as if they were attached directly to the CO.

Switching

The switching systems, resident in the carriers' COs, set up a dedicated transmission path that connects the calling and called parties for two-way communication. When the conversation is finished and one of the two parties hangs up, the path is torn down, freeing the network resources to handle another call.

Path setup and teardown through the long-distance portion of the PSTN are handled by a separate data network known as Signaling System 7 (SS7). This high-speed data network transfers network control and routing information such that telephone network resources are never committed to handling a call unless SS7 determines that the call actually can go through to completion.

Trunking

Trunking refers to the high-capacity links between switches in the PSTN. This interoffice trunking is accomplished primarily over fiberoptic links. At a minimum, interoffice trunks operate at the DS3 level, or 44.736 Mbps. Many of today's interoffice trunks operate at the SONET OC-3 and

OC-12 levels, or 155 and 622 Mbps, respectively. Along high-traffic corridors, higher-capacity interoffice trunks in the gigabits-per-second range are used.

Summary

While most carriers view the PSTN as voice-oriented and have preferred to run data over a separate overlay network, the trend is clearly toward merging voice and data over a single network. The two choices for accomplishing this convergence hinge around ATM and IP platforms, with some carriers committing to one or the other, and some committing to multiservice platforms that handle both ATM and IP. Still other carriers have not made any commitment, preferring instead to continue investing in their circuit-switched voice networks. Under this course, however, the carrier risks being saddled with an obsolete voice network that is not capable of supporting future market opportunities, which may very well be data-oriented.

See also

Central Office Switches

Internet

Local Loop

Q

QUALITY OF SERVICE

Quality of service (QoS) refers to attempts to ensure the delivery of traffic across packet data networks based on the differing performance requirements of the various applications that share the network. There are two approaches to implementing QoS. Packet-by-packet solutions seek to improve the delivery of each individual packet through differentiated treatment at a router, while application-centric QoS solutions focus on the delivery of applications as experienced by users.

With the availability of increasingly cheap bandwidth, there is the temptation to simply add more of it to solve application performance problems. If an application exhibits poor response time, for example, the easiest way for some companies to solve the problem is to purchase more bandwidth in an attempt to ease congestion somewhere in the network. This also saves the up-front cost of purchasing expensive tools or upgrading the routers with software and memory to classify and mark different traffic types to regulate flow in an effort to meet the specific performance requirements of all applications. However, this stopgap approach may not yield the desired results because the faster more bandwidth is made available, the faster it gets used. As a result, most companies never really get ahead of the performance curve.

Another problem with this "fat pipe" approach is that it ignores the need for prioritization schemes to implement service guarantees. Under this approach, certain applications simply would have more bandwidth to hog, still leaving other applications gasping for more. Without careful bandwidth management, routine HyperText Transfer Protocol (HTTP) traffic can make it impossible to implement voice over Internet Protocol (IP), for example. To get ahead of the performance curve requires a combination of intelligence and bandwidth. Intelligence comes in the form of tools that facilitate bandwidth management through such means as partitioning and policy setting.

With partitioning, a certain amount of bandwidth is allocated to a class of traffic. When a traffic class is partitioned, in essence, a separate, exclusive channel is created for it within the access link. While partitions control the traffic aggregate for a class, they do not influence individual flows within that aggregate. For example, a partition for all File Transfer Protocol (FTP) traffic can be created, thereby limiting how much of the link all FTP traffic will be allowed to consume. If the bandwidth within that partition is not needed for an FTP session, it can be used by other applications that have traffic to send.

The advantage that bandwidth management tools provide is the ability to configure the network to support many more users and applications than it would otherwise support. All applications are given a set amount of bandwidth in accordance with their priority, which is determined by each application's performance requirements. Not only does this save on the cost of bandwidth, however cheap, it also eliminates the need to buy more equipment and manage more boxes on the edges of the enterprise network.

Regardless of the particular bandwidth management tools used, the ability to assign a QoS to each type of traffic ensures the optimal performance of all applications on the network. Bandwidth management tools have the added advantage of making the enterprise network easier to manage and administer. They also make the network easier to scale without necessarily having to add network equipment and bandwidth.

Policy-Based QoS

Frame relay and IP do not inherently support QoS with the same granularity as Asynchronous Transfer Mode (ATM), so protocols must be added in order to prioritize different types of traffic for appropriate handling through the network. Most routers today can add prioritization schemes to expedite the delivery of real-time traffic over frame relay. Some of these prioritization schemes, implemented in the router's operating system, are summarized in Table Q-1.

Likewise, there are prioritization schemes that can be added to routers to expedite the delivery of real-time traffic over IP networks. Some of these prioritization schemes are summarized in Table Q-2.

With so many devices on these types of networks, however, IT staff easily can get bogged down performing manual QoS configurations to fully optimize the enterprise network. This task can be less tedious and error prone by using policy-based network management solutions.

To address the QoS challenges of IP networks, policy-based bandwidth management solutions are available from a growing number of vendors. These tools allow network administrators to create assured service levels and deploy security features across enterprise networks, including intranets and virtual private networks (VPNs) based on IP. These tools allow network administrators to set traffic policies designed to guarantee that both mission-critical and routine data traffic are delivered to the network in a timely and consistent manner.

Some of these policy-based tools for IP are available as software solutions that are installed on existing routers

Table Q-1 Prioritization Schemes Used for Frame Relay Service, as Implemented in the Router's Operating System

QoS Mechanism	Description	Applications
Rate enforcement on a per-VC basis	The peak rate value for outbound traffic can be set to match the committed information rate (CIR) to provide a constant bit rate.	Suited for real-time applications such as voice, data streaming, and large file transfers.
Dynamic traffic throttling on a per-VC basis	When backward-explicit congestion notification (BECN) packets indicate congestion on the network, the outbound traffic rate is automatically stepped down by 25%; when congestion eases, the outbound traffic rate is allowed to increase.	This network function ensures that all traffic gets the minimum acceptable incoming or outgoing CIR during times of congestion Routers that do not respond to BECN risk having their traffic discarded.
Enhanced queue support on a per-VC basis	Either custom queue or priority output queue can be configured for individual virtual circuits (VC).	Custom queuing is used in environments that need to guarantee a minimal level of service to all applications.Priority output queuing is used to give mission-critical data the highest priority and hold back less critical traffic during periods of congestion.

Table Q-2 Select Prioritization Schemes Used for IP Service, as Implemented in the Router's Operating System

QoS Mechanism	Description	Applications
Resource ReSerVation Protocol (RSVP)	Sets up resources through the network to deliver the data stream to each router on the network that has attached subscribers who have preregistered to receive it.	Suited for real-time applications such as scheduled audio/video multicasts, computer-based training (CBT), and distance learning.
Protocol Independent Multicast (PIM)	Sends the data stream only once from the server, which is replicated at a rendezvous point (RP) only as many times as necessary to reach the nearest subscribers who have registered to receive it.	Handles same applications as above but is more bandwidth-efficient than RSVP. PIM also con serves processing resources at the server, since the stream goes out to the network only once.
Real-Time Protocol (RTP)	Sequentially tags IP packets to enable proper reassembly of the packet stream at the receiving end point before conversion to the real-time application.	Handles real-time, multicast and simulation applications but does not set up network resources, as do RSVP and PIM. RTP is augmented by a control protocol (RTCP) to allow monitoring of data delivery and provide minimal control and identification functionality.
IP Precedence	Expedites the handling of IP packets based on the partitioning of the traffic into as many as six classes that can be indicated in the type of service (ToS) field of the IPv4 header.	Handles a range of real-time and non-time- sensitive applications on the basis of the class of service they are assigned.

Table Q-2 *(Continued)*

QoS Mechanism	Description	Applications
Differentiated Services (Diffserv)	Supersedes the original IP Precedence specification for defining packet priority. Diffserv first prioritizes traffic by class and then differentiates and prioritizes same-class traffic, offering finer priority granularity.	Satisfies differing performance needs of services and applications on the basis of the QoS specified by each packet.

located on the edges of the network, while others are implemented in hardware, requiring the purchase of dedicated devices that are also deployed at the edges of the network. The tool vendors typically specialize in IP because intranets based on the TCP/IP protocol suite are very economical and globally available.

Summary

Policy-based QoS management tools make it easier to configure and control network resources to accommodate network changes and new applications while staying ahead of the performance curve—things that cannot always be achieved merely by adding cheap bandwidth. Smaller companies that appreciate the value of bandwidth management but that do not have the resources to do it themselves can opt for the services of an integrated communications provider (ICP). The ICP configures, installs, and manages the customer premises equipment as well as the access and transport links end to end for the optimal performance of all applications in keeping with a service-level agreement that is verified with performance reports and backed up with lifecycle services.

See also

Asynchronous Transfer Mode

Bandwidth Management Systems

Integrated Access Devices

Inverse Multiplexers

Multiservice Networking

Voice-Data Convergence

R

REPEATERS

A repeater is a device that extends the inherent distance limitations of various transmission media, including wireless links, by boosting signal power so that it stays at the same level regardless of the distance the signal must travel. This type of device is particularly useful for extending the coverage area of Wi-Fi networks throughout a community for Internet access. As such, the repeater operates at the lowest level of the Open Systems Interconnection (OSI) reference model—the Physical Layer (Figure R-1).

Repeaters are necessary because signal strength weakens with distance: The longer the path a signal must travel, the weaker it gets. This condition is known as "signal attenuation." On a telephone call, a weak signal will cause low volume, interfering with the parties' ability to hear each other. In cellular networks, when a mobile user moves beyond the range of a cell site, the signal fades to the point of disconnecting the call. In the local area network (LAN) environment, a weak signal can result in corrupt data, which can substantially reduce throughput by forcing retransmissions when errors are detected. When the signal level drops low enough, the chances of interference from external noise increase, rendering the signal unusable.

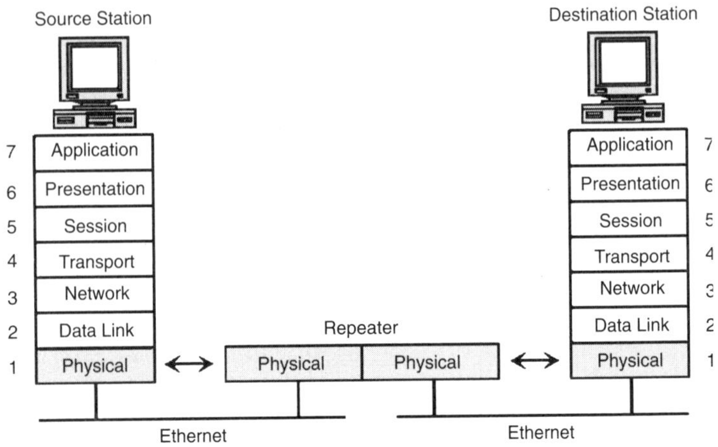

Figure R-1 Repeaters operate at the Physical Layer of the OSI Reference Model.

Repeaters also can be used to link different types of network media—fiber to coaxial cable, for example. Often LANs are interconnected in a campus environment by means of repeaters that form the LANs into connected network segments. The segments may employ different transmission media—thick or thin coaxial cable, twisted-pair wiring, or optical fiber. The cost of media converters is significantly less than full repeaters and can be used whenever media distance limitations will not be exceeded in the network.

Hubs or switches usually are equipped with appropriate modules that perform the repeater and media conversion functions on sprawling LANs. But the use of hubs or switches also can eliminate the need for repeaters, since most cable segments in office buildings will not run more than 100 feet (about 30 meters), which is well within the distance limitation of most LAN standards, including 1000BaseT Gigabit Ethernet running over Category 5 cable.

Regenerators

Often the terms *repeater* and *regenerator* are used interchangeably, but there is a subtle difference between the two. In an analog system, a repeater boosts the desired signal strength but also boosts the noise level as well. Consequently, the signal-to-noise ratio on the output side of the repeater remains the same as on the input side. This means that once noise is introduced into the desired signal, it is impossible to get the signal back into its original form again on the output side of the repeater.

In a digital system, regenerators are used instead of repeaters. The regenerator determines whether the information-carrying bits are 1s or 0s on the basis of the received signal on the input side. Once the decision of 1 or 0 is made, a fresh signal representing that bit is transmitted on the output side of the regenerator. Because the quality of the output signal is a perfect replication of the input signal, it is possible to maintain a very high level of performance over a range of transmission impairments. Noise, for instance, is filtered out because it is not represented as a 1 or 0.

Summary

Stand-alone repeaters have transceiver interface modules that provide connections to various media. There are fiberoptic transceivers, coaxial transceivers, and twisted-pair transceivers. Some repeaters contain the intelligence to detect packet collisions and will not repeat collision fragments to other cable segments. Some repeaters also can "deinsert" themselves from a hub or switch when there are excessive errors on the cable segment, and they can submit performance information to a central management station.

See also

Bridges

Gateways

Routers

ROUTERS

A router operates at layer 3 of the OSI Reference Model, the Network Layer. The device distinguishes among Network Layer protocols—such as Internet Protocol (IP), Internet Packet Exchange (IPX), AppleTalk, and DEC Local Area Telephony (LAT)—and makes intelligent packet delivery decisions using an appropriate routing protocol. It can be used to segment a network with the goals of limiting broadcast traffic and providing security, control, and redundant paths. A router also can provide multiple types of interfaces, including those for T1, frame relay, Integrated Services Digital Network (ISDN), Asynchronous Transfer Mode (ATM), cable networks, and Digital Subscriber Line (DSL) services, among others. Some routers can perform simple packet filtering to control the kind of traffic that is allowed to pass through them, providing a rudimentary firewall service. Larger routers can perform advanced firewall functions.

A router is similar to a bridge in that both provide filtering and bridging functions across the network. But while bridges operate at the Physical and Data Link Layers of the OSI Reference Model, routers join LANs at the Network Layer (Figure R-2). Routers convert LAN protocols into wide area network (WAN) protocols and perform the process in reverse at the remote location. They may be deployed in mesh as well as point-to-point networks and, in certain situations, can be used in combination with bridges.

Although routers include the functionality of bridges, they differ from bridges in the following ways: They generally offer more embedded intelligence and, consequently, more sophisticated network management and traffic control capabilities than bridges. Another distinction—perhaps the

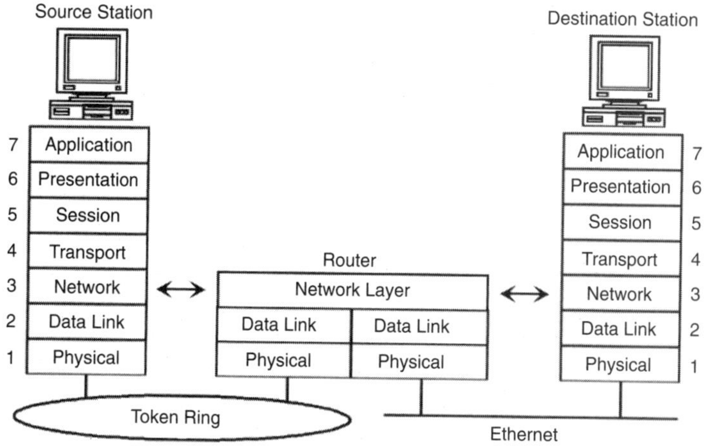

Figure R-2 Routers operate at the Network Layer of the OSI Reference Model.

most significant one—between a router and a bridge is that a bridge delivers packets of data on a "best effort" basis, specifically by discarding packets it does not recognize onto an adjacent network. Through a continual process of discarding unfamiliar packets, data get to their proper destination—on a network where the bridge recognizes the packets as belonging to a device attached to its network. By contrast, a router takes a more intelligent approach to getting packets to their destination—by selecting the most economical path (i.e., least number of hops) on the basis of its knowledge of the overall network topology, as defined by its internal routing table. Routers also have flow control and error protection capabilities.

Types of Routing

There are two types of routing: static and dynamic. In static routing, the network manager configures the routing table to set fixed paths between two routers. Unless reconfigured, the paths on the network never change. Although a static router will recognize that a link has gone down and issue an alarm, it will not automatically reroute traffic. A dynamic router, on the other hand, reconfigures the routing table automatically and recalculates the most efficient path in terms of load, line delay, or bandwidth.

Some routers balance the traffic load across multiple access links, providing an $N \times$ T1 inverse multiplexer function. This allows multiple T1 access lines operating at 1.544 Mbps each to be used as a single higher-bandwidth facility. If one of the links fails, the other links remain in place to handle the offered traffic. As soon as the failed link is restored to service, traffic is spread across the entire group of lines as in the original configuration.

Routing Protocols

Each router on the network keeps a routing table and moves data along the network from one router to the next using such protocols as Open Shortest Path First (OSPF) and Routing Information Protocol (RIP).

Although still supported by many vendors, RIP does not perform well in today's increasingly complex networks. As the network expands, routing updates grow larger under RIP and consume more bandwidth to route the information. When a link fails, the RIP update procedure slows route discovery, increases network traffic and bandwidth usage, and may cause temporary looping of data traffic. Also, RIP cannot base route selection on such factors as delay and bandwidth, and its line selection facility is capable of choosing only one path to each destination.

The newer routing standard, OSPF, overcomes the limitations of RIP and even provides capabilities not found in RIP. The update procedure of OSPF requires that each router on the network transmit a packet with a description of its local links to all other routers. On receiving each packet, the other routers acknowledge it, and in the process, distributed routing tables are built from the collected descriptions. Since these description packets are relatively small, they produce a minimum of overhead. When a link fails, updated information floods the network, allowing all the routers to simultaneously calculate new tables.

Types of Routers

Multiprotocol nodal, or hub, routers are used for building highly meshed internetworks. In addition to allowing several protocols to share the same logical network, these devices pick the shortest path to the end node, balance the load across multiple physical links, reroute traffic around points of failure or congestion, and implement flow control in conjunction with the end nodes. They also provide the means to tie remote branch offices into the corporate backbone, which might use such WAN services as TCP/IP, T1, ISDN, and ATM.

Access routers are typically used at branch offices. These are usually fixed-configuration devices available in Ethernet and Token Ring versions, which support a limited number of protocols and physical interfaces. They provide connectivity to high-end multiprotocol routers, allowing large and small nodes to be managed as a single logical enterprise network. Although low-cost, plug-and-play bridges can meet the need for branch office connectivity, low-end routers can offer more intelligence and configuration flexibility at comparable cost.

The newest access routers are multiservice devices, which are designed to handle a mix of data, voice, and video traffic. They support a variety of WAN connections through built-in interfaces that include dual ISDN BRI interfaces, dual analog ports, a T1/frame relay port, and an ISDN interface for videoconferencing. Such routers can run software that provides Internet Protocol Security (IPSec) VPN, firewall, and encryption services.

Midrange routers provide network connectivity between corporate locations in support of workgroups or the corporate intranet, for example. These routers can be stand-alone devices or packaged as modules that occupy slots in an intelligent wiring hub or LAN switch. In fact, this type of router is often used to provide connectivity between multiple wiring hubs or LAN switches over high-speed ATM or Ethernet backbones.

There is a consumer class of router that provides shared access to the Internet over such broadband technologies as cable and DSL. They are used to connect a small group of PCs to a high-speed Internet connection or to an Ethernet backbone. Configurable through any networked PC's Web browser, the router can be set up as a firewall and Dynamic Host Configuration Protocol (DHCP) server, allowing it to act as an externally recognized Internet device with its own IP address for the home LAN.

These routers also support Network Address Translation (NAT), a feature that translates one public IP address, given by the cable or DSL Internet provider, and assigns automatically up to 253 private IP addresses to users on the LAN. All the users given an IP address by the router are safe behind the firewall, so incoming and outgoing requests are filtered, keeping unwanted requests off the LAN.

At the same time, the router supports a feature called DMZ/Expose Host, which disassembles 1 of the 253 private IP addresses to become a public IP address so that outside users can access that PC without getting blocked by the firewall. An example would be a gamer playing another gamer via the Internet. They want to access each other's computers so that they can play the game.

Summary

Routers fulfill a vital role in implementing complex mesh networks such as the Internet and private intranets using Layer 3 protocols, usually IP. They also have become an economical means of tying branch offices into the enterprise network and providing PCs tied together on a home network with shared access to broadband Internet services such as cable and DSL. Like other interconnection devices, routers are manageable via Simple Network Management Protocol (SNMP), as well as the proprietary management systems of vendors. Just as bridging and routing functions made their way into a single device, routing and switching functions are being combined in the same way, and even add firewall, DHCP, and NAT capabilities.

See also

Bridges
Firewalls
Inverse Multiplexers

S

SIMPLE NETWORK MANAGEMENT PROTOCOL

Since 1988, the Simple Network Management Protocol (SNMP) has been the de facto standard for the management of multivendor Transmission Control Protocol/Internet Protocol (TCP/IP)–based networks. SNMP specifies a structure for formatting messages and for transmitting information between reporting devices and data-collection programs on the network. The SNMP-compliant devices on the network are polled for performance-related information, which is passed to a network management console. Alarms are also passed to the console. There, the gathered information can be viewed to pinpoint problems on the network or stored for later analysis.

SNMP runs on top of TCP/IP's datagram protocol—the User Datagram Protocol (UDP)—a transport protocol that offers a connectionless-mode service. This means that a session need not be established before network management information can be passed to the central control point. Although SNMP messages can be exchanged across any protocol, UDP is well suited to the brief request/response message exchanges characteristic of network management communications.

SNMP is a very flexible network management protocol that can be used to manage virtually any object, even Open System Interconnection (OSI) objects. An "object" refers to hardware, software, or a logical association, such as a connection or virtual circuit. An object's definition is written by the equipment vendor and is held in a management information base (MIB). The MIB is simply a list of switch settings, hardware counters, in-memory variables, or files that are used by the network management system to determine the alarm and reporting characteristics of each device on the network, including those connected over local area networks (LANs).

As noted, SNMP is basically a request/response protocol. The management system retrieves information from the agents through SNMP's "get" and "get-next" commands. The "get" request retrieves the values of specific objects from the MIB. The MIB lists the network objects for which an agent can return values. These values may include the number of input packets, the number of input errors, and routing information. The "get-next" request permits navigation of the MIB, enabling the next MIB object to be retrieved, relative to its current position. A "set" request is used to request a logically remote agent to alter the values of variables. In addition to these message types, there are "trap" messages, which are unsolicited messages conveyed from management agent to management stations. Other commands are available that allow the network manager to take specific actions to control the network. Although these commands look like SNMP commands, they are really vendor-specific implementations. For example, some vendors use a "stat" command to determine the status of network connections.

All the major network management platforms support SNMP. In addition, many of the third-party systems and network management applications that plug into these platforms support SNMP. The advantage of using such products is that they take advantage of SNMP's capabilities while providing a graphical user interface (GUI) to make SNMP easier to use. Even MIBs can be selected for display and navigation through the GUI.

Another advantage of commercial products is that they can use SNMP to provide additional functionality. For example, Hewlett-Packard's OpenView is used to manage network devices that are IP addressable and run SNMP. Their automatic discovery capability finds and identifies all IP nodes on the network, including those of other vendors that support SNMP. On the basis of discovered information, the management system automatically draws a network topology map. Nodes that cannot be discovered automatically can be represented in either of two ways: first, by manually adding custom or standard icons to the appropriate map views; second, by using SNMP-based Application Program Interfaces (APIs) for building map applications without having to manually modify the configuration to accommodate non-SNMP devices.

Architectural Components

SNMP is one of three components constituting a total network management system (Figure S-1). The other two are the MIB and the network manager (NM). The MIB defines the controls embedded in network components, while the NM contains the tools that enable network administrators to comprehend the state of the network from the gathered information.

Network Manager

The network manager is a program that may run on one host or more than one host, with each responsible for a particular subnet. SNMP communicates network management data to a single site, called a "network management station" (NMS). Under SNMP, each network segment must have a device, called an "agent," that can monitor devices (called "objects") on that segment and report the information to the NMS. The

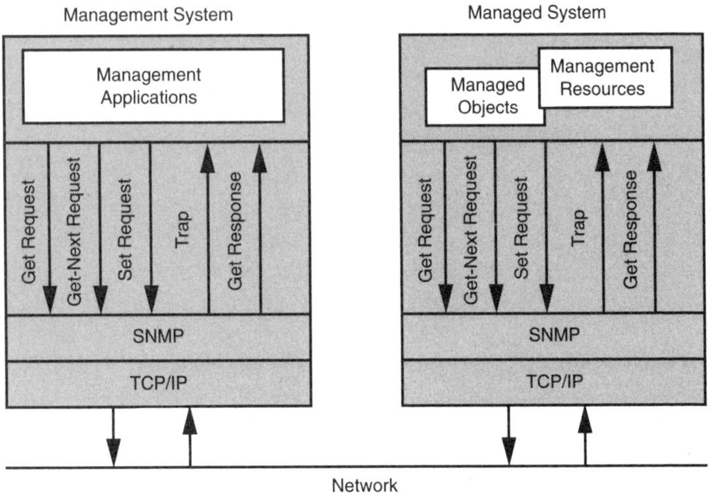

Figure S-1 The SNMP architecture.

agent may be a passive monitoring device whose sole purpose is to read the network, or it may be an active device that performs other functions as well, such as bridging, routing, and switching. Devices that are non-SNMP-compliant must be linked to the NMS via a proxy agent.

The NMS provides the information display, communication with agents, information filtering, and control capabilities. The agents and their appropriate information are displayed in a graphical format, often against a network map. Network technicians and administrators can query the agents and read the responses on the NMS display. The NMS also periodically polls the agents, searching for anomalies. Detection of an anomaly results in an alarm at the NMS.

Management Information Base

The MIB is a list of objects necessary to manage an entity on the network. As noted, an object refers to hardware, software, or a logical association such as a connection or virtual circuit. The attributes of an object might include such things as the number of packets sent, routing table entries, and protocol-specific variables for IP routing. A basic object of any MIB is sysDescr, which is a textual description of the entity. This value includes the full name and version identification of the system's hardware type, software operating system, and networking software. This object should contain only printable ASCII characters.

The first MIB was primarily concerned with IP routing variables used for interconnecting different networks. There are 110 objects that form the core of the standard SNMP MIB. The latest generation MIB, known as MIB II, defines over 160 objects. It extends SNMP capabilities to a variety of media and network devices, marking a shift from Ethernets and TCP/IP wide area networks (WANs) to all media types used on LANs and WANs. Many vendors want to add value to their products by making them more manageable, so they create private extensions to the standard MIB, which can include 200 or more additional objects.

Many vendors of SNMP-compliant products include MIB tool kits that generally include two types of utilities. One, an MIB compiler, acts as a translator that converts ASCII text files of MIBs for use by an SNMP management station. The second type of MIB tool converts the translator's output into a format that can be used by the management station's applications or graphics. These output handlers, also known as "MIB editors" or "MIB walkers," let users view the MIB and select the variables to be included in the management system. Some vendors of SNMP management stations do not offer MIB tool kits but rather an optional service whereby they will integrate into the management system any MIB a user requires for a given network. This service includes debugging and technical support.

There are also MIB browsers that allow network managers, technicians, and engineers to query a remote device for software and hardware configurations via SNMP and make changes to the remote device. The remote device could be a router, switch hub, server, firewall, or any other device that supports SNMP. Another common use for an MIB browser is to find out what MIBs and object IDs (OIDs) are supported on a particular device.

Summary

SNMP's popularity stems from the fact that it works, it is reliable, and it is widely supported. The protocol itself is in the public domain. SNMP capabilities have been integrated into just about every conceivable device that is used on today's LANs and WANs. MIBs contain a list of objects that can be monitored by the SNMP.

See also

Network Statistics
Ping
Protocol Analyzers

SMART BUILDINGS

So-called smart buildings are multitenant office or residential buildings that provide the latest telecommunications technologies. Usually, one or more carriers enter into an agreement with the building owner to provide a variety of services to building tenants at very attractive rates. Using cutting-edge technologies, including fixed wireless and fiber networks, providers offer tenants discounted rates on bundles of services that may include

- Local and long-distance calling
- High-speed Internet access
- Broadband data services via metro Ethernet services
- Paging
- Faxing

- Voice messaging
- Videoconferencing
- Television programming and interactive television

Tenants get one-stop shopping for all their telecommunications needs, and their services cost less—up to 50 percent less than traditional phone companies. Previously, customers had one company for local phone service, another for long distance, and still another for services such as Internet access. Under the smart-building concept, customers can choose to have all these services bundled together, all under a single monthly bill from one company.

Service Installation

Upgrading an existing building to smart building standards can be done in various ways. One upgrade method is to deploy a fixed wireless technology called Local Multipoint Distribution Service (LMDS). With LMDS, antennas the size of a dinner plate are mounted on the rooftop. The installation process is simple, taking 1 or 2 days, and requires little wiring because the system depends on over-the-air transmissions. Only rarely can these receivers be viewed from street level.

The antenna is connected to an indoor unit at the tenant locations via a single coaxial cable. This unit has Ethernet and T-carrier ports, allowing customers to plug in existing routers, channel banks, private branch exchanges (PBXs), and videoconference systems. Once the LMDS system is installed, the service provider can configure it and implement future changes to the indoor unit right over the air link. A serial port on the indoor unit connects to a land line, allowing the carrier's network operation center to access the system in the event of an air link failure.

In most cases, LMDS provides an Asynchronous Transfer Mode (ATM)–based service that can support mixed types of traffic—voice, data, and video. Despite this promising convergence capability, LMDS has been deployed primarily to meet customers' needs for high-speed Internet access. The reputation of LMDS has been tarnished by continued controversy in the industry about the technology's susceptibility to rain fade—the disruption of the signal during heavy rain.

Another way to upgrade an existing building is through wireline technology. This entails running fiber from the basement of the building to the top floor inside the building's vertical utility shaft. A junction box installed on each floor enables the connection of a discrete pair of fibers from the vertical riser in the utility shaft to the customer's LAN. The junction boxes are connected to an intermediate frame distribution panel on every sixth or seventh floor. Depending on the number of tenants in a building, the service provider may pull 48 to 96 strands of fiber through the vertical utility shaft.

Inside the building, usually in the basement, the service provider also establishes a building point of presence (POP), where it installs routers and other equipment that enables the transmission of data and video traffic and aggregates and distributes traffic to and from the fiber infrastructure. The service provider typically obtains the right to use a small amount of space in the basement of the buildings to establish the POP.

Within each metropolitan area, there is a POP at which the service provider aggregates and distributes traffic to and from all its interconnected buildings. The buildings are connected to the metropolitan POP via broadband data circuits leased from carriers that provide local transport capacity. Some service providers also lease lines to offer branch office connections so that their customers can exchange data with their remote locations.

Each POP is connected to the service provider's network operations center (NOC), which manages and monitors network traffic on a 24 × 7 basis. From the NOC, technical staff provision, activate, maintain, and troubleshoot circuits and equipment. Traffic and service-level statistics are gathered to report performance, plan additional capacity, and communicate changes to customers.

Summary

The smart building industry points out that buildings offering advanced communications services are higher-valued buildings. These communications services create work and living spaces that are more competitive and sought after in the marketplace. It is believed that people who work and live in smart buildings are more satisfied and less likely to leave. In turn, owners face less churn and less exposure to financial loss. The carriers assume all installation costs and are responsible for repairs and payments for damages. In many cases, building owners and service providers jointly market telecommunications offerings to tenants. The owners may receive up-front payments for access to their buildings and may even share in the ongoing revenues from monthly usage.

See also

Building Local Exchange Carriers
Local Multipoint Distribution Services

STORAGE AREA NETWORKS

A storage area network (SAN) is a specialized network that enables fast, reliable access among servers and external or independent storage resources regardless of physical location. Fibre Channel or Gigabit Ethernet links can provide high-speed transfers of data between systems distributed within a building, campus, or metropolitan area. For longer distances, ATM and IP technologies can be used to transport data over the WAN.

In a SAN, a storage device is not the exclusive property of any one server. Rather, storage devices are shared among all networked servers as peer resources. Just as a LAN can be used to connect clients to servers, a SAN can be used to connect servers to storage, servers to each other, and storage to storage for load balancing and protection.

SAN Advantages

Redundancy is an inherent part of the SAN architecture, making for high availability. The "pluggable" nature of SAN resources—storage, nodes, and clients—enables much easier scalability while preserving ubiquitous data access. And under centralized management, there is more efficiency in carrying out tasks such as optimization, reconfiguration, and backup/restore.

SANs are particularly useful for backups. Previously, there were only two choices: Either a tape drive had to be installed on every server and someone went around changing the tapes, or a backup server was created and the data moved across the network, which consumed bandwidth. Performing backup over the LAN can be excruciatingly disruptive and slow. A daily backup can introduce gigabytes of data into the normal LAN traffic. With SANs, organizations can have the best of both worlds: high-speed backups managed from a central location.

Instead of dedicating a specific kind of storage to one or more servers, a SAN allows different kinds of storage—mainframe disk, tape, and Redundant Array of Inexpensive Disk (RAID)—to be shared by different kinds of servers, such as Windows NT/2000, UNIX, and OS/390. With this shared capacity, organizations can acquire, deploy, and use storage devices more efficiently and cost-effectively. ATM would be adept at connecting heterogeneous storage resources over the WAN because it slices and dices different protocol traffic into standardized packets called "cells" for high-speed, jitter-free transmission between distributed storage nodes.

With a SAN, there is no need for a physically separate network to handle storage and archival traffic. This is so because the SAN can function as a virtual subnet that operates on a shared network infrastructure. For this to work, however, different priorities or classes of service must be established. Fortunately, both Fibre Channel and ATM provide the means to set different classes of service, and this capability can be added to IP.

All of this makes SANs highly suited for data-intensive environments like those used for video editing, prepress, online transaction processing (OLTP), data warehousing, storage management, and server clustering applications.

Fibre Channel

SANs have existed for years in the mainframe environment in the form of Enterprise Systems Connection (ESCON). In midrange environments, the high-speed data connection was primarily Small Computer System Interface (SCSI)—a point-to-point connection that is severely limited in terms of the number of connected devices it can support as well as the distance between devices. Fibre Channel overcomes these limitations.

While traditional SCSI allows only a 25-meter distance (about 82 feet) between machines and Ultra2 SCSI allows only a 12-meter distance (about 40 feet), Fibre Channel supports spans of 10 kilometers (about 6.2 miles), making it suited to building campus-wide storage networks. SCSI can only connect up to 16 devices, whereas Fibre Channel can link as many as 127. By combining LAN networking models with the core building blocks of server performance and mass storage capacity, SANs eliminate the bandwidth bottlenecks and scalability limitations imposed by previous SCSI bus-based architectures.

More recently, vendors have pushed the speed of Fibre Channel from 1 to 2 Gbps and increased the distance beyond the original 6.2 miles to about 75 miles. As the SAN concept has evolved, it has moved beyond association with any single technology. In fact, just as LANs and WANs use a diverse mix of technologies, so can SANs. This mix can include Fiber Distributed Data Interface (FDDI), ATM, and IBM's Serial Storage Architecture (SSA), as well as Fibre Channel. Synchronous Optical Network (SONET) and Dense Wave Division Multiplexing (DWDM) have been added to the mix to extend the operating range of storage networks. Even the TCP/IP suite of Internet protocols is being used for a more economic implementation of storage networks.

Although early implementations of SANs have been local or campus-based, there is no technological reason why they cannot be extended much farther with such proven technologies such as SONET and ATM. With its 50-millisecond recovery time, SONET also offers the benefit of extremely high resiliency, which has yet to be matched by any other transport technology, including Fibre Channel. Under SONET, data travel to their destination in opposite directions over a dual-ring architecture (Figure S-2). If one of the fibers is cut or a node fails, protection mechanisms kick in to ensure that data get to their destination with little or no loss. ATM's quality-of-service (QoS) capabilities and priority queuing techniques allow the SAN to be extended over a much wider area—perhaps globally—with little or no performance fatigue

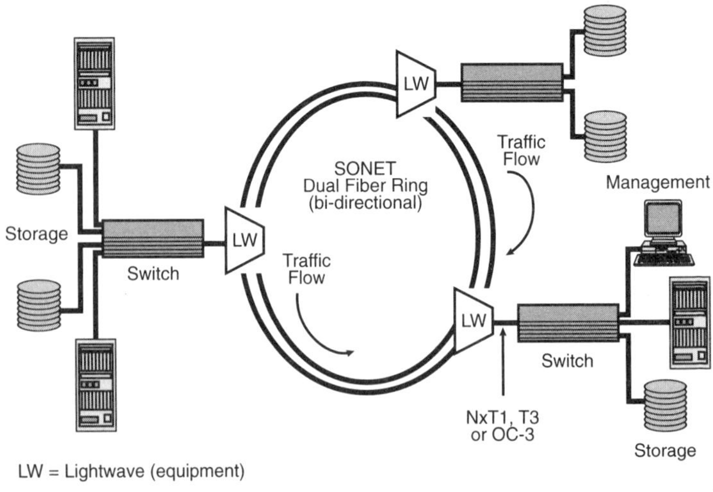

Figure S-2 For metropolitan SANs, SONET offers the highest resiliency of any transport technology. This ensures that data get to their proper destination with little or no loss, even if a fiber gets cut or a node on the ring fails.

SAN Components

Several components are required to implement a SAN. A Fibre Channel adapter is installed in each server. These are connected via the server's PC interface (PCI) bus to the server's operating system and applications. Because Fibre Channel's transport-level protocol wraps easily around SCSI frames, the adapter appears to be a SCSI device.

The adapters are connected to a single Fibre Channel hub running over fiberoptic cable or copper coaxial cable. Category 5, the unshielded twisted-pair wiring rated for 100-Mbps Fast Ethernet and 155-Mbps ATM, also can be used.

A LAN-free backup architecture may include some type of automated tape library that attaches to the hub via Fibre Channel. This machine typically includes a mechanism capable of feeding data to multiple tape drives and may be bundled with a front-end Fibre Channel controller. Existing SCSI-based tape drives can be used through the addition of a Fibre Channel-to-SCSI bridge.

Storage management software running in the servers performs contention management by communicating with other servers via a control protocol to synchronize access to the tape library. The control protocol maintains a master index and uses data maps and time stamps to establish the server-to-hub connections.

Currently, control protocols are specific to the software vendors. Eventually, the storage industry likely will standardize on one of the several protocols now in proposal status before the Storage Network Industry Association (SNIA).

From the hub, a standard Fibre Channel protocol, Fibre Channel–Arbitrated Loop (FC-AL), functions similarly to Token Ring to ensure collision-free data transfers to the storage devices. The hub also contains an embedded SNMP agent for reporting to network management software.

Emerging Role of IP

One of the hottest new trends in building SANs is the use of the ubiquitous IP, which can be used to connect storage devices and servers more economically than Fibre Channel and ATM.

One protocol, IP SAN, also known as iSCSI (Internet Protocol Small Computer System Interface), uses the IP networking infrastructure to transport large amounts of block storage data over existing LANs and WANs. Among the companies with IP SAN solutions is IBM. The company's IP Storage 200i provides storage that is directly attachable to an Ethernet LAN. This solution supports heterogeneous Windows NT, Windows 2000, and Linux clients, enabling users to take advantage of many SAN-like capabilities without the infrastructure and support cost of Fibre Channel SAN environments.

With the potential to support all major networking protocols, IP SAN can unify network architecture across an entire enterprise, reducing the overall network cost and complexity while ensuring widespread availability. To facilitate administration, IP SAN can use known network management tools and utilities that have been developed for IP networks. IBM's IP Storage 200i, for example, comes with a browser-based interface that allows system administrators to easily configure the system, set permissions, and implement changes from anywhere on the network.

To meet an organization's diverse connectivity requirements, there are switches that address the challenges of connecting multiple SAN islands across a variety of network topologies. Such switches feature an assortment of network connection options that include T3 for today's ATM-based WANs, OC-3 and higher feeds for WAN/metropolitan area network (MAN) networks, and Gigabit Ethernet for implementing SANs over existing high-speed IP networks. In supporting connectivity for the most commonly used WAN services, these all-in-one solutions meet today's current and emerging storage needs while allowing for future technology migration, bandwidth scalability, and convergence.

Summary

Companies faced with a continuous bombardment of information are turning to SANs to house, manage, and protect this vital asset. While NAS is intended for data access at the file level, SANs are optimized for high-volume block-oriented data transfers. Although both solutions address the need to remove direct storage-to-server connections to facilitate more flexible storage access, SANs provide a higher-performance and more scalable storage environment. They achieve this by enabling many direct connections between servers and storage devices—such as disk storage systems and tape libraries—over a variety of transports. The choice between Fibre Channel, Gigabit Ethernet, IP, or ATM will hinge on such factors as the distance between storage locations, the presence of other types of traffic, and the organization's budget constraints.

See also

Asynchronous Transfer Mode
Ethernet

STREAMING CONTENT

Streaming is a method of delivering content to subscribers over a network. The content can be in the form of a stock ticker, video program, news feed, movie preview, audio track, or DVD segment. Usually the stream is sent over the most economical transmission medium possible, which is an IP network such as the public Internet or private intranet. But streaming content may be sent over frame relay and ATM networks as well.

In an IP environment, the stream may be sent in either of two ways: unicast or multicast. Unicast delivery involves sending a separate data stream to each recipient. Multicast delivery involves sending only one data stream into the network, which is replicated only as many times as necessary to distribute the stream to the nodes (i.e., routers) with registered subscribers attached.

Multicast

Multicast has a number of applications. It is ideal for content providers with real-time applications, such as news and entertainment events, and for the distribution of dynamic content, such as financial information and sports scores. The application itself can be audio, video, or text—or any combination of these. For content providers, IP multicast is a low-cost way to supplement current broadcast feeds. In fact, the major news networks are among the biggest users of IP multicast.

Corporations can use IP multicast to deliver training to employees and keep them informed of internal news, benefits programs, and employment opportunities within the organization. They also can use IP multicast to broadcast annual meetings to shareholders or introduce new products to their sales channels. Associations can use IP multicast to broadcast conference sessions and seminars to members who would not otherwise be able to attend in person.

Political parties and issue advocacy groups can use IP multicast to keep their members informed of late-breaking developments and call them to action. Entrepreneurs can use IP multicast to offer alternative programming to the growing base of Internet users.

Performance

There is concern among potential users of IP multicast about the effects of delay on performance. After all, they have experienced the long delays accessing multimedia content on the Web. They see video that is slow and jerky and hear audio that pauses periodically until something called a “buffer” has a chance to fill. When video and audio run together, often the two are out of synchronization. Thus they wonder how the Internet can handle a real-time multicast with acceptable quality.

Currently, multicast works best on a managed IP backbone network, where a single company or carrier has control of all the equipment, protocols, and bandwidth end to end. This is not possible on the public Internet because there is no central management authority. While simple real-time applications may work well enough, such as stock tickers, the performance of a sophisticated graphically enriched real-time multimedia application suffers.

Not only can the performance of a private IP network be controlled to eliminate potential points of congestion and minimize delay, the company or carrier can place dedicated multicast routers throughout its network. This type of router replicates and distributes the content stream in a highly efficient way that does not require massive amounts of bandwidth.

With Protocol Independent Multicast (PIM), for example, instead of sending out 100 information streams to 100 subscribers, only one information stream is sent from the source server. The multicast routers replicate and distribute the stream within the network to only the nodes that have subscribers (Figure S-3) who requested the stream through a registration process.

When subscribers join a multicast group, the directly connected routers send PIM "join" messages to the rendezvous point (RP). The RP keeps track of multicast groups. Servers that send multicast packets are registered with the RP by the first-hop router. The RP then sends "join" messages toward the source. At this point, packets are forwarded on a shared distribution tree. The result is that content providers no longer need to purchase enormous amounts of bandwidth to accommodate a large number of subscribers or buy multiple high-capacity servers to send out all the data streams. Instead, a single data stream is sent, the size of which is based on the type of content.

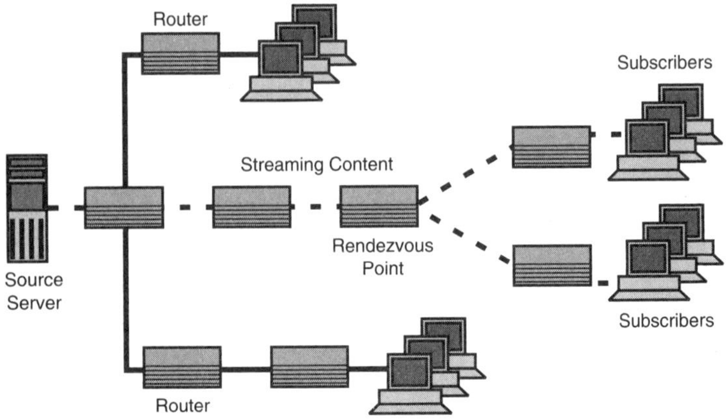

Figure S-3 In Protocol Independent Multicast (PIM), streaming content goes out the server one time and is replicated at the rendezvous point (RP) to reach the nearest subscribers who have specifically requested the stream. This method of content delivery reduces the processing burden of the source server and conserves network bandwidth.

Registration

A multicast can reach potentially anyone who specifically subscribes to the session—whether they have a dedicated connection or use a dial-up modem connection. Of course, the content originator can put distance limits on the transmission and restrict the number of subscribers that will be accepted for any given program.

A variety of methods can be used to advertise a multicast. A program guide can be sent to employees and other appropriate parties via e-mail, or it can be posted on a Web site. If the company already has an information channel on the Web that delivers content to subscribers, the program guide can be one of the items "pushed" to users when they access the channel.

When a person wants to receive a program, he or she enrolls through an automated registration procedure. The request is sent to the server running the multicast, which adds the subscriber's IP address to its subscriber list. In this way, only users who want to participate will receive packets from the server.

The user also selects a multicast node from those listed in the program guide. Usually, this will be the router closest to the user's location. The user becomes a member of this particular node. Group membership information is distributed among neighboring routers so that multicast traffic gets routed only along paths that have subscribers at the end nodes. From the end node, the data stream is delivered right to the user's computer.

Once the session is started, users can join and leave the multicast group at any time. The multicast routers adapt to the addition or deletion of network addresses dynamically, so the data stream gets to new destinations when users join and is stopped from going to destinations that no longer want to receive the session.

Multicast Services

For companies that understand the value of multicast but prefer not to handle it themselves, multicast host services are available from such sources as UUNET, which offers a multicast hosting service called UUCast. As summarized in Table S-1, the company offers six data streams of varying size to accommodate virtually any real-time data transmission.

Table S-1 Data Streams Offered under UUNET's Multicast Hosting Service Called UUcast

Data Stream Size. Kbps	Application Example	End-User Access Speed
5	Ticker banner	Dial-up modem up to 56 kbps
10	Ticker banner or audio message	Dial-up modem up to 56 kbps
25	Audio/video applications	Dial-up modem up to 56 kbps
35	Audio/video applications	Dial-up modem up to 56 kbps
64	High-quality audio/video	ISDN up to 128 kbps
128	High-quality audio/video	ISDN up to 128 kbps

UUCast requires a dedicated UUNET connection. The subscribing organization supplies its own content and equipment. On installation, the organization is given a unique multicast group address for each of the data streams. The organization's router is configured with a virtual point-to-point connection to the multicast router located in a local UUNET point of presence (POP). Since these multicast routers only transmit data streams to the corresponding multicast group address, there is no interference with other traffic sources.

UUNET has equipped all domestic POPs with multicast routers, so any of its dial-up customers also can receive the data stream, if it is made available to them. A dial-access router located within each POP recognizes the request for a particular multicast data stream and begins transmitting the appropriate content to the end-user's desktop.

Costs

Telecommunications costs are always an important consideration. Companies are continually looking for ways to save money in this area and are understandably interested in the cost of multicasting.

The cost to implement multicast on an internal IP network is minimal because an existing infrastructure is simply being leveraged. There might be some upgrades to hardware and software and possibly the need for management tools to monitor the multicasts. Organizations that are serious about IP multicast often have a dedicated full-time administrator.

An overlooked cost comes in the form of continuously developing the multicast content. This could take a whole staff of creative people with specialized skills—writers, editors, graphic artists, audio/video production people, and a multimedia server administrator—not to mention all the expensive equipment and facilities they will need.

The cost of production will hinge on the type of content to be developed. Obviously, it will be much cheaper to use text only, but not very many multicast applications will attract viewers if only text is involved. Production costs jump dramatically as audio and video components are added because special equipment and expertise are required. An alternative to the do-it-yourself approach is to outsource production to specialized firms. This can cut development costs by as much as 60 percent.

For companies that prefer to outsource IP multicast, companies such as UUNET offer a predictable price for a large-scale Internet broadcast, and they take care of all the server and router management. UUNET multicast hosting is available in dedicated configurations. The monthly fee is based on a server component, starting at $3000, and a multicast stream size, starting at $2200 per month for a 5-kbps stream, $10,000 per month for a 25-kbps stream, and $15,000 for a 35-kbps stream.

Table S-2 compares the cost of a unicast transmission to 1000 users on a private IP network (coast-to-coast) with a multicast transmission to 1000 users over the managed Internet backbone of UUNET.

The cost savings of multicasting over traditional unicasting is quite compelling. Of course, this scenario assumes

Table S-2 Comparison of Costs for Traditional Unicast versus Multicast via UUcast

Traditional Unicast Transmission	UUCast Transmission (Multicast)
1000 dial users ¥ 28.8 kbps per user = 28.8 Mbps of bandwidth required	One 35-kbps data stream required
One T3 connection (45 Mbps) (This unicast transmission will use approximately 50% of the T3.)	Tremendous amounts of bandwidth are no longer needed to support a single broadcast.
Monthly cost for full T3 = $54,000	Monthly cost: $15,000
Total savings = $39,000 per month, or $468,000 per year	

that the private T3 link is dedicated to unicasting and that the carrier does not yet offer fractional T3, in which case the monthly cost could be cut in half, to about $27,000. Even in this case, the monthly savings is $12,000, or $144,000 per year, with outsourcing.

Summary

As noted, the Internet is not centrally managed, making it difficult to convey sophisticated streaming content in real time with any consistency in performance. Congestion and delays are still obstacles that must be overcome. Other real-time applications, such as IP telephony, experience the same problem when run over the public Internet. The next big step toward full deployment of IP multicast is convincing more Internet service providers to offer the service to customers. The argument is not hard to make, especially since multimedia applications are obviously running on the Internet anyway, and multicasting actually conserves bandwidth. With only modest costs, mostly for router upgrades, wider support of multicast would improve the Internet's performance for everyone. Until then, multicasting will be most effective on private intranets and carrier-managed backbones where performance can be controlled end to end.

See also

Content Delivery Networks
Internet
Multiservice Networking
Quality of Service
Routers
Transmission Control Protocol/Internet Protocol
Voice-Data Convergence

T

T1 LINES

T1 lines are a type of T-carrier facility that provides a transmission rate of up to 1.544 Mbps using digital signal level 1 (DS1) signaling. Two pairs of wires are used to achieve full-duplex transmission—one pair for the send path and one pair for the receive path. The available bandwidth is divided into 24 channels operating at 64 kbps each, plus an 8-kbps channel for basic supervision and control.

Voice is sampled and digitized via Pulse Code Modulation (PCM) before being placed on the line. T1 digital lines are used for more economic and efficient transport of voice, data, and Internet traffic over the wide area network (WAN). Some channels can be designated to carry voice, while others can be designated to carry data or Internet traffic. The different traffic types are split out in the carrier's network via a digital cross-connect system (DCS), at which point they take different paths to their destinations.

Cost savings is the result of consolidating multiple lower-speed voice, data, and Internet channels via a multiplexer or channel bank and sending the traffic out over the higher-speed T1 line. This is more cost-effective than dedicating a separate lower-speed line to each terminal device connected to the WAN. The economics are such that only five to eight analog lines are needed to cost-justify the move to T1.

Greater bandwidth efficiency can be obtained by compressing voice and data to make room for even more channels over the available bandwidth. This can result in even more cost savings. Individual channels also can be dropped or inserted at various destinations along the line's route. Network management information can be embedded in each channel for enhanced levels of supervision and control.

Usually a T1 multiplexer provides the means for companies to realize the full benefits of T1 lines, while channel banks offer a lower-cost alternative. The difference between the two devices is that T1 multiplexers offer higher line capacity, support more types of modules and interfaces, and provide more management features than channel banks.

Summary

T1 lines are the basic building blocks of digital networks. They can support voice, data, and Internet traffic. The individual channels can be added to or dropped from the aggregate bandwidth to improve the efficiency and economy of a private network. In addition, these channels can traverse the public switched telephone network (PSTN) to bring off-net locations into the private network. Alternatively, the channels can go through a carrier's frame relay network or Asynchronous Transfer Mode (ATM) network, allowing even greater efficiencies and economies for certain applications. The channels of a T1 can even be bonded together to support bandwidth-intensive applications on private networks.

See also

Inverse Multiplexers

Voice Compression

TOKEN RING

Token Ring is a type of local area network (LAN) that was introduced by IBM in 1985. It had a top speed of 4 Mbps and was developed as a response to the commercial availability of Ethernet, which was developed jointly by DEC, Intel, and Xerox. When Ethernet was introduced, IBM did not endorse it, mainly because its equipment would not work in that environment. Later, in 1989, the speed of Token Ring was boosted to 16 Mbps. More recently, the speed of Token Ring has been boosted to 100 Mbps, but the technology is not as scalable as Ethernet, which now has a speed of 10 Gbps. Both types of networks support Internet Protocol (IP) traffic to the Internet, as well as a range of business applications.

The ring is essentially a closed loop, although various wiring configurations that employ a multistation access unit (MAU)[1] and patch panel may cause it to resemble a star topology (Figure T-1). In addition, today's intelligent wiring hubs and Token Ring switches can be used to create dedicated pipes between rings and provide switched connectivity between users on different rings.

The cable distance of a 4-Mbps Token Ring is limited to 1600 feet between stations, while the cable distance of a 16-Mbps Token Ring is 800 feet between stations. Because each node acts as a repeater in that data packets and the tokens are regenerated at their original signal strength, Token Ring networks are not as limited by distance, as are bus-type networks. Like its nearest rival, Ethernet, Token Ring networks normally use twisted-pair wiring, shielded or unshielded.

Advantages of Token Ring

The ring topology offers several advantages:

[1] This is a nonintelligent concentrator that can be used as the basis for implementing Token Ring LANs.

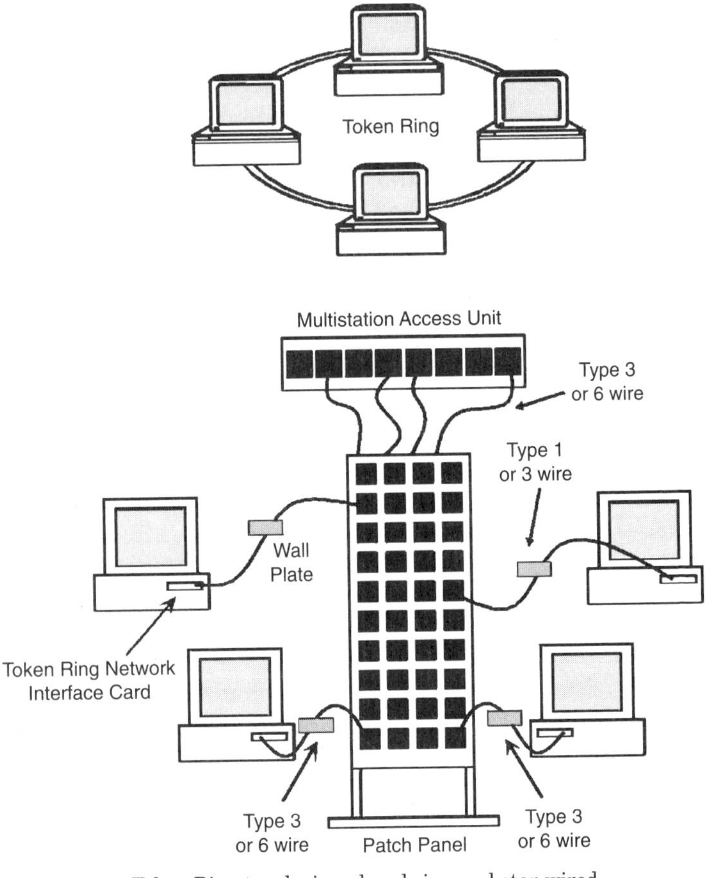

Figure T-1 Token Ring topologies: closed ring and star-wired.

- Since access to the network is not determined by a contention scheme, as is Ethernet, a higher throughput rate is possible in heavily loaded situations, limited only by the slowest element—sender, receiver, or link speed.

- With all messages following the same path, there are no routing problems to contend with. Logical addressing may be accommodated to permit message broadcasting to selected nodes.

- Adding terminals is easily accomplished—one connector is unplugged, the new node is inserted, and both nodes are plugged into the network. Other nodes are updated with the new address automatically.
- Control is simple, requiring little in the way of additional hardware or software to implement.
- The cost of network expansion is proportional to the number of nodes.

Another advantage of Token Ring is that the network can be configured to give high-priority traffic precedence over lower-priority traffic. Only if a station has traffic equal to or higher in priority than the priority indicator embedded in the token can it transmit data onto the network.

The Token Ring in its pure configuration is not without liabilities, however. Failed nodes and links can break the ring, preventing all the other terminals from using the network. At extra cost, a dual-ring configuration with redundant hardware and bypass circuitry is effective in isolating faulty nodes from the rest of the network, thereby increasing reliability. Through the use of bypass circuitry, physically adding or deleting terminals to the Token Ring network is accomplished without breaking the ring. Specific procedures must be used to ensure that the new station is recognized by the others and is granted a proportionate share of network time. The process for obtaining this identity is referred to as "neighbor notification." This situation is handled quite efficiently, since each station becomes acquainted with the address of its predecessor and successor on the network on initialization (power-up) or at periodic intervals thereafter.

Frame Format

The frame size used on 4-Mbps Token Rings is 4048 bytes, while the frame size used on 16-Mbps Token Rings is 16192 bytes. The IEEE 802.5 Standard defines two data formats—tokens and frames (Figure T-2). The token, three octets in length, is the means by which the right to access the medium

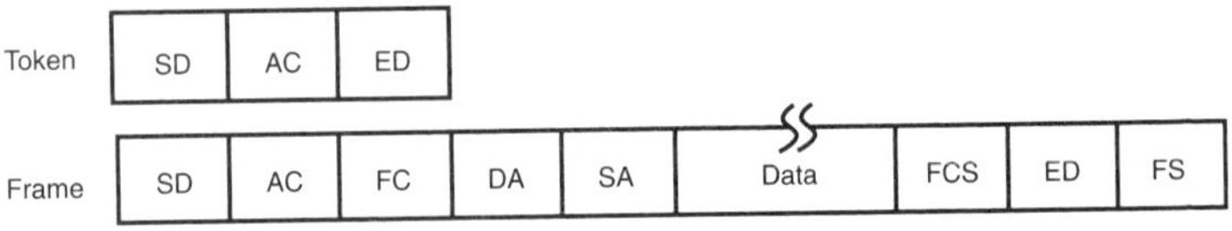

AC Access control
DA Destination address
ED Ending delimiter
FC Frame control
FCS Frame check sequence
FS Frame status
SA Source address
SD Starting delimiter

Figure T-2 Format of IEEE 802.5 token and frame.

is passed from one station to another. The frame format of Token Ring differs only slightly from that of Ethernet. The following fields are specified for IEEE 802.5 Token Ring frames:

- *Start Delimiter (SD)* Indicates the start of the frame.
- *Access Control (AC)* Contains information about the priority of the frame and a need to reserve future tokens, which other stations will grant if they have a lower priority.
- *Frame Control (FC)* Defines the type of frame, either Media Access Control (MAC) information or information for an end station. If the frame is a MAC frame, all stations on the ring read the address, but only the destination station can read the user data.
- *Destination Address (DA)* Contains the address of the station that is to receive the frame. The frame can be addressed to all stations on the ring.
- *Source Address (SA)* Contains the address of the station that sent the frame.
- *Data* Contains the data "payload." If the frame is a MAC frame, this field may contain additional control information.
- *Frame Check Sequence (FCS)* Contains error-checking information to ensure the integrity of the frame to the recipient.

- *End Delimiter (ED)* Indicates the end of the frame.
- *Frame Status (FS)* Provides indications of whether one or more stations on the ring recognized the frame, whether the frame was copied, or whether the destination station is not available.

Operation

A token is circulated around the ring, giving each station in sequence a chance to put information on the network. The station seizes the token, replacing it with an information frame. Only the addressee can claim the message. At the completion of the information transfer, the station reinserts the token on the ring. A token-holding timer controls the maximum amount of time a station can occupy the network before passing the token to the next station.

A variation of this token-passing scheme allows devices to send data only during specified time intervals. The ability to determine the time interval between messages is a major advantage over the contention-based access method used by Ethernet. This time-slot approach can support voice transmission and videoconferencing, since latency is controllable.

To protect the Token Ring from potential disaster, one terminal is typically designated as the control station. This terminal supervises network operations and does important housecleaning chores, such as reinserting lost tokens, taking extra tokens off the network, and disposing of "lost" packets. To guard against the failure of the control station, every station is equipped with control circuitry so that the first station detecting the failure of the control station assumes responsibility for network supervision.

Dedicated Token Ring

Dedicated Token Ring (DTR), also known as "full-duplex Token Ring," lets devices directly connected to a Token

Ring switch transmit and receive data simultaneously at 16 Mbps, effectively providing each station with 32 Mbps of throughput.

Under the IEEE 802.5r Standard for DTR, which defines the requirements for end stations and concentrators that operate in full-duplex mode, all new devices will coexist with existing Token Ring equipment and will adhere to the token-passing access protocol. The DTR concentrator consists of C-Ports and a data transfer unit (DTU). The C-Ports provide basic connectivity from the device to Token Ring stations, traditional concentrators, or other DTR concentrators. The DTU is the switching fabric that connects the C-Ports within a DTR concentrator. In addition, DTR concentrators can be linked to each other over a LAN or WAN via data transfer services such as ATM.

High-Speed Token Ring

With 16-Mbps Token Ring, connections between switches easily become congested at busy times, and high-performance servers become less able to deliver their full bandwidth potential. The need for a high-speed solution for Token Ring has become readily apparent in recent years. Other high-speed technologies exist—Fiber Distributed Data Interface (FDDI), Fast Ethernet, and ATM—but they are inadequate for the Token Ring environment.

In 1997, several Token Ring vendors teamed up to address this situation by forming the High Speed Token Ring Alliance (HSTRA). A year later, the alliance issued a specification for High Speed Token Ring (HSTR), which offers 100 Mbps and preserves the native Token Ring architecture. However, to keep costs to a minimum and to shorten its development time, HSTR is based on the IEEE 802.5r Standard for Dedicated Token Ring, adapted to run over the same 100-Mbps physical transmission scheme used by dedicated Fast Ethernet. HSTR links can be run in either half- or full-duplex mode, just like Dedicated Token Ring.

HSTR uses existing switches, hubs, bridges, routers, network interface cards (NICs), and cabling. This introduces greater throughput where the enterprise needs it most—at the server and backbone. Upgrading these connections with HSTR requires only that an HSTR uplink be plugged into a Token Ring switch and that the existing 16-Mbps server network NIC be replaced with a 100-Mbps HSTR NIC. To complete the upgrade, the two devices are connected with appropriate cabling. The 100-Mbps HSTR operates over both Category 5 UTP and IBM Type 1 STP cable, as well as multimode fiberoptic cabling.

It is also possible to connect desktop systems to Token Ring switches on dedicated 100-Mbps HSTR connections. Token Ring vendors offer 4/16/100-Mbps adapter cards that enable companies to standardize on a single network adapter and prepare their infrastructure for the eventual move to HSTR. While the HSTR standard does not define an autonegotiation algorithm, individual vendors have a number of ways to implement the feature while adhering to the standard. With this feature, HSTR products operate at the maximum connection speed, automatically determining whether to transmit at 4, 16, or 100 Mbps. Many corporations install autonegotiating 4/16/100 Mbps NICs in today's desktops, even if there is no immediate need for 100-Mbps throughput to the desktop. When the hub or switch at the other end of the connection is later upgraded to 100-Mbps HSTR, the Token Ring desktop will automatically adjust transmission to 100 Mbps.

Since Ethernet packets can be carried over Token Ring links, HSTR makes a good backbone medium for the mixed-technology LAN. With support for the maximum Token Ring frame size, an HSTR backbone segment is able to handle Ethernet and Token Ring frames on the same virtual LAN (VLAN) connection, which Fast Ethernet would not be able to do without a lot of processing to break down the larger Token Ring frames.

Summary

Token Ring is a stable technology with proven capacity for handling today's applications. At the same time, network managers can protect their current investments in Token Ring by understanding application performance and the capacity of the network and tuning it accordingly. The DTR standard prolongs the useful life of Token Ring networks while meeting the increased bandwidth requirements of emerging applications such as document imaging, desktop videoconferencing, and multimedia. Nevertheless, Token Ring has been overtaken by Ethernet, in terms of both technology and market share. Not only is Ethernet cheap to implement, it offers a migration path to higher speeds that Token Ring standards lack. While Ethernet has reached gigabit-per-second speeds, Token Ring has not been standardized beyond 100 Mbps.[2]

See also

Ethernet

TRANSCEIVERS

A transmitter-receiver (transceiver) connects a computer, printer, or other device to a LAN (Figure T-3). The transceiver may be a component integrated on the network interface card (NIC), or it can be a separate device that connects to the NIC with a drop cable. In the latter case, the transceiver cable and connectors form the attachment unit inter-

[2] Although some vendors such as Cisco Systems offer a form of Gigabit Token Ring, critics claim that they are proprietary products and not authentic Token Ring.

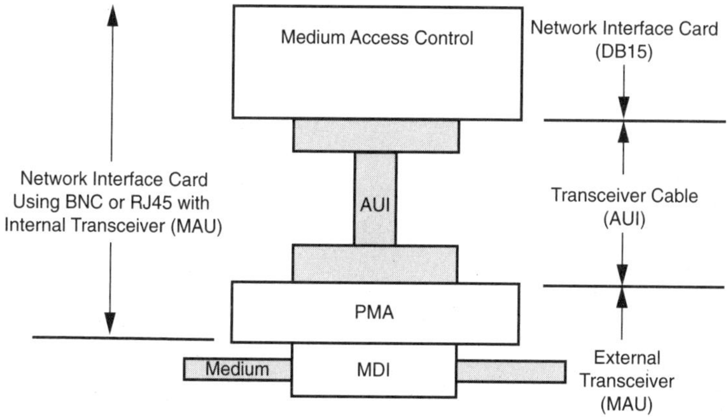

Figure T-3 Transceiver architecture.

face (AUI), and the transceiver is the medium attachment unit (MAU).

In the Ethernet environment, the MAU has four basic functions:

- *Transmit* Transmits serial data onto the medium.
- *Receive* Receives serial transmission and passes these signals to the attached station.
- *Collision detection* Detects the presence of simultaneous signals on the network and alerts the station.
- *Jabber function* Automatically interrupts the transmit function to inhibit abnormally long data stream output.

The MAU consists of the physical medium attachment (PMA), which provides the functions and two connectors. On the network side, the MAU attaches to the medium dependent interface (MDI). The specific interface depends on the type of media used. For example, 10BaseT (twisted-pair) uses an RJ-45 connector and 10Base2 (thin coax) uses a BNC connector, while the older 10Base5 (thick coax) implements a special "vampire" tap that pierces the coaxial cable and makes contact with both the center conductor and the shield.

Status Indicators

Transceivers offer several indicators to keep the user informed of performance status at any given time:

- *Transmit* Indicates packets are being transmitted onto the medium.
- *Receive* Indicates packets are being received from the medium.
- *SQE* Indicates a signal quality error (SQE) test signal is present.
- *Collision* Indicates a collision has occurred.

A user-selectable switch is provided that permits the network manager to choose between enabling and disabling the SQE test function. This feature permits the transceiver to be used with repeaters that cannot support the heartbeat function.

Summary

Transceivers are available in a variety of configurations to support different LAN types and media. There are transceivers for all versions of Ethernet, as well as FDDI and ATM networks. There are transceivers for coaxial cable (thick and thin), twisted-pair wiring (shielded and unshielded), optical fiber (single- and multimode), and wireless (spread spectrum and infrared).

See also

Network Interface Cards

TRANSMISSION CONTROL PROTOCOL/INTERNET PROTOCOL (TCP/IP)

The Transmission Control Protocol/Internet Protocol (TCP/IP) is a suite of networking protocols that is valued for its ability to interconnect diverse computing platforms—from PCs, Macintoshes, and UNIX systems to mainframes and supercomputers. The protocol suite originated from the work done by four key individuals over 30 years ago: Vinton Cerf, Robert Kahn, Leonard Kleinrock, and Lawrence Roberts (Figure T-4). Each disagrees on who deserves the lion's share of credit in the development of the Internet. Although the early experiments of Kleinrock made his computer the first node on the early Advanced Research Projects Agency Network (ARPANET), Cerf and Roberts generally get the credit for designing the network architecture that eventually became known as the Internet.

The U.S. government's Advanced Research Projects Agency (ARPA) funded the further development of the TCP/IP suite in the 1970s. As noted, the protocol suite was developed to enable different networks to be joined to form a virtual network known as an "internetwork." The original Internet was formed by converting an existing conglomeration of networks belonging to ARPANET over to TCP/IP, which evolved to become the backbone of today's Internet. Today, the Internet Engineering Task Force (IETF) oversees the development of the TCP/IP protocol suite and related protocols.

Several factors have driven the acceptance of TCP/IP for mainstream business and consumer use over the years. These include the technology's ability to support local and wide area connections, its open architecture, and a set of specifications that are freely available in the public domain. Although not the most functional or robust transport available, TCP/IP offers a mature, dependable environment for corporate users who need a common denominator for their diverse and sprawling networks.

Figure T-4 Vinton Cerf (*top left*), Robert Kahn (*top right*), Leonard Kleinrock (*bottom left*), and Lawrence Roberts (*bottom right*) are the four individuals generally credited with the initial development work that led to today's Internet.

Key Protocols

The key protocols in the suite include the Transmission Control Protocol (TCP), the Internet Protocol (IP), and the User Datagram Protocol (UDP). There are also application services that include the Telnet protocol, providing virtual terminal service, the File Transfer Protocol (FTP), the Simple Mail Transfer Protocol (SMTP), and the Simple Network Management Protocol (SNMP).

Transmission Control Protocol (TCP) TCP forwards data delivered by IP to the appropriate process at the receiving host. Among other things, TCP defines the procedures for breaking up the data stream into packets and reassembling them in the proper order to reconstruct the original data stream at the receiving end. Since the packets typically take different routes to their destination, they arrive at different times and out of sequence. All packets are temporarily stored until the missing packets arrive so that they can be put in the correct order. If a packet arrives damaged, it is simply discarded, and another one is resent.

To accomplish these and other tasks, TCP breaks the messages or data stream down into a manageable size and adds a header to form a packet. The packet's header (Figure T-5) consists of

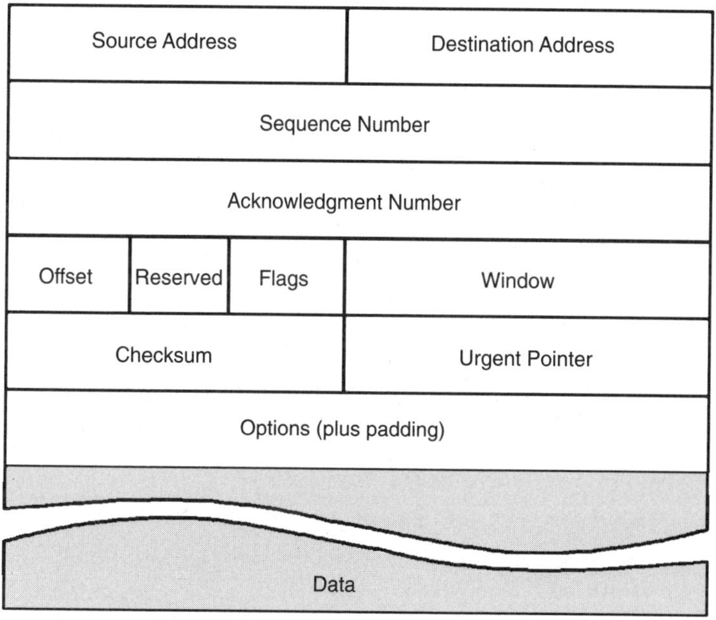

Figure T-5 TCP packet header.

- *Source port (16 bits)/destination port (16 bits) address* The source and destination ports correspond to the calling and called TCP applications. The port number is usually assigned by TCP whenever an application makes a connection. There are well-known ports associated with standard services such as Telnet, FTP, and SMTP.
- *Sequence number (32 bits)* Each packet is assigned a unique sequence number that lets the receiving device reassemble the packets in sequence to form the original data stream.
- *Acknowledgment number (32 bits)* The acknowledgment number indicates the identifier or sequence number of the next expected packet. Its value is used to acknowledge all packets transmitted in the data stream up to that point. If a packet is lost or corrupted, the receiver will not "acknowledge" that particular packet. This negative acknowledgment triggers a retransmission of the missing or corrupted packet.
- *Offset (4 bits)* The offset field indicates the number of 32-bit words in the TCP header. This is required because the TCP header may vary in length, according to the options that are selected.
- *Reserved (6 bits)* This field is not currently used but may accommodate some future enhancement of TCP.
- *Flags (6 bits)* The flags field serves to indicate the initiation or termination of a TCP session, reset a TCP connection, or indicate the desired type of service.
- *Window (16 bits)* The window field, also called the "receive window size," indicates the number of 8-bit bytes that the host is prepared to receive on a TCP connection. This provides precise flow control.
- *Checksum (16 bits)* The checksum is used to determine whether the received packet has been corrupted in any way during transmission.
- *Urgent pointer (16 bits)* The urgent pointer indicates the location in the TCP byte stream where urgent data end.

- *Options (0 or more 32-bit words)* The options field is typically used by TCP software at one host to communicate with TCP software at the other end of the connection. It passes such information as the maximum TCP segment size that the remote machine is willing to receive.

The bandwidth and delay of the underlying network impose limits on throughput. Poor transmission quality causes packets to be discarded, which in turn results in retransmissions and causes poor throughput.

Internet Protocol (IP) The Internet is composed of a series of autonomous systems, or subnetworks, each of which is locally administered and managed. These subnets may consist of Ethernet LANs, Integrated Services Digital Network (ISDN) networks, frame relay networks, and ATM networks over which IP runs. IP delivers data between these different networks through routers that process packets from one autonomous system (AS) to another.

Each node in the AS has a unique IP address. The IP adds its own header and checksum to make sure the data are routed properly (Figure T-6). This process is aided by the presence of routing update messages that keep the address tables in each router current. Several different types of update messages are used, depending on the collection of subnets involved in a management domain. The routing tables list the various nodes on the subnets as well as the paths between the nodes. If the data packet is too large for the destination node to accept, it will be segmented into smaller packets.

The IP header consists of the following fields:

- *IP version (4 bits)* The current version of IP is 4; the next generation of IP is 6.
- *IP header length (4 bits)* Indicates header length; if options are included, the header may have to be padded with extra 0s so that it can end at a 32-bit-word boundary. This is necessary because header length is measured in 32-bit words.

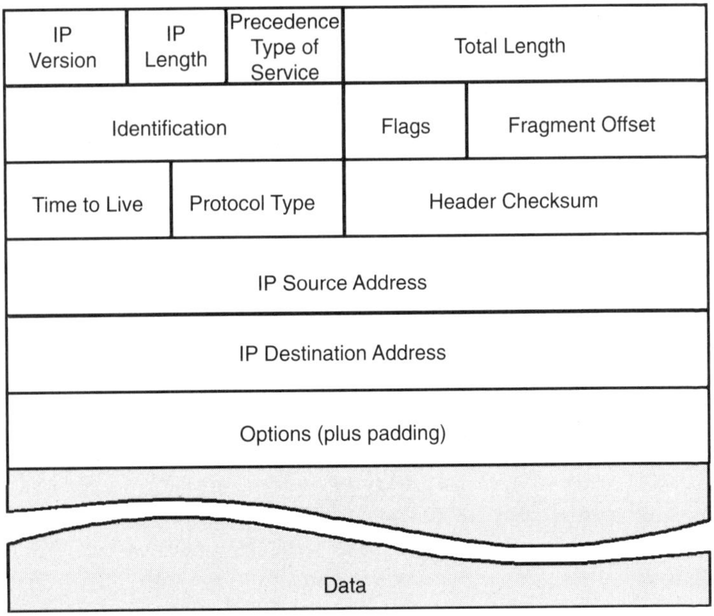

Figure T-6 IP packet header.

- *Precedence and type of service (8 bits)* Precedence indicates the priority of data packet delivery, which may range from 0 (lowest priority) for normal data to 7 (highest priority) for time-critical data (i.e., multimedia applications). Type of service contains quality-of-service (QoS) information that determines how the packet is handled over the network. Packets can be assigned values that maximize throughput, reliability, or security and minimize monetary cost or delay. This field will play a larger role in the future as the Internet evolves to handle more multimedia applications.[3]

[3] Differentiated Services (Diffserv) supersedes the original IP Precedence/Type of Service specification for defining packet priority. Using this field, Diffserv first prioritizes traffic by class and then differentiates and prioritizes same-class traffic, offering finer priority granularity.

- *Total packet length (16 bits)* Total length of the header plus the total length of the data field of the packet.
- *Identification (16 bits)* A unique ID for a message that is used by the destination host to recognize packet fragments that belong together.
- *Flags (3 bits)* Indicates whether or not the packets can be fragmented for delivery; if a packet cannot be delivered without being fragmented, it will be discarded and an error message will be returned to the sender.
- *Fragmentation offset (13 bits)* If fragmentation is allowed, this field indicates how IP packets are to be fragmented. Each fragment has the same ID. Flags are used to indicate that more fragments are to follow, as well as indicate the last fragment in the series.
- *Time to live (8 bits)* This field indicates how long the packet is allowed to exist on the network in its undelivered state. The hop counter in each host or gateway that receives the packet decrements the value of the time-to-deliver field by one. If a gateway receives a packet with the hop count decremented to zero, it will be discarded. This prevents the network from becoming congested by undeliverable packets.
- *Protocol type (8 bits)* Specifies the appropriate service to which IP delivers the packets, such as TCP or UDP.
- *Header checksum (16 bits)* This field is used to determine whether the received packet has been corrupted in any way during transmission. The checksum is updated as the packet is forwarded because the time-to-live field changes at each router.
- *IP source address (32 bits)* The address of the source host (e.g., 130.132.9.55).
- *IP destination address (32 bits)* The address of the destination host (e.g., 128.34.6.87).
- *Options (up to 40 bytes)* Although seldom used for routine data, this field allows one or more options to be specified. Option 4, for example, time stamps all stops that the

packet made on the way to its destination. This allows measurement of overall network performance in terms of average delay and nodal processing time.

Internet performance depends on the resources available at the various hosts and routers—transmission bandwidth, buffer memory, and processor speed—and how efficiently these resources are used. Although each type of resource is manageable, there are always tradeoffs between cost and performance.

User Datagram Protocol (UDP) While TCP offers assured delivery, it does so at the price of overhead. UDP, on the other hand, functions with minimum overhead; it merely passes individual messages to IP for transmission. Since IP is not reliable without TCP, there is no guarantee of delivery.

Nevertheless, UDP is very useful for certain types of communications, such as quick database lookups. For example, the Domain Name System (DNS) consists of a set of distributed databases that provide a service that translates between system names and their IP addresses. For simple messaging between applications and these network resources, UDP does the job. UDP is also well suited to the brief request/response message exchanges characteristic of the Simple Network Management Protocol (SNMP). The UDP header consists of the following fields:

- *Source port (16 bits)* This field identifies the source port number.
- *Destination port (16 bits)* This field identifies the destination port number.
- *Length (16 bits)* Indicates the total length of the UDP header and data portion of the message.
- *Checksum (16 bits)* Validates the contents of a UDP message. Use of this field is optional. If it is not computed for the request, it can still be included in the response.

Applications using UDP communicate through a specified numbered port that can support multiple virtual connections, which are called "sockets." A socket is an IP address and port, and a pair of sockets (source and destination) forms a TCP connection. One socket can be involved in multiple connections.

Some ports are registered ("well known") and can be found on many TCP/IP implementations. Well-known ports are numbered from 0 to 1023. Telnet, for example, always uses port 23 for communications, while File Transfer Protocol (FTP) uses port 21. The well-known ports are assigned by the Internet Assigned Numbers Authority (IANA) and on most systems can only be used by system (or root) processes or by programs executed by privileged users. Other examples of UDP well-known ports are listed in Table T-1.

In addition to the well-known ports, there are also registered ports numbered from 1024 to 49151 and private ports numbered from 49152 to 65535.

Table T-1 Examples of UDP Well-Known Ports

Service	Port	Description
Users	11	Shows all users on a remote system.
Quote	17	Returns a "quote of the day."
Mail	25	Used for electronic mail via SMTP.
Domain Name Server	53	Translates system names and their IP addresses.
BOOTpc	68	Client port used to receive configuration information.
TFTP	69	Trivial File Transfer Protocol used for initializing diskless workstations.
World Wide Web	80	Provides access to the Web via the HyperText Transfer Protocol (HTTP).
Snagas	108	Provides access to an SNA Gateway Access Server.
nntp	119	Provides access to a newsgroup via the Network News Transfer Protocol (NNTP).
SNMP	161	Used to receive network management queries via the Simple Network Management Protocol.

High-Level TCP/IP Services

The TCP/IP model includes three simple types of services for file transfers, electronic mail, and virtual terminal sessions.

- *File Transfer Protocol (FTP)* A protocol used for the bulk transfer of data from one remote device to another. Usually implemented as application-level programs, FTP uses the Telnet and TCP protocols. Most FTP offerings have options to support the unique aspects of each vendor's file structures. Data in the FTP environment consist of a stream of data followed by an end-of-file marker, allowing only entire files to be transferred—not selected records within a file. Sending a file via FTP to a user on another TCP/IP network requires a valid user ID and password for a host on that network.
- *Simple Mail Transfer Protocol (SMTP)* A protocol for exchanging mail messages between systems without regard for the type of user interface or the functionality that is available locally. SMTP sessions consist of a series of commands, starting with both ends exchanging "handshake" messages to identify themselves. This is followed by a series of commands that indicate that a message is to be sent and receipts are needed and by commands that actually transfer the data. Separating the data message from the address field allows a single message to be delivered to multiple users and to verify that there is at least one deliverable addressee before sending the contents. SMTP modifies every message that it receives by adding a time stamp and a reverse path indicator. This means that a mail message in the SMTP environment usually consists of a fairly long header with information from each node that handled the message. Many user interfaces are able to automatically filter out this kind of information, however.
- *Telnet Virtual Terminal Service* The Telnet protocol defines a network-independent virtual terminal through which a user can log in to remote TCP/IP hosts. The user goes through the standard logon procedure on the remote

TCP/IP host and must know the characteristics of the remote operating system to execute host-resident commands. Telnet enables remote terminals to access different hosts by fooling an operating system into thinking that a remote terminal is locally connected. Most Telnets operate in the full-duplex mode, meaning that they are capable of sending and receiving at the same time. There is a half-duplex mode to accommodate IBM hosts, however. In this case, a turnaround signal switches the sending of data to the other side of the connection.

Summary

In its early years of development and implementation, TCP/IP was considered of interest only to research institutions, academia, and defense contractors. Today, corporations have embraced TCP/IP as a platform that can meet their needs for multivendor, multinetwork connectivity. Because it was developed in large part with government funding, TCP/IP code is in the public domain; this availability has encouraged its use by thousands of vendors worldwide, who apply it to support nearly all types of computers and network devices. Because of its flexibility, comprehensiveness, and nonproprietary nature, TCP/IP has captured a considerable and growing share of the commercial internetworking market.

See also

Internet

Simple Network Management Protocol

TWISTED-PAIR WIRING

Twisted-pair wiring is the most common transmission medium; it is currently installed in most office buildings and residences. Twisted-pair wiring consists of one or more pairs of copper wires. To reduce cross-talk or electromagnetic induction between pairs of wires, two insulated copper wires are twisted around each other. For some business locations, twisted pair is enclosed in a shield that functions as a ground. This is known as "shielded twisted pair" (STP). Ordinary wire to the home is unshielded twisted pair (UTP).

In the local loop, hundreds of insulated wires are bundled into larger cables, with each pair color-coded for easy identification. Most telephone lines between central offices and local subscribers consist of this type of cabling, which is mounted on poles or laid underground. Bundling facilitates installation and reduces costs. Special sheathing offers protection from natural elements. In the business environment, Category 5 unshielded twisted-pair cable is usually delivered as a bundle of 25 pairs. Each color-coded wire is solid conductor, 24 AWG (American Wire Gauge).

The same unshielded twisted-pair wiring has become the most popular transmission medium for LANs. The pairs of wires in UTP cable are color-coded so that they can be identified easily at each end. The most common color scheme is the one that corresponds to the Electronic Industry Association/Telecommunications Industry Association's

Table T-2 Color Scheme Specified by Electronic Industry Association/ Telecommunications Industry Association's Standard 568B for Category 5 UTP Cable

Wire Pair	Color Code
#1	White/blue
	Blue
#2	White/orange
	Orange
#3	White/green
	Green
#4	White/brown
	Brown

Standard 568B. Table T-2 summarizes the proper color scheme.

The cable connectors and jacks that are used most commonly with Category 5 UTP cables are RJ45. The RJ simply means "Registered Jack," and the number 45 designates the pin numbering scheme. The connector is attached to the cable, and the jack is the device that the connector plugs into, whether it is in the wall, the NIC in the computer, or the hub.

In response to the growing demand for data applications, cable has been categorized into various levels of transmission performance, as summarized in Table T-3. The levels are hierarchical in that a higher category can be substituted for any lower category.

The use of unshielded twisted-pair wiring has several advantages. The technology and standards are mature and stable for voice and data communications. Telephone systems, which use twisted-pair wiring, are present in most buildings, and unused pairs usually are available for LAN connections. When required, additional twisted pair can be installed relatively easily, and the cost of Category 5 cabling is relatively inexpensive.

Of course, unshielded twisted-pair wiring has a few disadvantages as well. It is sensitive to electromechanical interference, so new installations must be planned to route around sources of EMI. Unshielded twisted pair is also more susceptible to eavesdropping, which makes encryption and other security precautions necessary to safeguard sensitive information. An additional requirement is a wiring hub. Although this involves another expense, the hub actually facilitates the installation of new wiring, keeps all wires organized, and makes it easier to implement moves, adds, and changes to the network. Over the long term, a hub saves much more than it costs.

Summary

UTP cable has evolved over the years, and different varieties are available for different needs. Improvements over the

Table T-3 Categories of UTP Cable

Category	Maximum Bandwidth	Application	Standards
7	600 MHz	1000BaseT and faster	Standard under development*
6	250 MHz	1000BaseT	TIA/EIA 568-B (Category 6)
5E	100 MHz	Same as CAT 5 plus 1000BaseT	ANSI/TIA/EIA-568-A-5 (Category 5E)
5	100 MHz	10/100BaseT	TIA/EIA 568-A (Category 5)
		100-Mbps TPDDI (ANSI X.319.5)	NEMA (extended frequency)
	155-Mbps ATM	ANSI/ICEA S-91-661	
4	20 MHz	10-Mbps Ethernet (IEEE 802.3)	TIA/EIA 568-A (Category 4)
		16-Mbps Token Ring (IEEE 802.5)	NEMA (extended distance)
		ANSI/ICEA S-91-661	
3	16 MHz	10-Mbps Ethernet (IEEE 802.3)	TIA/EIA 568-A (Category 3) NEMA (standard loss) ANSI/ICEA S-91-661
2	4 MHz	IBM Type 3 1.544-Mbps T1 1 Base 5 (IEEE 802.3) 4-Mbps Token Ring (IEEE 802.5) ANSI/ICEA S-91-661 ANSI/ICEA S-80-576	IBM Type 3
1	Less than 1 MHz	POTS (plain old telephone service)	ANSI/ICEA S-80-576
		RS 232 and RS 422 ISDN basic rate	ANSI/ICEA S-91-661

*In new installations, fiber to the desk may be less expensive than installing Category 7 cable.

years, such as variations in the twists or in individual wire sheaths or overall cable jackets, have led to the development of EIA/TIA-568 Standard-compliant categories of cable that have different specifications on signal bandwidth. Because UTP cable is lightweight, thin, and flexible, as well as versatile, reliable, and inexpensive, millions of nodes have been and continue to be wired with UTP cable, even for high-data-rate applications. For the best performance, UTP cable should be used as part of a well-engineered structured cabling system. However, businesses that require reliable gigabit-per-second data transmission speeds should give serious consideration to moving to optical fiber rather than Category 7 UTP.

See also

Hybrid Fiber/Coax

T1 Lines

U

UNIFIED MESSAGING

While the methods of communicating have diversified, so too has the number of devices people must use to receive all the messages—desk phones, cellular phones, fax machines, alphanumeric pagers, and e-mail systems, to name a few. Unified messaging brings order to this communications chaos by consolidating the reception, notification, presentation, and management of what have until now been standalone messaging systems. The goal of unified messaging is to make individuals, workgroups, and organizations more efficient and responsive.

The unified messaging capability can be provided through message servers connected to a corporate private branch exchange (PBX) (Figure U-1) or carrier switch. These servers can have a distributed architecture, allowing unified messaging services to be added incrementally throughout the carrier or corporate network as demand warrants. More recently, unified messaging services have become available over the Internet, allowing users to view communication activity through their Web browser.

A unified messaging service deposits each subscriber's e-mail, fax, and voice messages into a universal messaging

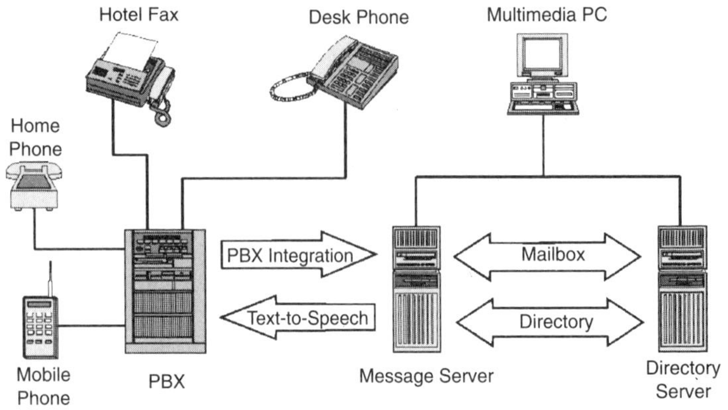

Figure U-1 Unified messaging application implemented through the integration of a corporate LAN and PBX.

inbox so that the subscriber can find all messages in a single place and through a single interface, such as a Web browser or desktop application.

Notification Options

The unified messaging systems notify subscribers whenever a new message arrives. Depending on the type of system or service that is used, notification methods include an e-mail message, message-waiting indicator light, stutter dial tone, pager, and out-dial.

When a service is used, for example, it is common for voice-messaging subscribers to hear a stutter dial tone on the arrival of a new voice message but not when an e-mail or fax message has arrived. With the advent of unified messaging, however, this changes, and the system notifies the subscriber of an incoming message whenever a message arrives, regardless of whether it is a voice, fax, or e-mail message. A stutter dial tone is only one of several potential notification options. The service provider may choose which notification methods it offers to subscribers.

If the specified notification method is e-mail, for example, the unified messaging system sends an e-mail notification message to a subscriber-specified e-mail address. In the body of the notification message, the system embeds a hypertext link that points to the voice or fax message in the system's message store. When the subscriber clicks on the hypertext link, the mail application passes the Uniform Resource Locator (URL) to the subscriber's Web browser, which opens up a window to the subscriber's universal inbox, where the voice or fax message appears.

If a voice message comes in and the subscriber's workstation supports multimedia, the subscriber can listen to the voice message over the workstation's audio system. If the message is a fax, the system presents the fax as a graphic image on the workstation's screen. If the message is itself an e-mail message, the system simply passes it along to the subscriber.

Let's say the subscriber wants to be notified of incoming messages via a telephone's message-waiting indicator. When a new voice, fax, or e-mail message arrives and the subscriber checks for messages, the system indicates how many of each type of message have arrived. The system's synthesized voice might tell the subscriber, "You have two new voice messages, three new fax messages, and five new e-mail messages." Using buttons on the handset, the subscriber can choose to listen to the voice messages, output the fax messages to the nearest fax machine, and save the e-mail messages for viewing at a more convenient time and place.

The Role of Browsers

Because the Web can support text, graphics, and audio efficiently, it provides an effective medium for the consolidation and presentation of e-mail, fax, and voice messages. Some unified messaging systems can take advantage of Web technology to securely handle these as binary objects for large numbers of subscribers.

With a Web browser, unified messaging system subscribers can view fax and e-mail messages on their computer screens and listen to voice mail messages over headphones or speakers attached to their workstation. They also can use the message compose and reply features of the browser interface to respond to incoming messages.

For example, a subscriber can send a voice-mail message to another recognized unified messaging subscriber in response to a voice, fax, or e-mail message. By clicking the "Compose Voice" button in the browser window and picking the recipient of the message from a directory window, the subscriber records a response and, when finished, clicks the "Send" button to deliver the voice mail.

A subscriber also can send an e-mail message in response to an incoming voice-mail, fax, or e-mail message. In response to an incoming voice-mail or fax message, the subscriber clicks the "Compose E-mail" button in the browser window. If the recipient is a recognized unified messaging subscriber, his or her address can be selected from the directory. If the recipient is not a recognized subscriber, the e-mail address must be entered manually. After composing the e-mail address, the subscriber clicks the "Send" button to deliver the e-mail message. If the original message was an e-mail message instead of a voice-mail or fax message, the return address is automatically entered into the reply whether or not the addressee is a unified messaging subscriber.

Messages also can be forwarded from the browser interface. A "Forward" button enables subscribers to send an incoming e-mail message to another subscriber. If the subscriber wants to forward a voice-mail or fax message, the subscriber saves the voice-mail message in an audio file (or the fax in a graphic file) format and sends it as an attachment to an e-mail message. Subscribers can use the print capabilities of their Web browsers to output hard copies of received fax and e-mail messages.

While certain levels of access are not yet available—such as a viewing a fax or e-mail message over a standard telephone handset—subscribers can use the handset to manipulate even these types of messages. With a unified messaging system between the PBX and local area network (LAN), the system can report who sent the fax or e-mail message and what time it arrived.

Via the handset, subscribers also can redirect fax messages for output on whatever fax machine they designate. The user simply dials the phone number of the desired fax machine and presses the “Enter” key. The fax machine could be in a hotel lobby, at a remote office, at a customer site, or at the subscriber’s own home. Incoming e-mail messages can be output in a similar manner, except that an image of the e-mail message arrives at the specified fax machine.

By pressing appropriate buttons in response to system prompts, a subscriber can record and send a voice-mail response to a voice- or e-mail message sent from another subscriber within the unified messaging environment. The unified messaging system addresses the reply automatically, so the subscriber does not have to remember (or even know) the address of the person to whom he or she is sending the reply. To reply to an externally originated message, the subscriber must dial the external number directly and rely on the recording capabilities attached to the phone service at the other end of the line.

Although there is usually no support for sending a voice message from the handset in direct response to an incoming fax message, a subscriber can forward voice-mail and fax messages to another recognized unified messaging subscriber by entering that subscriber’s mailbox number at the prompt. Subscribers also can attach voice annotations to messages they forward from the handset.

Subscribers also can forward fax and e-mail messages to systems outside the unified messaging environment by using the print capabilities of the unified messaging system. The subscriber forwards the fax by designating a remote recipient’s fax machine or fax mailbox as the target output device. This feature also enables handset users to forward e-mail messages to recipients outside the system: As noted, the unified messaging system actually faxes an image of the e-mail message to a remote recipient's fax machine. Beyond providing a message to be faxed, the e-mail component of the unified messaging system does not play a role in forwarding messages outside the environment from the handset.

The unified messaging system ensures that actions initiated via the browser and handset interfaces are kept closely synchronized. If a subscriber listens to a new voice-mail message through the browser interface, it is flagged as "read" and is not announced as a "new" message when the subscriber later accesses the in-box via the handset interface. And as noted earlier, if a subscriber deletes a message via one interface, it is deleted from the list of messages accessible via any other interface.

Summary

While unified messaging solutions are available as a service or dedicated system, they are also available over the Internet. Numerous companies offer services that consolidate voice mail, e-mail, and faxes in one mailbox. Ranging in price from $10 to $20 a month, these services provide individuals and companies with a telephone number for incoming voice messages and faxes and an e-mail address. Despite the benefits of unified messaging, a strong market has failed to materialize. Part of the problem may be generational—younger people seem more inclined to accept unified messaging because they have grown up with the technology and seem more adept at using it, whereas older people are less inclined to accept unified messaging because they have grown up using separate communications tools and are less trusting of "all in one" solutions. Further study in this area would be required to draw definitive conclusions.

See also

Call Centers
Electronic Mail
Internet Facsimile

V

VIRTUAL PRIVATE NETWORKS

Carrier-provided networks that function like private networks are referred to as virtual private networks, or VPNs, which are oriented toward either voice or data. Traditionally, VPNs have been voice oriented, allowing companies to interconnect their PBXs over a carrier's wide area network. In early 1997, a new trend emerged in which private data is routed between corporate locations worldwide over carrier-provided Internet Protocol (IP) networks.

The carrier is responsible for network management, security, and quality of service issues. In many cases, service-level guarantees are available from the carriers, which provide users with credits for poor performance. Although frame relay and asynchronous transfer mode (ATM) continue to be used for VPNs—supporting voice in packet form as well as data—IP is now the most popular type of VPN. Sometimes a mix of services can be used to implement VPNs. For example, companies can use their current frame relay service for remote access, eliminating the need for additional VPN access circuits, routers, and modem pools for IP traffic.

IP-Based VPNs

IP-based VPNs are an increasingly popular option for interconnecting corporate locations over the Internet, including branch offices and telecommuters. They also can be used for electronic commerce and making enterprise applications available to customers and strategic partners worldwide. Basically, this type of data service lets business users carve out their own IP-based WANs within the carrier's high-speed Internet backbone. Security functions are performed on IP packets, which are then encapsulated, or tunneled, inside other IP packets for routing across the Internet. By drawing on the economies of transmission and switching that the larger Internet provides, VPNs offer substantial cost savings.

Several protocols are used to implement IP-based VPNs: the Layer 2 Forwarding Protocol (L2F), the Layer 2 Tunneling Protocol (L2TP), the Point-to-Point Tunneling Protocol (PPTP), and the IP Security (IPSec). The first three are vendor implementations of tunneling protocols that transport Layer 3 packets such as AppleTalk, IP, and IPX over the Internet by encasing them in Layer 2 PPP (Point-to-Point Protocol) frames.

A tunnel is an end-to-end connection similar to a virtual circuit (VC) used to provision frame relay and ATM services. Since a tunnel goes only to a designated destination, it provides the ability to securely transport one or more protocols. The tunnel can handle both IP and non-IP traffic, which can be assigned various priorities to provide quality of service (QoS). By themselves, however, many of the tunneling protocols do not provide native support for data security. To enhance data security, software-controlled encryption, authentication, and integrity functions can be applied to the packets for transport through the tunnel.

The Internet Engineering Task Force's (IETF) IPSec is one tunneling protocol that does provide for packet-by-packet authentication, integrity, and encryption. Authentication positively identifies the sender. Integrity guarantees that no third party has tampered with the packet stream. Encryption allows only the intended receiver to read the data. These security functions must be applied to every IP packet because Layer 3 protocols such as IP are stateless; that is, there is no way to be sure whether a packet is really associated with a particular connection. Higher-layer protocols such as TCP are stateful, but their connection-tracing mechanisms can be easily duplicated or "spoofed" by knowledgeable hackers. The key limitation of IPSec is that it can carry only IP packets.

The latest approach to achieving secure IP-based VPNs makes use of Multiprotocol Label Switching (MPLS), another IETF standard. With MPLS, a sending device knows a receiving device by a tag or label. If VPN traffic is assigned a tag or label, it will flow onto a specific set of virtual circuits in the core network, separate from those used for public service, such as the Internet. The use of different virtual circuits makes for real security and provides the potential to control QoS.

The major carriers offer some compelling features with their IP-based VPN offerings. To encourage customers to consider their services, carriers also are making promises about trouble response time, network uptime, and dial port availability. The overriding concern of corporate IS managers, however, is end-to-end-latency. If the VPN cannot get the packets through, then it is of little importance if the network is available 100 percent of the time. Accordingly, most VPN service providers offer a latency guarantee.

Classes of Service

Carriers are beginning to offer classes of service for their IP VPN offerings—high priority, low priority, and best effort. Network managers can assign these classes of service in any configuration to address the specific bandwidth demands of their network traffic.

Classes of service allow businesses to differentiate one application running on their network from another. The capability helps network managers use their network bandwidth more efficiently by letting them map applications into "classes," assign a priority to each class, and based upon the priority, treat the classes differently.

For example, Voice over IP (VoIP) and video conferencing applications have stricter performance requirements than Internet browsing traffic. When both VoIP and Internet browsing traffic are on the same network, classes of service allow network managers to assign a higher priority to the VoIP traffic. That way, voice traffic is always serviced first so the voice quality can be maintained.

Summary

IP VPNs are available as fully managed, comanaged, or customer-managed services. They provide the means for companies to extend the reach of their networks globally. Selected segments of the VPN can be securely opened to business partners, suppliers, and clients, depending on unique needs. Customers of a managed VPN service pay a single, all-inclusive service price, which includes network access, customer premises equipment, software and hardware maintenance, VPN installation project management support, and 24 × 7 proactive management and monitoring. Local loop access charges are additional.

See Also

Asynchronous Transfer Mode
Frame Relay
Multiprotocol Label Switching

VOICE COMPRESSION

Voice compression entails the application of various algorithms to the voice stream to reduce bandwidth requirements, while preserving the quality or audibility of the voice transmission. Numerous compression standards for voice have emerged over the years, which allow businesses to achieve substantial savings on leased lines with only a modest cost for additional hardware. Using these standards, the normal 64-kbps voice channel can be reduced to 32, 16, or 8 kbps, or even as little as 6.3 and 5.3 kbps. As the compression ratio increases, however, voice quality tends to diminish.

In the 1960s, Pulse Code Modulation (PCM) was the internationally accepted coding standard (G.711) for toll-quality voice transmission. Under this standard, a single voice channel requires 64 kbps when transmitted over the telephone network, which is based on time division multiplexing (TDM). The 64-kbps PCM time slot—or payload bit rate—forms the basic building block for today's public telephone services and equipment, such that 24 time slots, or channels of 64 kbps each, can be supported on a T1 line.

Pulse Code Modulation

A voice signal takes the shape of a wave, with the top and the bottom of the wave constituting the signal's frequency level, or amplitude. The voice is converted into digital form by an encoding technique called Pulse Code Modulation (PCM). Under PCM, voice signals are sampled at the minimum rate of 2 times the highest voice frequency level of 4000 Hertz (Hz), which equates to 8000 times per second. The amplitudes of the samples are encoded into binary form using enough bits per sample to maintain a high signal-to-noise ratio. For quality reproduction, the required digital transmission speed for 4-kHz voice signals works out to: 8000 samples per second × 8 bits per sample = 64,000 bps (64 kbps).

The conversion of analog voice signals to and from digital is performed by a coder-decoder, or codec, which is a key component of D4 channel banks and multiplexers. The codec translates amplitudes into binary values and performs mu-law quantizing. The mu-law process (North America only) is an encoding-decoding scheme for improving the signal-to-noise ratio. This is similar in concept to Dolby noise reduction, which ensures quality sound reproduction.

Other components in the channel bank or multiplexer interleave the digital signals representing as many as 24 channels to form a 1.544-Mbps bit stream (including 8 kbps for control), suitable for transmission over a T1 line. PCM exhibits high quality, is robust enough for switching through the public network without suffering noticeable degradation, and is simple to implement. But PCM allows for only 24 voice channels over a T1 line. Digital compression techniques can be applied to multiply the number of channels on a T1 line, several of which are described in Table V-1.

Compression Basics

Among the most popular compression methods is Adaptive Differential Pulse Code Modulation (ADPCM), which has been a worldwide standard since 1984. It is used primarily on private T-carrier networks to double the channel capacity of the available bandwidth from 24 channels to 48 channels.

The ADPCM device accepts the 8000-sample-per-second rate of PCM and uses a special algorithm to reduce the 8-bit samples to 4-bit words. These 4-bit words, however, no longer represent sample amplitudes, but only the difference between successive samples. This is all that is necessary for a like device at the other end of the line to reconstruct the original amplitudes.

Integral to the ADPCM device is circuitry called the adaptive predictor, which predicts the value of the next signal based only on the level of the previously sampled signal. Since the human voice does not usually change significantly from one sampling interval to the next, prediction accuracy can be very high. A feedback loop used by the predictor ensures that voice variations are followed with minimal deviation. Consequently, the high accuracy of the prediction means that the difference in the predicted and actual signal

Table V-1 Description of Commonly Used Digital Compression Techniques

Digital encoding method	Standard	Bit rate, Kbps	Mean opinion score
Pulse Code Modulation (PCM)	G.711	64	4.4
Adaptive Differential Pulse Code Modulation (ADPCM)	G.726	40, 32, 24, 16	4.2
Low Delay Code Excited Linear Prediction (LDCELP)	G.728	16	4.2
Conjugate Structure Algebraic Code Excited Linear Prediction (CSACELP)	G.729A	8	4.2
Multipulse Maximum Likelihood Quantization (MPMLQ)	G.723	6.3	3.9
Algebraic Code Excited Linear Prediction (ACELP)	G.723	5.3	3.5

The Mean Opinion Score (MOS) is the accepted measure of voice quality, determined through a statistical sample of user opinions.

is very small and can be encoded with only four bits, rather than the eight bits used in PCM. In the event that successive samples vary widely, the algorithm adapts by increasing the range represented by the four bits. However, this adaptation will decrease the signal-to-noise ratio and reduce the accuracy of voice frequency reproduction.

At the other end of the digital facility is another compression device, where an identical predictor performs the process in reverse to reinsert the predicted signal and restore the original 8-bit code.

By halving the number of bits to accurately encode a voice signal, T1 transmission capacity is doubled from the original 24 channels to 48 channels, providing the user with a 2-for-1 cost savings on monthly charges for leased T1 lines.

It is also possible for ADPCM to compress voice to 16 kbps by encoding voice signals with only two bits, instead of four bits as discussed above. This 4:1 level of compression provides 96 channels on a T1 line without significantly reducing signal quality.

Although other compression techniques are available for use on wire networks, ADPCM offers several advantages. ADPCM holds up well in the multinode environment, where it may undergo compression and decompression several times before arriving at its final destination. And, unlike many other compression methods, ADPCM does not distort the distinguishing characteristics of a person's voice during transmission.

Variable Rate ADPCM

Some vendors have designed ADPCM processors that not only compress voice, but accommodate 64-kbps pass-through as well. The use of very compact codes allows several different algorithms to be handled by the same ADPCM processor. The selection of algorithm is controlled in software and is done by the network manager. Variable-rate ADPCM offers several advantages.

Compressed voice is more susceptible to distortion than uncompressed voice—16 kbps more so than 32 kbps. When line conditions deteriorate to the point where voice compression is not possible without seriously disrupting communication, a lesser compression ratio may be invoked to compensate for the distortion. If line conditions do not permit compression even at 32 kbps, 64-kbps pass-through may be invoked to maintain quality voice communication. Of course, channel availability is greatly reduced, but the ability to communicate with the outside world becomes the overriding concern at this point, rather than the number of channels.

Variable-rate ADPCM provides opportunities to allocate channel quality according to the needs of different classes of user. For example, all intracompany voice links may operate at 16 kbps, while those used to communicate externally may be configured to operate at 32 kbps.

The number of channels may be increased temporarily by compressing voice to 16 kbps instead of 32 kbps, until new facilities can be ordered, installed, and put into service. As new links are added to keep up with the demand for more channels, the other links may be returned to operation at 32 kbps. Variable-rate ADPCM, then, offers much more channel configuration flexibility than products that offer voice compression at only 32 kbps.

Other compression schemes can be used over T-carrier facilities, such as Continuously Variable Slope Delta (CVSD) modulation and Time Assigned Speech Interpolation (TASI). There are still other voice compression techniques used on cellular communications networks and for carrying voice over IP networks.

Summary

Adding lines and equipment is one way that organizations can keep pace with increases in traffic. But even when funds are immediately available for such network upgrades, communications managers must contend with the delays inherent in ordering, installing, and putting new facilities into service. To accommodate the demand for bandwidth in a timely manner, communications managers can apply an appropriate level of voice compression to obtain more channels out of the available bandwidth. Depending on the compression technique selected, there need not be a noticeable decrease in voice quality.

See Also

Data Compression

Voice over IP

VOICE-DATA CONVERGENCE

The combination of voice, video, and data signals over the same equipment, line, and protocol is often referred to as voice-data convergence. Examples of voice-data convergence include:

- *Ethernet* Combines voice, broadband data, messaging, and video for transmission over Category 5 wiring.
- *Digital Subscriber Line (DSL)* Combines voice, Internet access, messaging, and broadband data over the same twisted-pair wires used for Plain Old Telephone Service (POTS).
- *Cable* Combines voice, broadband data, Internet access, messaging, television programming, and interactive features over the same coaxial cable connected to homes and businesses.
- *Local Multipoint Distribution Service* Combines voice, broadband data, Internet access, and television programming over the same short-haul wireless link operating in the 28- to 31-GHz range.
- *Optical fiber* Carries voice, broadband data, Internet access, video, and images over separate wavelengths within a single fiber strand.

- *Integrated Services Digital Network (ISDN)* Carries voice, data, and video within separate bearer channels provisioned over a digital line, which are switched through the public telephone network.
- *Asynchronous Transfer Mode (ATM)* Interleaves voice, data, and video for transmission through a fiber-based broadband network.
- *Internet Protocol (IP) networks* Interleave voice, data, and video for transmission through the Internet or private intranet.

IP is often viewed as the optimal convergence platform. Regardless of the specific transmission medium or service, voice-data convergence can be done more efficiently and economically on networks that support the Internet Protocol. With IP, the distinct traffic types are reduced to a single stream of binary ones and zeros that are organized into packets, which can be carried by any delivery platform and traverse any environment from LAN to WAN. Although IP can be carried over ISDN and ATM, voice-data convergence over IP can also be done at much less cost than either ISDN or ATM, which were specifically designed to handle multimedia traffic, but are too costly to extend to every desktop.

As part of the TCP/IP protocol suite, IP can be used to bridge the gap between dissimilar computers, operating systems, and networks. It is supported on nearly every computing platform—from PCs, Macintoshes, and Unix systems to thin clients and servers, legacy mainframes, and the newest supercomputers. In supporting both local and wide area connections, IP provides seamless interconnectivity between these traditionally separate environments.

Another protocol often considered for convergence is ATM. Some carriers, like AT&T, use ATM on the access links as well as the backbone. Instead of forcing customers to use separate access lines for different types of traffic, this approach entails the use of a customer premises ATM switch, also known as an Integrated Access Device (IAD), which consolidates voice, frame, and IP traffic on the same access line or group of access lines that are bonded together to achieve higher-speed channels. At the IAD, the various traffic types are assigned an appropriate class of service by special prioritization algorithms. The traffic is then fed into the nearest network switch in the ATM cell format without any performance degradation.

Summary

IP is flexible enough to overcome the traditional boundaries between voice and data services. Unlike other protocols, application developers can come up with innovative new services like unified messaging and multimedia call centers that combine different content formats and immediately load them onto the existing IP infrastructure. Because particular services are no longer locked into specific forms of infrastructure, voice-data convergence over IP creates new markets and new efficiencies as no other protocol can. The competition spawned from these new markets and efficiencies lowers the cost of communications and fuels the continuous cycle of innovation.

The reduction in long distance phone costs, however, is stalling progress toward voice-data convergence over packet networks. With long distance rates so low, many companies do not see immediate benefits from replacing traditional PBXs with Voice over IP (VoIP) systems because the two technologies currently offer similar performance levels. No business is going to pull out a PBX that works perfectly well to put in a VoIP system to provide what they were already getting from a traditional PBX. Instead of a wholesale adoption of voice-data convergence, there will be a slow migration.

See Also

Asynchronous Transfer Mode
Cable Telephony
Digital Subscriber Line Technologies

Integrated Access Devices
Multiservice Networking
Unified Messaging
Voice over IP

VOICE MAIL

Over the years, voice mail has become an effective communications tool that can enhance productivity and permit personal mobility without the risk of being out of touch with friends, family, or colleagues. Businesses, especially, need to ensure that calls are correctly routed and messages are reliably delivered—even during busy times and off-hours. Voice mail provides the convenience of allowing callers to leave messages for playback by the called party at a more convenient time.

Voice mail may be implemented in a variety of ways—by a machine connected to a telephone line, by a PBX or add-on messaging system, or as a service provided by a telephone company or firm that specializes in voice messaging. Voice mail offers many advantages to callers as well, including:

- Provides the opportunity to leave detailed messages.
- Ensures that the message is accurate.
- Enables information to be delivered in the caller's own speaking style.
- Provides privacy over other message delivery methods such as operators, receptionists, and secretaries.

A business-class voice mail system for the small to midsize organization typically comes with multiple ports—expandable from two to eight ports or more—for interfacing with the telephone system, storage for 35 to 50 hours of messages, and a virtually unlimited number of password-protected voice mailboxes. Custom call routing lets the system administrator tailor options and menus to the specific needs of the organization.

Users can access the system to customize personal greetings and use dial codes to record, check, and send outgoing messages—with or without a confirmation receipt. Other dial codes allow users to play, skip, save, or delete incoming messages locally or remotely. Users can also recover deleted messages. A copy of received messages can be forwarded to another voice mailbox, or broadcast to a list of mailboxes—with or without annotation. Other features that can be implemented by the user include:

- Multiple delivery options including private, urgent, future delivery, and confirmation receipt.
- Messages can be played faster, slower, louder, or softer.
- Attach date/time stamp and sending mailbox ID on every outgoing message.
- Create multiple greetings per mailbox, including standard, temporary, busy, and time sensitive.
- Pager notification allows the voice mailbox to contact the user's pager when a message has been left.
- Call screening conveys the calling party name to the mailbox owner who can then accept, reject, or forward the call to another extension.

Automated Attendant

Many business-class voice mail systems can manage communications without a live operator. The system will greet callers with an introduction, ask them to enter an extension or select from options, and even ask for the caller's name before transferring so the recipient can screen the call.

The system can park a call in a hold location and announce the call over the in-house paging system. It can also inform a caller to a busy station of his or her position in the station's queue ("third in line"). Plus, the system can provide frequently requested information such as directions to the company's headquarters.

Administration

Today's voice mail systems typically employ Windows-based setup screens to facilitate installation and administration from a laptop or desktop computer. In some cases, programming can even be done from a telephone set from an off-site location. With the installation wizards, integration with popular telephone systems is easily implemented. In such cases, mailboxes, message-waiting strings, transfer strings, and time of day greetings are automatically implemented.

Voice Mail Service

Instead of purchasing and managing their own voice mail equipment, self-employed individuals and small businesses may prefer to subscribe to the voice mail services of a carrier. An added advantage of carrier-provided service is that it works during a local power outage. Users access all the features available through the service right from the telephone's keypad. There is usually a choice of plans based on the number of voice mailboxes and control options required. For individuals, basic service usually starts at $5 per month, plus a one-time mailbox setup fee. For businesses with multiple users who need extra recording time, pager notification, added security, and more message-handling features, the service is priced based on the number of mailboxes and the specific features associated with each mailbox. Many carriers have Web pages that offer customers a convenient way to order voice mail services.

Internet Voice Mail

The most economical way to send voice messages is over the Internet as an attachment to electronic mail. There are several ways to record and send messages. With encoder software loaded on a multimedia-equipped notebook or desktop computer, the user can record a voice message through the microphone and store it on the hard disk as an audio file. The file is then sent as an attachment using any e-mail software.

Voice messaging capabilities are also offered in some Internet phone products. Another solution is to use the voice mail capabilities offered by some e-mail programs. This approach allows users to open the voice encoder from within the e-mail program, record a message, and send it just like any other e-mail message.

Summary

Voice mail has been in use by major corporations almost 20 years. It is implemented in the business environment through a PBX, add-on messaging system, or voice messaging switch. Voice mail also is a service offered by telephone companies and numerous third-party messaging firms. Similar technology is available for use over the Internet, enabling anyone with a multimedia PC to record, send, and playback voice messages. Since the Internet spans the globe, voice mail offers an economical way to send personalized messages to friends, family, or colleagues. Regardless of the method used, voice mail provides the opportunity to leave detailed messages in the caller's own speaking style and provides privacy over other message delivery methods such as operators, receptionists, and secretaries.

See Also

Electronic Mail
Unified Messaging
Voice-Data Convergence

VOICE OVER IP

IP (Internet Protocol) telephony has been around since the 1980s when it was referred to as a "Voice Funnel." In 1983, both the ARPANET and Internet were being run from the

Network Operations Center facility at the offices of Bolt Beranek and Newman (BBN) in Cambridge, Massachusetts. There, among the workstations dedicated to special projects, was one labeled "Voice Funnel" that digitized voice, arranged the resulting bits into packets, and sent them through the Internet between sites on the east and west coasts. The Voice Funnel was part of an ARPA research project dealing with packetized speech. ARPA and its contractors used the Voice Funnel, and related video facilities, to do three- and four-way conferencing, saving travel time and money.

The technology was rediscovered in the 1990s, but did not become popular until 1995 when improvements in microprocessors, digital signal processing (DSP) technology, and routing protocols all came together to make feasible products for mainstream use. Since then, IP telephony has been adapted for commercial telecommunications service. Some service providers compete directly with established long distance carriers and regional Bell operating companies (RBOCs), allowing consumers to use IP telephony services with a look and feel identical to today's phone service. Corporations are adding IP telephony service to leverage investments in private intranets and save on long distance call charges. They also see the potential of IP telephony for adding value to existing network applications such as call centers, customer support, help desks, and electronic commerce.

First Generation Technology

First generation IP telephony focused on establishing calls over the public Internet between similarly equipped multimedia PCs. Placing calls involved logging onto the Internet and starting up the "phoneware," which provided several ways of establishing a voice link. Users could connect to the vendor's directory server to check the "white pages" for other phoneware users who also were logged on to the Internet. The public directories were organized by user name and topic of interest to make it easy for everyone to strike up a conversation with like-minded and willing participants. The directory was periodically updated, reflecting changes as people entered and left the network.

Alternatively, users could click on a name in a locally stored private phone book. Of course, users could simply enter the IP address or e-mail address to establish a direct user-to-user connection from the start without having to first go through a directory server (Figure V-1). Whether a public or private listing, the names corresponded with the static IP addresses of other users. As Internet service providers (ISPs) increasingly adopted dynamic IP addresses,[1] the names in public directories and private address books corresponded with e-mail addresses instead.

System Requirements

To make calls over the Internet, users required a computer equipped with a modem, sound card, speakers (or headset), and a microphone. Sound cards came in two types—half duplex and full duplex. Half-duplex works like a citizen's band (CB) radio, where one person can talk at a time and says "over," indicating when he or she is finished talking. With full-duplex audio cards, both parties can talk at once, just like an ordinary telephone call. However, if a full-duplex user connected to a half-duplex user, the conversation defaulted to the half-duplex mode.

[1] With dynamic IP addressing, an address is assigned by the Internet service provider each time the user dials into the server. The Internet service provider has a pool of IP addresses for this purpose. With static IP addressing, the same address is used each time the user connects to the Internet. The proliferation of TCP/IP-based networks, coupled with the growing demand for Internet addresses, makes it necessary to conserve IP addresses. Issuing IP addresses on a dynamic basis provides a way to recycle this finite resource. Even companies with private intranets are increasingly using dynamic IP addresses, instead of issuing unique IP addresses to every machine. The standard that addresses this issue is the Dynamic Host Configuration Protocol (DHCP), developed by the Internet Engineering Task Force (IETF). From a pool of IP addresses, a DHCP server doles them out to users as they establish Internet connections. When they log off the net, the IP addresses become available to other users.

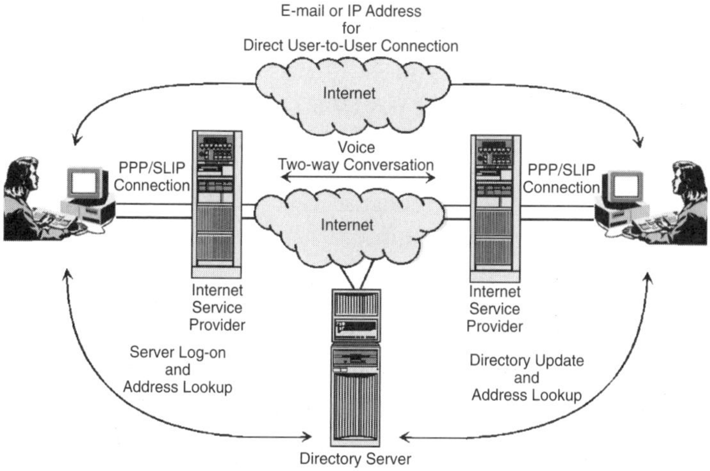

Figure V-1 Early phoneware products logged users onto a directory server, which enabled them to receive a list of other registered phoneware users. The connection was then user to user, bypassing the vendor's directory server.

In addition to the hardware, three software components were typically required—a TCP/IP dialer program (since most users dialed into the Internet with a modem), Web browser, and the phone software itself. The critical component is the phoneware, which provides algorithms that compress the recorded speech obtained from the sound card and apply optimization techniques to ensure its efficient delivery over the Internet in the form of data packets.

Phoneware vendors use a variety of compression algorithms to minimize bandwidth over the Internet. A software coder/decoder (codec) uses mathematical strategies to reduce the bit rate requirements as much as possible and yet still provide acceptable reproduction of the original content. Numerous codecs have appeared on the scene in recent years. For example, NetSpeak's WebPhone started out in 1995 using two audio compression algorithms—GSM and TrueSpeech. GSM is the Global System for Mobile communications and is a worldwide standard for digital cellular communications. It provides close to a 5:1 compression of raw audio with an acceptable loss of audio quality on decompression. TrueSpeech, a product of the DSP Group, provides compression ratios as high as 18:1 with an imperceptible loss of audio quality on decompression. NetSpeak's WebPhone used GSM compression when it was installed on a 486-based computer and TrueSpeech when it is installed on a Pentium-based computer. Offering a high compression ratio, TrueSpeech is more CPU-intensive than GSM, so it requires a faster processor to compress the same audio signal in real time.

The strategy of NetSpeak has been to deal with codecs in a "plug-and-play" fashion to allow new codecs to be incorporated into the client application with little or no effort on the part of the user. As NetSpeak adds new codecs to its repertoire, they are easily incorporated into the clients. The clients can then negotiate between themselves the best codec to be used for an individual session based on available session bandwidth limitations, network delay characteristics, or individual PC resource limitations.

Other phoneware vendors offer proprietary compression algorithms and support one or more accepted industry-standard algorithms. Most now support the G.7xx international standards as well, which guarantee various levels of speech quality and facilitate interoperability between various H.323-compliant products. H.323 is the umbrella standard which includes the G.7xx audio standards. Recommended by the International Telecommunication Union (ITU), H.323 defines how audio and visual conferencing data is transmitted across networks.

In addition to the algorithms that compress/decompress sampled voice, some phoneware products include optimization techniques to deal with the inherent delay of the Internet. The packets may take different paths to their destination and may not all arrive in time to be reassembled in the proper sequence. If this was ordinary data, late or bad packets would simply be dropped and the host's error checking protocols would request a retransmission of those packets. But this concept cannot be applied to packets containing compressed audio without causing major disruption to real-time voice conversations.

If only a small percentage of the packets are dropped, say 2 to 5 percent, the users at each end may not notice the gaps in their conversation. When packet loss approaches 20 percent, however, the quality of the conversation begins to deteriorate. Some products, such as VocalTec's Internet Phone, employ predictive analysis techniques to reconstruct lost packets, thereby minimizing this problem.

Occasionally, the Internet can become overloaded or congested, resulting in lost packets and choppy sound quality. This problem can be overcome by introducing artificial delay into the signal while regulating the flow of voice to the receiver to smooth out any gaps. This scheme affords extra time to retransmit lost packets using a proprietary algorithm. The end result is better quality, at the expense of increased but predictable delay.

Operation

Once the call is placed, either by IP address or e-mail address, the users at each end speak into the microphones connected to the sound cards in their respective computers. The phone software samples the speech, digitizes and compresses the audio signal, and transmits the packets via TCP/IP over the Internet to the remote party. At the other end, the packets of are received and pieced together in the right order. The audio is then decompressed and sent to the sound card's speaker for the other party to hear. The compression algorithm compensates for much of the Internet's inherent delay. As the packets are decompressed and the audio signals are being played, more compressed packets are arriving. This process approximates real-time conversation.

To improve overall sound quality, early users of IP telephony software often found it necessary to fine-tune the sampling rate and compression level to suit their modem's speed. For example, to overcome the annoying problem of clipped speech, the user could reduce the sampling rate until smooth speech resumed.

Some products allow users to adjust recording and playback quality in response to the speed of the modem connection. The user can start this manual tuning process by connecting at the default sampling rate, and then incrementally increasing the sampling rate. With the new setting, the connection is renegotiated at the higher sampling rate. The user can continue to increment the sampling rate until the other party's speech begins to break up, and then back it down until it is clear again.

The sampling rate can be set from 4000 to 44,000 bytes per second, depending on the capabilities of the sound card and the speed of the Internet connection. In general, the higher the speed of the modem connection, the higher the sampling rate can be set. In conjunction with the sampling rate, users can set the compression level. With a lower-speed modem, a higher compression level can be selected to improve performance, but with some loss in sound quality. With a higher-speed modem, users can select a lower compression level for better sound quality.

Some products provide real-time statistics that can help users determine the quality of the Internet connection at any given time. The network statistics window provides a count of incoming and outgoing packets, the average round-trip delay of packets, and the number of lost packets in both directions—incoming and outgoing. This information helps the user pinpoint the source of the problem as originating from the network or locally (e.g., sound card, modem, or software) so corrective action can be taken.

Features

Internet phone products offer many features and new ones are being added all the time. The benefit of using a computer for telephone calls—rather than an ordinary phone—is that the user can take advantage of integrated voice-data features. The following list provides the most common features found across a broad range of products.

- *Adjustable volume control* Allows the volume of the microphone and speakers to be adjusted during the conversation.
- *Advanced caller ID* Not only is the calling party identified by name, but some phoneware products offer a brief introduction message about what callers want to talk about, which is displayed as the call comes through. This information can help users decide whether or not to answer the call.
- *Advanced phone book* Not only holds contact information, but also offers a search capability by name, e-mail, country, company, or any other parameter that can identify a particular person.
- *Audio date/time stamp* Notifies unavailable users of call attempts by date and time.
- *Automatic notification* With this feature, the phoneware automatically looks for and provides notification of when specific other users come on line so they can be called.
- *Busy notification* If a call is placed to someone who is busy with another call, an appropriate "busy" message is returned. Some products allow callers to send an e-mail message, voice mail, or other notification to the busy party indicating that they have tried to call.
- *Call blocking* Annoying or unwanted incoming calls can be blocked by fixed IP or e-mail address.
- *Call conferencing* The ability to converse with three or more people at the same time.
- *Call duration timer* Provides an indication of the amount of time spent on each call.
- *Call hold* Allows an initial call to be put on hold while the user answers another incoming call. The user may continue

the first conversation after holding or hanging up the second call.

- *Call log* Records information about incoming and outgoing calls, allowing the user to keep track of incoming and outgoing calls.
- *Call queue* A place where incoming calls are held until they can be answered.
- *Caller ID* Identifies the caller by name, nickname, e-mail address, or phone number so a user can see who is calling before deciding whether to take the call.
- *Configuration utility* Scans the computer system to determine if the proper hardware is installed to use the phoneware and offers the user advice on configuring various operating parameters such as IRQs, DMAs, and I/O base address settings to prevent conflicts with other communications applications.
- *Database repair utility* The phoneware maintains one or more databases to hold such things as private phone books and configuration data. If a database gets corrupted or destroyed, the user will be notified and have the option of running the database repair utility to restore it.
- *Dedicated server* For those who receive a large volume of incoming calls, some phoneware vendors offer special servers to facilitate call handling.
- *Directory assistance* A searchable directory of users currently on line is automatically maintained. Users can initiate a phone call simply by mouse-clicking on a person's name, or by typing in the first few characters of a name.
- *Dynamic, on-screen directory* Provides the latest information on users who have registered with the server, indicating that they are on line and ready to take or initiate calls. This display is periodically refreshed with new information.
- *Encryption* To ensure secure voice communication over the Internet, a public-key encryption technology such as

PGP (pretty good privacy) can be applied. Depending on vendor, PGP is integrated into the phoneware or may be licensed separately for use with the phoneware. Other products may not accommodate encryption at all.

- *Event message system* Allows users to view the ongoing status of the phoneware to determine what features and functions are active at any given moment.
- *File transfer* Allows the user to transmit a file to the other party during a conversation. The file transfer process takes place in the background and does not interfere with the conversation.
- *Greeting message* When a user is not available or too busy to take a call, a recorded message can be played to callers.
- *H.323 compliance* A worldwide standard for audiovisual communication over packet data networks, such as the Internet. Users of different H.323-compliant products are capable of conversing with each other over the Internet.
- *Last-party redial* Allows the user to redial the last party called without having to look up the address in a directory.
- *Map* Displays the connection of the call against the background of a U.S. or world map, showing the points of origination and destination.
- *Multiple calling mechanisms* Some phoneware products offer multiple methods of initiating calls, including by fixed IP address, domain name, e-mail address, saved addresses, and online directory of registered users.
- *Multiple lines* Some phoneware products allow users to carry on a conversation on one line and take an incoming call on another line, or put one call on hold while another call is initiated.
- *Multiple user configurations* If several people share the same computer, some phoneware products allow each of them to have a private configuration, including caller ID information and address books.

- *Music on hold* Plays music to a caller on hold, until the call can be answered.
- *Mute* A mute button allows private, "offline" conversations.
- *Online help* Offers help on the proper use of various phoneware features without having to resort to a manual or opening a separate read-me file.
- *Picture compression* Some phoneware allows the user to call up a photo of the person they are talking to (if the remote user supplies one). Compression allows fast photo loading over the Internet of the commonly supported file formats.
- *Programmable buttons* Allows users to configure quick-dial buttons for the people they call most frequently. In some cases, buttons are added automatically and written over on the basis of the most recent calls.
- *Redial* If a person is not reachable on the first call attempt, the phoneware can be configured to automatically redial at designated time intervals until the connection is established.
- *Remote information display* Displays the operating system and sound card information of the remote user.
- *Remote time display* Displays the remote time of the called party.
- *Selectable codec* Provides a choice of codecs, depending on the processing power of the computer. A high-compression codec can be used for Pentium-class machines and a lower-compression codec can be used for 486-based machines. The choice is made during phoneware installation.
- *Silence detection* Detects periods of silence during the conversation to avoid unnecessary transmission.
- *Statistics window* Allows the user to monitor system performance and the quality of the Internet link.
- *Text chat* Some phoneware products offer an interactive text or "chat" capability to augment voice conversations.

The chat feature can be used before, during, or apart from voice mode.

- *Toolbar* Icons provide quick access to frequently performed tasks such as hang up, mute, chat, view settings, and help.
- *Toolbox mode* The interface can be collapsed into a compact toolbox to save desktop space. This makes it easier to work in other applications until the phoneware is used for calls.
- *User Location Service (ULS) compliance* ULS technology enables Internet phone users to find each other through existing Internet Directory Services.
- *User-defined groups* Allows users to set up private "calling circles" for calls among members only or establish new topic groups for public access.
- *Video* Some phoneware products allow the calling and called parties to see each other as they converse. This requires a video camera connection to the computer.
- *Voice mail* Allows users to record and playback greeting messages as well as send voice mail messages for playback by recipients. Depending on vendor, this might include the ability to give specific messages to callers when they enter a personal code.
- *Voice mail screening* Allows users to delete voice messages before they can be downloaded to their computer.
- *Web links* Allows users to put links into their Web pages that, when activated, establish a call with the visitor.
- *Whiteboard* Some multifunction products allow participants to draw or annotate shared text and images while conversing.

While early phoneware products focused on PC-to-PC communication, they had the advantage of combining audio, video, and text capabilities. Such products are still available and are continually being enhanced with new capabilities and features. The business version of NetSpeak's WebPhone, for example, offers four lines with call holding, muting, do not disturb, and call blocking options. It also offers a large video display area with self and remote views.

Use of Conventional Phones

While early VoIP products required the use of a multimedia PC, the technology has progressed to the point where conventional phones can be used to place calls over IP networks. In this case, the user dials an 800 number to access the nearest gateway of a service provider. After receiving dial tone, the user enters the domestic or international long distance number. The local gateway digitizes and compresses the incoming voice signals into packets, which travel over the Internet or the service provider's privately managed intranet to a remote gateway that is nearest to the dialed location. At that gateway, the packets are decompressed and reconstructed into their original form, making them suitable for transmission over the Public Switched Telephone Network (PSTN). From there, the call is then routed over the local loop connection to the destination telephone or PBX.

IP/PSTN gateways offer many features that expedite administration, call monitoring, security, and billing. They are even equipped with an Interactive Voice Response (IVR) application, which acts as the interface between the PSTN/PBX and the IP network. The application includes an auto attendant, which guides users through the calling process if they need assistant using the technology. Service level guarantees are becoming available with some gateways, with vendors offering a 100 percent guarantee that IP calls will go through to completion the first time. The gateway continually monitors IP network quality and if it detects congestion, automatically switches a call from the data network to the PSTN, thus ensuring the call will be completed.

Summary

Many of the issues that plagued users of first-generation Internet telephony products have been addressed by hardware and software vendors with the goal of facilitating the growth of commercial, carrier-grade VoIP services. There are now scalable IP switches that compete with traditional central office switches in terms of call-processing capacity and call-handling features. There are roaming agreements between service providers that help ensure the broadest possible coverage for commercial IP telephony service.

Mechanisms have been developed to accurately meter usage and charge for voice calls over IP networks. Administrative tools allow individual users or groups of users to be assigned a class of service. There are even tools that allow network managers to monitor the performance of the IP network in real time and check on the quality of service being delivered to each user at any given time. These and other developments have prompted carriers to adopt this once-spurned technology to help reduce operating costs. Businesses are adopting the technology to reduce the cost of telecommunications. ISPs use the technology to offer value-added services to their subscribers.

See Also

Cable Telephony
Internet Telephony
LAN Telephony
Multiservice Networking
Transmission Control Protocol/Internet Protocol (TCP/IP)
Voice Compression
Voice-Data Convergence

W

WIRELESS LANS

Wireless local area network (WLAN) technologies are implemented as an extension—or as an alternative—to wired LANs. Using a variety of technologies, including narrowband radio, spread spectrum, and infrared, wireless LANs transmit and receive data through the air, minimizing the need for wired connections. Wireless links to access points connected to wired LANs offer users the means to reach corporate information on servers, peripherals such as printers, and routers for reaching resources on the Internet.

Applications

Wireless LANs have become popular in a number of vertical markets, including health care, retail, manufacturing, and warehousing. These industries have profited from the productivity gains of using hand-held terminals and notebook computers to transmit real-time information to centralized hosts for processing.

Wireless LANs allow users to go where wires cannot go. While the initial investment required for wireless LAN hardware can be higher than the cost of conventional LAN hardware, overall installation expenses and life-cycle costs can be significantly lower. Long-term cost savings are greatest in dynamic environments requiring frequent moves, adds, and changes. Wireless LANs can be configured in a variety of topologies to meet the needs of specific applications and installations. They can grow by adding access points and extension points to accommodate virtually any number of users.

Technologies

There are several technologies to choose from when selecting a wireless LAN solution, each with advantages and limitations. Most wireless LANs use spread spectrum, a wideband radiofrequency technique developed by the military for use in reliable, secure, mission-critical communications systems. To achieve these advantages, the signal is spread out over the available bandwidth and resembles background noise that is virtually immune from interception.

There are two types of spread spectrum radio: frequency hopping and direct sequence. Frequency-hopping spread spectrum (FHSS) uses a narrowband carrier that changes frequency in a pattern known only to the transmitter and receiver. Properly synchronized, the net effect is to maintain a single logical channel. To an unintended receiver, FHSS appears to be short-duration impulse noise.

Direct-sequence spread spectrum (DSSS) generates a redundant bit pattern for each bit to be transmitted and requires more bandwidth for implementation. This bit pattern, called a “chip” (or “chipping code”), is used by the receiver to recover the original signal. Even if one or more bits in the chip are damaged during transmission, statistical techniques embedded in the radio can recover the original data without the need for retransmission. To an unintended receiver, DSSS appears as low-power wideband noise.

Another technology used for wireless LANs is infrared (IR), which uses very high frequencies that are just below visible light in the electromagnetic spectrum. Like light, IR cannot penetrate opaque objects—to reach the target system, the waves carrying data are sent in either directed (line-of-sight) or diffuse (reflected) fashion. Inexpensive directed systems provide very limited range of not more than 3 feet and typically are used for personal area networks but occasionally are used in specific wireless LAN applications. High-performance directed IR is impractical for mobile users and is therefore used only to implement fixed subnetworks. Diffuse IR wireless LAN systems do not require line-of-sight transmission, but cells are limited to individual rooms. As with spread-spectrum LANs, infrared LANs can be extended by connecting the wireless access points to a conventional wired LAN.

Operation

As noted, wireless LANs use electromagnetic waves (radio or infrared) to communicate information from one point to another without relying on a wired connection. Radio waves are often referred to as "radio carriers" because they simply perform the function of delivering energy to a remote receiver. The data being transmitted are superimposed on the radio carrier so that they can be extracted accurately at the receiving end. This process is generally referred to as "carrier modulation." Once data are modulated onto the radio carrier, the radio signal occupies more than a single frequency, since the frequency or bit rate of the modulating information adds to the carrier.

Multiple radio carriers can exist in the same space at the same time without interfering with each other if the radio waves are transmitted on different frequencies. To extract data, a radio receiver tunes into one radio frequency while rejecting all other frequencies.

In a typical wireless LAN configuration, a transmitter/receiver (transceiver) device, called an "access point," connects to the wired network from a fixed location using standard cabling. At a minimum, the access point receives, buffers, and transmits data between the wireless LAN and the wired network infrastructure. A single access point can support a small group of users and can function within a range of less than 100 to several hundred feet. The access point (or the antenna attached to the access point) is usually mounted high but may be mounted essentially anywhere that is practical as long as the desired radio coverage is obtained.

Users access the wireless LAN through wireless LAN adapters. These adapters provide an interface between the client network operating system (NOS) and the airwaves via an antenna. The nature of the wireless connection is transparent to the NOS.

Configurations

Wireless LANs can be simple or complex. The simplest configuration consists of two PCs equipped with wireless adapter cards that form a network whenever they are within range of one another (Figure W-1). This peer-to-peer network

Figure W-1 A wireless peer-to-peer network created between two laptops equipped with wireless adapter cards.

requires no administration. In this case, each client would only have access to the resources of the other client and not to a central server.

Installing an access point can extend the operating range of the wireless network, effectively doubling the range at which the devices can communicate. Since the access point is connected to the wired network, each client would have access to the server's resources as well as to other clients (Figure W-2). Each access point can support many clients—the specific number depends on the nature of the transmissions involved. In some cases, a single access point can support up to 50 clients.

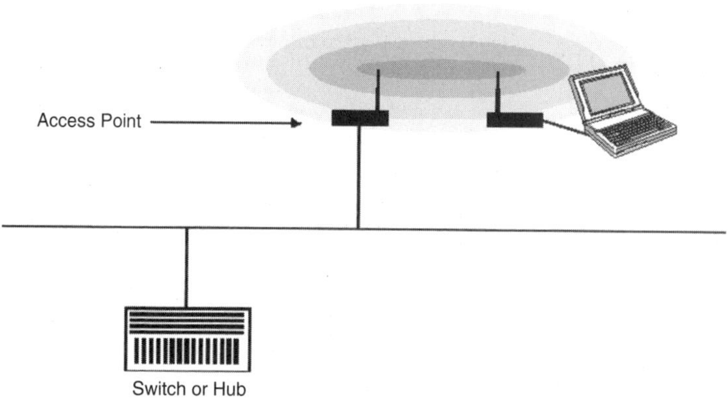

Figure W-2 A wireless client connected to the wired LAN via an access point.

Access points have an operating range of about 500 feet indoors and 1000 feet outdoors. In a very large facility such as a warehouse or on a college campus, it probably will be necessary to install more than one access point. Access point positioning is determined by a site survey. The goal is to blanket the coverage area with overlapping coverage cells so that clients can roam throughout the area without ever losing network contact. Access points hand the client off from one to another in a way that is invisible to the client, ensuring uninterrupted connectivity.

To solve particular problems of topology, the network designer might choose to use extension points (EPs) to augment the network of access points. These devices look and function like access points (APs), but they are not tethered to the wired network, as are APs. EPs function as repeaters by boosting signal strength to extend the range of the network by relaying signals from a client to an AP or another EP.

Another component of wireless LANs is the directional antenna. If a wireless LAN in one building must be connected to a wireless LAN in another building a mile away, one solution might be to install a directional antenna on the two buildings—each antenna targeting the other and connected to its own wired network via an AP (Figure W-3).

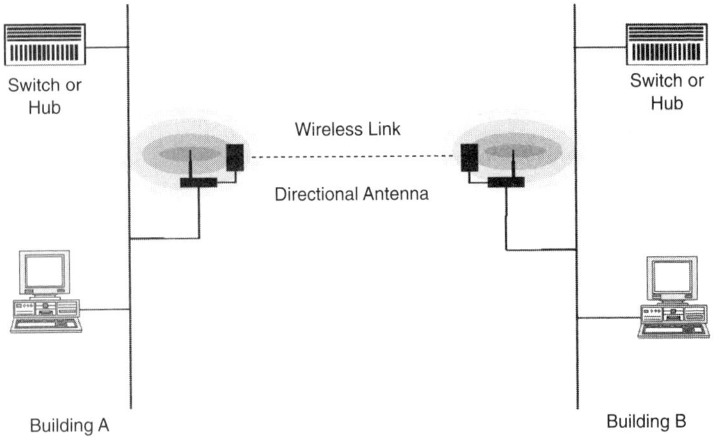

Figure W-3 A directional antenna can be used to interconnect wireless LANs in different buildings.

Network Management

The setup and management of wireless LANs typically are done with Windows-based tools, which facilitate configuration, remote management, and troubleshooting. Such tools also can be used to verify building coverage, identify coverage patterns, select alternate frequencies, locate and tune around RF interference, and customize network access security. Beyond that, they make it easy for system administrators to monitor the quality of communications at multiple stations in a wireless network. Wireless management systems offer five basic functions:

- *Communications indicator* Located on the Windows taskbar, it provides mobile users graphical, real-time information on the level of communication quality between a wireless station and the nearest access point.
- *Link test diagnostics* Verifies the communications path between neighboring wireless stations, as well as between stations and access points within a wireless cell. Link test diagnostics measure signal quality, signal-to-noise ratio, and the number of successfully received packets.
- *Site monitor* Ensures optimal placement of access points. While carrying a wireless-equipped notebook computer through the facility, the management tool graphically displays changing communication quality levels with the various access points installed in the building. This tool makes it easy to locate radio dead spots or sources of interference.
- *Frequency select* Manages RF channel selection. It enables the user to choose from up to eight different channels (in the 2.4-GHz frequency band).
- *Access control table manager* Enables the system administrator to provide extra levels of security by restricting access to individual computers in a facility.

Wireless LAN Standards

There are several wireless LAN standards, each suited for a particular environment: IEEE 802.11b (Wi-Fi), HomeRF, and Bluetooth.

For the corporate environment, IEEE 802.11b (referred to as Wireless Fidelity, or simply Wi-Fi) offers a data transfer rate of up to 11 Mbps at a range of up to 300 feet from the base station. It operates in the 2.4-GHz band and transmits via the Direct Sequence Spread Spectrum method. Multiple base stations can be linked to increase that distance as needed, with support for multiple clients per access point. Products designed for corporate use feature better security and management than Wi-Fi products for the home. Another version of the technology is IEEE 802.a, which operates in the 5-GHz band. This is called "Wi-Fi5" and is capable of reaching speeds of up to 54 Mbps and higher with proprietary "turbo mode" features.

The HomeRF 2.0 standard draws from IEEE 802.11b and Digital Enhanced Cordless Telecommunication (DECT), a popular standard for portable phones worldwide. Operating in the 2.4-GHz band, HomeRF was designed from the ground up for the home market for both voice and data. It offers throughput rates comparable to IEEE 802.11b and supports the same kinds of terminal devices in both point-to-point and multipoint configurations. HomeRF transmits at up to 10 Mbps over a range of about 150 feet from the base station, which makes it suitable for the average home. HomeRF transmits using frequency hopping; that is, it hops around constantly within its prescribed bandwidth. When it encounters interference, such as a microwave oven or the wireless LAN in the next apartment, it adapts by moving to another frequency.

The key advantage that HomeRF has over IEEE 802.11b in the home environment is its superior ability to adapt to interference from such devices as portable phones and microwaves. As a frequency hopper, it coexists well with other frequency-hopping devices that proliferate in the home. Another advantage of HomeRF is that it continuously reserves a chunk of bandwidth via "isochronous channels" for voice services. Speech quality is high; there is no clipping while the protocol deals with interference.

The IEEE 802.11b Standard does not include frequency hopping. In response to interference, IEEE 802.11b simply retransmits or waits for the higher-level TCP/IP protocol to sort out signal from noise. This works well for data but can result in voice transmissions sounding choppy. Voice and data are treated the same way, converting voice into data packets but offering no priority to voice. This results in unacceptable voice quality. Another problem with IEEE 802.11b is that its Wired Equivalent Privacy (WEP) encryption, designed to safeguard privacy, has had problems living up to its claim.

Bluetooth also operates in the 2.4-GHz band but was not originally created to support wireless LANs; it was intended as a replacement for cable between desktop computers, peripherals, and hand-held devices. Operating at the comparatively slow rate of 30 to 400 kbps across a range of only 30 feet, Bluetooth supports "piconets" that link laptops, personal digital assistants, mobile phones, and other portable devices on an as-needed basis. It improves on infrared in that it does not require a line of sight between the devices and has greater range than infrared's 3 to 10 feet. Bluetooth also supports voice channels.

While Bluetooth does not have the power and range of a full-fledged LAN, its master-slave architecture does permit the devices to face different piconets, in effect extending the range of the signals beyond 30 feet. Like HomeRF, Bluetooth is a frequency hopper, so devices that use these two standards can coexist by hopping out of each other's way. Bluetooth has the faster hop rate, so it will be the first to sense problems and act to steer clear of interference from HomeRF devices.

The three standards each have particular strengths that make them ideal for certain situations, as well as specific shortcomings that render them inadequate for use beyond their intended purpose:

- While suited for the office environment, IEEE 802.11b is not designed to provide adequate interference adaptation and voice quality for the home. Data collisions force packet retransmissions, which is fine for file transfers and print jobs but not for voice or multimedia that cannot tolerate the resulting delay.
- HomeRF delivers an adequate range for the home market but not for many small businesses. It is better suited than IEEE 802.11b for streaming multimedia and telephony,

applications that may become more important for home users as convergence devices become popular.

- Bluetooth does not provide the bandwidth and range required for wireless LAN applications but instead is suited for desktop cable replacement and ad-hoc networking for both voice and data within the narrow 30-foot range of a piconet.

Wireless LAN technology is continually improving. The IEEE 802.11b Standard developers seek to improve encryption (IEEE 802.11i), make the standard more multimedia-friendly (IEEE 802.11e), and increase the speed to 54-Mbps (IEEE 802.11a) operating in the 5-GHz band—more than enough to move full-motion video through the home. Dozens of vendors are shipping IEEE 802.11b products, and the standard's proliferation in corporate and public environments is a distinct advantage. An office worker who already has an IEEE 802.11b–equipped notebook will not likely want to invest in a different network for the home.

Furthermore, the multimedia and telephony applications HomeRF advocates tout have not yet arrived to make the technology a compelling choice. Although HomeRF currently beats IEEE 802.11b in terms of security, this is not a big issue in the home. For these and other reasons, industry analysts predict that IEEE 802.11b will soon overtake HomeRF in the consumer marketplace, especially since the price difference between the two has just about reached parity.

Summary

Once expensive, slow, and proprietary, wireless LAN products are now reasonably fast, standardized, and priced for mainstream business and consumer use. Wireless LAN configurations range from simple peer-to-peer topologies to complex networks offering distributed data connectivity and roaming. To solve problems of vendor interoperability, the Wireless Ethernet Compatibility Alliance (WECA) offers a certification program that tests vendor-submitted products.

Those that pass WECA's battery of tests receive the right to bear the Wi-Fi logo of interoperability.

See also

Bluetooth

Ethernet

Token Ring

WORLD WIDE WEB

Since its development by Tim Berners-Lee in 1990 at the European Particle Physics Laboratory in Switzerland (CERN), the World Wide Web (WWW) has grown to become one of the most sophisticated and popular services on the Internet. The two main mechanisms that make the Web work are the HyperText Transfer Protocol (HTTP) and HyperText Markup Language (HTML).

HTTP is used to transfer hypertext documents among the Web servers on the Internet and, ultimately, to a client—the end user's browser-equipped computer (Figure W-4). Collectively, the hundreds of thousands of servers distributed worldwide that support HTTP are known as the "World Wide Web." HTML is used to structure information that resides on the servers in a way that can be readily rendered by browser software, such as Microsoft's Internet Explorer and Netscape Navigator, installed on the clients. HTML makes documents portable from one computer platform to another and is intended as a common medium for tying together information from widely different sources.

History

As noted, the Web was invented by Tim Berners-Lee, who was named by *Time* magazine as one of the 100 greatest minds of the twentieth century. The original vision of

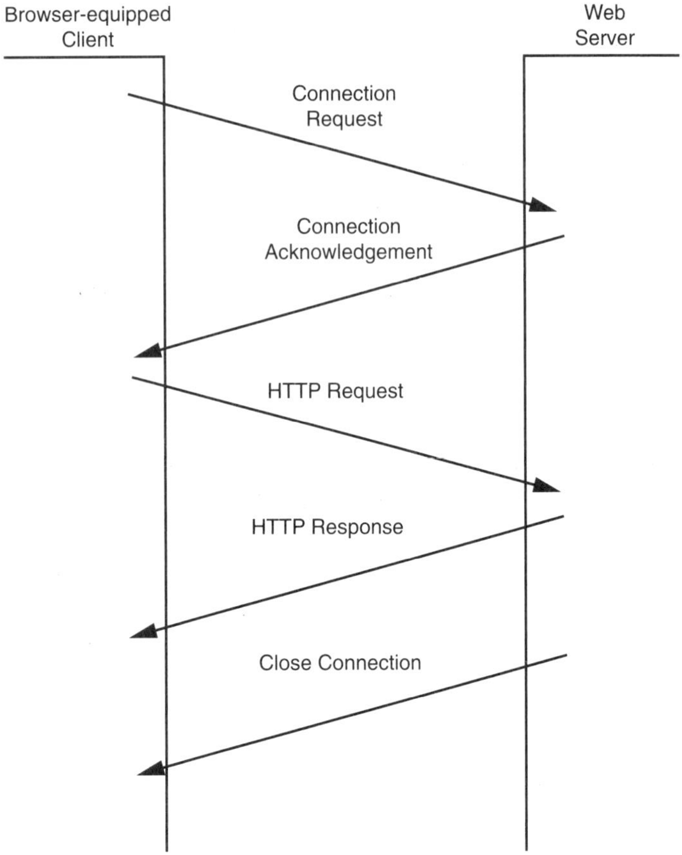

Figure W-4 The HyperText Transfer Protocol (HTTP) delivers documents from Web servers to browser-equipped clients in response to specific requests and then closes the connection until a new request is made from the client.

Berners-Lee has inspired the Web's further development as a powerful technological force for social change, commerce, and individual creativity. Berners-Lee is now the director of the World Wide Web Consortium (W3C) based at the Massachusetts Institute of Technology (MIT).

Since its establishment in 1994, the W3C has been focused on developing interoperable technologies (specifications, guidelines, software, and tools) to lead the Web to its full potential as a forum for information, commerce, communication, and collective understanding. The W3C has over 510 members and nearly 60 full-time staff around the world who contribute to the development of specifications and software.

Although hypertext systems had been around since the 1980s, they were limited to working across a single database. The contribution of Berners-Lee was to apply hypertext links to multiple databases distributed across a network, enabling the links to point to any document anywhere on the network.

The idea for the Web occurred to Berners-Lee while working at CERN. He noticed that many people came in and out of the facility with great ideas. They did some work, and when they left, there was no trace of what they had accomplished. He decided the organization needed a method of sharing all this information so that others could benefit from it rather than merely grabbing somebody at coffee hour for a one-time conversation that would soon be forgotten. The intent of the Web was to enable people to work together as a self-managing team in an ongoing collaborative way regardless of each participant's location or what computer platform they were using.

Web Characteristics

The Web itself can best be described as a dynamic, interactive, graphically oriented, distributed, platform-independent hypertext information system.

- The Web is dynamic because it changes daily. Web servers are continually being added to the Web. New information also is continually being added, as are new hypertext links and innovative services.

- The Web is interactive in that specific information can be requested through various search engines and returned moments later in the form of lists, with each item weighted according to how well it matched the search parameters. Another example of interactivity is text chat, whereby users communicate online in near real time via their keyboards. Even voice conversations and videoconferences can take place over the Web.
- The Web is graphics-oriented. The use of graphics makes the Web not only visually appealing but also easy to navigate. Graphical signposts direct users to other sources of information accessed via hypertext links. Sound, animation, and video capabilities may be added to Web pages as well.
- The Web is distributed, meaning that information resides on hundreds of thousands of individual Web servers around the world. If one server goes down, there is no significant impact on the Web as a whole, except that access to the failed server will be denied until it can be brought back into service. Some servers are mirrored—duplicated at other locations—to keep information available if the primary server crashes.
- The Web is platform-independent, which means that virtually any client can access the Web, whether it uses Windows, OS/2, Macintosh, or UNIX operating environment. This platform independence even applies to the Web servers. Although most Web servers are based on UNIX, Windows NT is growing in popularity.
- The Web makes extensive use of hypertext links. A hypertext link, usually identified by an underlined word or phrase or a graphical symbol, points the way to other information. That information may be found virtually anywhere: in the same document, in a different document on the same server, or in another document on a different server that may be located anywhere in the world. A hypertext link does not necessarily point to text docu-

ments; it can point to maps, forms, images, sound and video clips, or applications. The links can even point to other Internet resources such as File Transfer Protocol (FTP) and Gopher sites and Usenet newsgroups.

Summary

The Web has become a pervasive medium for information distribution, collaborative projects, electronic commerce, and the delivery of support services of all kinds. The capabilities of the Web are continually being expanded. In addition to text and images, the Web is being used for instant messaging, telephony, videoconferencing, faxing, remote printing, and paging. With the emergence of Java as a popular network programming language, applets can be embedded into Web pages, which enable users to take advantage of the capabilities of larger applications that may reside on a corporate Web site.

See also

Electronic Commerce
Internet
Intranets
Java

ACRONYMS

A

AAL	ATM Adaptation Layer
ABR	Available Bit Rate
AC	Access Control
AC	Authentication Center
ACP	Access Control Point
DPCM	Adaptive Differential Pulse Code Modulation
ADSL	Asymmetric Digital Subscriber Line
ANSI	American National Standards Institute
ANT	ADSL Network Terminator
AOL	America Online
API	Application Programming Interface
ARIN	American Registry of Internet Numbers
ARP	Address Resolution Protocol
ARPA	Advanced Research Projects Agency
ARQ	Automatic Repeat Request
AS	Autonomous System
ASCII	American Standard Code for Information Interchange

ASIC	Application-Specific Integrated Circuit
ASP	Application Service Provider
ATM	Asynchronous Transfer Mode
AUI	Attachment Unit Interface
AWG	American Wire Gauge

B

B2B	Business-to-Business
BACP	Bandwidth Allocation Control Protocol
BBS	Bulletin Board System
BCCH	Broadcast Control Channel
BECN	Backward Explicit Congestion Notification
Bellcore	Bell Communications Research, Inc.
BER	Bit Error Rate
BERT	Bit Error Rate Tester
BGP	Border Gateway Protocol
BHCA	Busy Hour Call Attempts
BIB	Backward Indicator Bit
BIOS	Basic Input-Output System
BLEC	Building Local Exchange Carrier
BMS-E	Bandwidth Management Service–Extended (AT&T)
BOC	Bell Operating Company
BONDING	Bandwidth on Demand Interoperability Group
BootP	Boot Protocol
BPDU	Bridge Protocol Data Unit
bps	Bits Per Second
BRI	Basic Rate Interface (ISDN)

BSC	Base Station Controller
BSN	Backward Sequence Number
BTS	Base Transceiver Station

C

CAN	Campus Area Network
CAP	Competitive Access Provider
CATV	Cable Television
CBR	Constant Bit Rate
CCC	Clear Channel Capability
CCCH	Common Control Channel
CD	Compact Disc
CDCS	Continuous Dynamic Channel Selection
CDMA	Code Division Multiple Access
CDN	Content Delivery Network
CDPD	Cellular Digital Packet Data
Centrex	Central Office Exchange
CHAP	Challenge Handshake Authentication Protocol
CIF	Common Intermediate Format
CIR	Committed Information Rate
CLEC	Competitive Local Exchange Carrier
CLP	Cell Loss Priority
CNR	Customer Network Reconfiguration
CO	Central Office
COS	Class of Service
CPE	Customer Premises Equipment
CPS	Cycles per Second (Hertz)
CPU	Central Processing Unit

CRC	Cyclic Redundancy Check
CRM	Customer Relationship Management
CSM	Communications Services Management
CSMA/CD	Carrier Sense Multiple Access with Collision Detection
CT	Cordless Telecommunications
CTI	Computer-Telephony Integration
CTIA	Cellular Telecommunications Industry Association
CVSD	Continuously Variable Slope Delta (Modulation)

D

DA	Destination Address
DACS	Digital Access and Cross-Connect System (AT&T)
DAP	Demand Access Protocol
DAT	Digital Audio Tape
dB	Decibel
DBMS	Database Management System
DBS	Direct Broadcast Satellite
DCCH	Digital Control Channel
DCE	Data Communications Equipment
DCE	Distributed Computing Environment
DCS	Digital Cross-Connect System
DDS	Digital Data Services
DDS/SC	Digital Data Service with Secondary Channel
DECT	Digital Enhanced (formerly European) Cordless Telecommunication

DES	Data Encryption Standard
DHCP	Dynamic Host Control Protocol
DLCS	Digital Loop Carrier System
DLEC	Data Local Exchange Carrier
DLL	Data Link Layer
DLL	Dynamic Link Library
DLSw	Data Link Switching (IBM Corp.)
DMT	Discrete Multitone
DNS	Domain Name Service
DOCSIS	Data over Cable Service Interface Specification
DoD	Department of Defense
DOS	Disk Operating System
DOV	Data over Voice
DQPSK	Differential Quadrature Phase-Shift Keying
DS0	Digital Signal Level 0 (64 kbps)
DS1	Digital Signal Level 1 (1.544 Mbps)
DS1C	Digital Signal Level 1C (3.152 Mbps)
DS2	Digital Signal Level 2 (6.312 Mbps)
DS3	Digital Signal Level 3 (44.736 Mbps)
DS4	Digital Signal Level 4 (274.176 Mbps)
DSL	Digital Subscriber Line
DSLAM	DSL Access Multiplexer
DSML	Directory Services Markup Language
DSN	Defense Switched Network
DSP	Digital Signal Processor
DSS	Decision Support System
DTE	Data Terminal Equipment
DTR	Dedicated Token Ring

DTU	Data Transfer Unit
DWMT	Discrete Wavelet Multitone

E

E-mail	Electronic Mail
EBCDIC	Extended Binary Coded Decimal Interexchange Code (IBM Corp.)
ED	Ending Delimiter
EDI	Electronic Data Interchange
EEROM	Electronically Erasable Read-Only Memory
EFRC	Enhanced Full Rate Codec
EFT	Electronic Funds Transfer
EGP	External Gateway Protocol
EISA	Extended Industry Standard Architecture
EMI	Electromagnetic Interference
EMS	Element Management System
EOC	Embedded Overhead Channel
EOT	End of Transmission
ESD	Electronic Software Distribution
ESF	Extended Superframe Format
ESP	Enhanced Service Provider

F

FACCH	Fast Associated Control Channel
FAT	File Allocation Table
FC	Frame Control
FC	Fibre Channel
FC-0	Fibre Channel—Layer 0
FC-1	Fibre Channel—Layer 1

FC-2	Fibre Channel—Layer 2
FC-3	Fibre Channel—Layer 3
FC-4	Fibre Channel—Layer 4
FCC	Federal Communications Commission
FCS	Frame Check Sequence
FDIC	Federal Deposit Insurance Corporation
FDL	Facilities Data Link
FECN	Forward Explicit Congestion Notification
FEP	Front-End Processor
FIB	Forward Indicator Bit
FIB	Forwarding Information Base
FIFO	First In, First Out
FITL	Fiber in the Loop
FRAD	Frame Relay Access Device
FS	Frame Status
FSN	Forward Sequence Number
FT1	Fractional T1
FTP	File Transfer Protocol
FTTB	Fiber to the Building
FTTC	Fiber to the Curb
FTTH	Fiber to the Home

G

GBIC	Gigabit Interface Converter
GFC	Generic Flow Control
GHz	Gigahertz (Billions of Cycles per Second)
GIS	Geographical Information System
GloBanD	Global Bandwidth on Demand
GPRS	General Packet Radio Services

GPS	Global Positioning System
GSM	Global System for Mobile (GSM) Telecommunications (formerly Groupe Spéciale Mobile)
GUI	Graphical User Interface

H

H0	High-Capacity ISDN Channel Operating at 384 kbps
H11	High-Capacity ISDN Channel Operating at 1.536 Mbps
HDSL	High-Bit-Rate Digital Subscriber Line
HEC	Header Error Check
HF	High Frequency (3–30 MHz)
HFC	Hybrid Fiber/Coax
HIC	Head-End Interface Converter
HPR	High-Performance Routing (IBM Corp.)
HSCSD	High-Speed Circuit-Switched Data
HTML	HyperText Markup Language
HTTP	HyperText Transfer Protocol
HVAC	Heating, Ventilation, and Air Conditioning
Hz	Hertz (Cycles per Second)

I

I/O	Input-Output
IAB	Internet Architecture Board
IANA	Internet Assigned Numbers Authority
ICANN	International Corporation for Assigned Names and Numbers
ICMP	Internet Control Message Protocol

ICP	Integrated Communications Provider
ID	Identification
IDLC	Integrated Digital Loop Carrier
IDPR	Interdomain Policy Routing
IDSL	ISDN Digital Subscriber Line
IEEE	Institute of Electrical and Electronics Engineers
IESG	Internet Engineering Steering Group
IETF	Internet Engineering Task Force
IFX	Interactive Financial Exchange
IGP	Interior Gateway Protocol
ILEC	Incumbent Local Exchange Carrier
IMA	Inverse Multiplexing over ATM
IMAP	Internet Mail Access Protocol
IOC	Interoffice Channel
IP	Internet Protocol
IPsec	Internet Protocol with Security
IPX	Internet Packet Exchange
IRQ	Interrupt Request
IS	Information System
IS	Industry Standard
IS-IS	Intraautonomous System to Intraautonomous System
ISA	Industry Standard Architecture
ISD	Independent Service Developer
ISDN	Integrated Services Digital Network
ISM	Industrial, Scientific, and Medical (Frequency Bands)
ISO	International Organization for Standardization
ISOC	Internet Society

ISP	Internet Service Provider
IT	Information Technology
ITU-TSS	International Telecommunication Union—Telecommunications Standardization Sector (formerly, CCITT)
IXC	Interexchange Carrier

J

JEPI	Joint Electronic Payments Initiative
JPEG	Joint Photographic Experts Group
JTAPI	Java Telephony Application Programming Interface

K

k (kilo)	One Thousand (e.g., kbps)
kB	Kilobyte
kHz	Kilohertz (Thousands of Cycles per Second)

L

L2F	Layer 2 Forwarding
L2TP	Layer 2 Tunneling Protocol
LAN	Local Area Network
LAPB	Link Access Procedure–Balanced
LAT	Local Area Transport (Digital Equipment Corp.)
LATA	Local Access and Transport Area
LCD	Liquid Crystal Display
LCN	Local Channel Number
LCP	Link Control Protocol

LD	Laser Diode
LDAP	Lightweight Directory Access Protocol
LEC	Local Exchange Carrier
LED	Light-Emitting Diode
LEO	Low Earth Orbit
LIB	Label Information Base
LIFO	Last In, First Out
LIPS	Lightweight Internet Person Schema
LLC	Logical Link Control
LMDS	Local Multipoint Distribution System
LSI	Large-Scale Integration
LSP	Label Switched Path
LSR	Label Switch Router

M

M	(Mega) One Million (e.g., Mbps)
MAC	Media Access Control
MAC	Moves, Adds, Changes
MAN	Metropolitan Area Network
MAPI	Messaging Applications Programming Interface (Microsoft Corp.)
MAU	Multistation Access Unit
MB	Megabyte
MCA	Microchannel Architecture (IBM Corp.)
MCU	Multipoint Control Unit
MD	Mediation Device
MDS	Multipoint Distribution Service
MF	Mediation Function
MF	Medium Frequency (300 kHz to 3 MHz)
MHz	Megahertz (Millions of Cycles per Second)

MIB	Management Information Base
MIME	Multipurpose Internet Mail Extensions
MIPS	Millions of Instructions per Second
MIS	Management Information Services
MISR	Multiprotocol Integrated Switch-Routing
MJU	Multipoint Junction Unit
MMDS	Multichannel Multipoint Distribution Service
Modem	Modulation/Demodulation
MOS	Multimedia Operating System
MPEG	Moving Pictures Experts Group
MPLS	Multiprotocol Label Switching
MPPP	Multilink Point-to-Point Protocol
ms	Millisecond (Thousandths of a Second)
MSN	Microsoft Network
MSRN	Mobile Station Roaming Number
MTBF	Mean Time Between Failure
MVC	Multicast Virtual Circuit

N

NAP	Network Access Point
NAT	Network Address Translation
NAUN	Nearest Active Upstream Neighbor
NC	Network Computer
NE	Network Element
NetBIOS	Network Basic Input-Output System
NEF	Network Element Function
NFS	Network File System (or Server)
NIC	Network Interface Card
NID	Network Interface Device

nm	Nanometer
NM	Network Manager
NNTP	Network News Transfer Protocol
NOC	Network Operations Center
NOS	Network Operating System
NSF	National Science Foundation
NTSC	National Television Standards Committee

O

OC	Optical Carrier
OC-1	Optical Carrier Signal Level 1 (51.84 Mbps)
OC-3	Optical Carrier Signal Level 3 (155.52 Mbps)
OC-9	Optical Carrier Signal Level 9 (466.56 Mbps)
OC-12	Optical Carrier Signal Level 12 (622.08 Mbps)
OC-18	Optical Carrier Signal Level 18 (933.12 Mbps)
OC-24	Optical Carrier Signal Level 24 (1.244 Gbps)
OC-36	Optical Carrier Signal Level 36 (1.866 Gbps)
OC-48	Optical Carrier Signal Level 48 (2.488 Gbps)
OC-96	Optical Carrier Signal Level 96 (4.976 Gbps)
OC-192	Optical Carrier Signal Level 192 (9.952 Gbps)
OC-256	Optical Carrier Signal Level 256 (13.271 Gbps)
OC-768	Optical Carrier Signal Level 768 (40 Gbps)
OC-1536	Optical Carrier Signal Level 1536 (80 Gbps)
OEM	Original Equipment Manufacturer
OFX	Open Financial Exchange
OLAP	Online Analytical Processing

OLE	Object Linking and Embedding
OS	Operating System
OSI	Open Systems Interconnection
OSN	Official Services Network
OTN	Optical Transport Network

P

PA	Preamble
PAD	Packet Assembler-Disassembler
PAL	Phase Alternating by Line
PAP	Password Authentication Protocol
PBX	Private Branch Exchange
PC	Personal Computer
PCB	Printed Circuit Board
PCM	Pulse Code Modulation
PCN	Personal Communications Networks
PCS	Personal Communication Services
PDA	Personal Digital Assistant
PDN	Packet Data Network
PDU	Payload Data Unit
PEM	Privacy Enhanced Mail
PGP	Pretty Good Privacy
PHY	Physical Layer
PIM	Protocol-Independent Multicast
PIN	Personal Identification Number
PMD	Physical Media Dependent
PnP	Plug and Play
POP	Point of Presence
POP	Post Office Protocol
POTS	Plain Old Telephone Service

PPP	Point-to-Point Protocol
PPTP	Point-to-Point Tunneling Protocol
PRI	Primary Rate Interface (ISDN)
PSN	Packet-Switched Network
PSTN	Public Switched Telephone Network
PT	Payload Type
PTT	Post, Telephone, and Telegraph
PVC	Permanent Virtual Circuit

Q

QAM	Quadrature Amplitude Modulation
QoS	Quality of Service
QPSK	Quadrature Phase Shift Keying

R

RAD	Rapid Application Development
RADSL	Rate Adaptive Digital Subscriber Line
RAID	Redundant Array of Inexpensive (or redundant) Disks
RAM	Random Access Memory
RAS	Remote Access Server
RDBMS	Relational Database Management System
RFC	Request for Comment
RF	Radiofrequency
RF	Routing Field
RFI	Radiofrequency Interference
RIP	Routing Information Protocol
RMON	Remote Monitoring
ROI	Return on Investment

ROM	Read Only Memory
RP	Rendezvous Point
RPC	Remote Procedure Call
RSVP	Resource ReSerVation Protocol
RT	Remote Terminal
RX	Receive

S

SA	Source Address
SAR	Segmentation and Reassembly
SAS	Single Attached Station
SCSI	Small Computer Systems Interface
SD	Starting Delimiter
SDSL	Symmetric Digital Subscriber Line
SET	Secure Electronic Transaction
SFD	Start Frame Delimiter
SHTTP	Secure HyperText Transfer Protocol
SIF	Signaling Information Field
SIG	Special Interest Group
SIIA	Software & Information Industry Association (formerly Software Publishers Association)
SLIC	Serial Line Interface Coupler (IBM Corp.)
SLIP	Serial Line Internet Protocol
SMS	Short Message Service
SMTP	Simple Mail Transfer Protocol
SNAL	Subscriber Network Access Line
SNI	Subscriber Network Interface
SNIA	Storage Network Industry Association
SNMP	Simple Network Management Protocol
SONET	Synchronous Optical Network

SPA	Software Publishers Association
SPC	Stored Program Control
SPI	Service Provider Interface
SPX	Synchronous Packet Exchange (Novell, Inc.)
SQL	Structured Query Language
SS	Switching System
SS7	Signaling System No. 7
SSL	Secure Sockets Layer
SSP	Service Switching Point
STDM	Statistical Time Division Multiplexing
STP	Shielded Twisted Pair
STP	Signal Transfer Point
STP	Spanning Tree Protocol
STS	Synchronous Transport Signal
STX	Start of Transmission
SVC	Switched Virtual Circuit
SWC	Serving Wire Center

T

T1	T-Carrier Service at the DS1 Rate of 1.544 Mbps
T3	T-Carrier Service at the DS3 Rate of 44.736 Mbps
TAPI	Telephony Application Programming Interface (Microsoft Corp.)
TASI	Time Assigned Speech Interpolation
TB	Terabyte (Trillion Bytes)
Tbps	Terabits per Second
TCP	Transmission Control Protocol
TDD	Time Division Duplexing

TDM	Time Division Multiplexer
TDMA	Time Division Multiple Access
TDMA/TDD	Time Division Multiple Access with Time Division Duplexing
TFTP	Trivial File Transfer Protocol
TIA	Telecommunications Industry Association
TIB	Tag Information Base
TSAPI	Telephony Services Application Programming Interface (Novell Inc.)
TSI	Time Slot Interchange
TSR	Terminal Stay Resident
TTRT	Target Token Rotation Time
TX	Transmit

U

UART	Universal Asynchronous Receiver/Transmitter
UBR	Unspecified Bit Rate
UDP	User Datagram Protocol
UHF	Ultrahigh Frequency (300 MHz to 3 GHz)
UMS	Universal Messaging System
UNI	User-Network Interface
UPS	Uninterruptible Power Supply
UTP	Unshielded Twisted-Pair

V

VAR	Value-Added Reseller
VBR	Variable Bit Rate

VC	Virtual Circuit
VCI	Virtual Channel Identifier
VDSL	Very-High-Speed Digital Subscriber Line
VF	Voice Frequency
VG	Voice Grade
VLAN	Virtual Local Area Network
VLSI	Very-Large-Scale Integration
VoFR	Voice over Frame Relay
VPI	Virtual Path Identifier
VP	Virtual Path
VPDN	Virtual Private Data Network
VPN	Virtual Private Network
VSAT	Very Small Aperture Terminal
VT	Virtual Terminal
VT	Virtual Tributary

W

W3C	World Wide Web Consortium
WAN	Wide Area Network
WAP	Wireless Application Protocol
WECA	Wireless Ethernet Compatibility Alliance
Wi-Fi	Wireless Fidelity
WLAN	Wireless Local Area Network
WLL	Wireless Local Loop
WWW	World Wide Web

INDEX

Boldface page range indicates a main entry.

U

About the Author

Nathan J. Muller is an independent consultant specializing in telecommunications technology marketing, research, and education. A resident of Sterling, VA, he serves on the Editorial Board of the *International Journal of Network Management,* the Advisory Board of the American Alliance of Service Providers, and the Advisory Panel of Faulkner Information Services. Among the 21 books he has authored are *The Desktop Encyclopedia of Telecommunications* and *Network Manager's Handbook*.